Recent Progress in Understanding the Mechanism and Consequences of Retrotransposon Movement

Special Issue Editors

David J. Garfinkel
Katarzyna J. Purzycka

MDPI • Basel • Beijing • Wuhan • Barcelona • Belgrade

MDPI

Special Issue Editors
David J. Garfinkel
University of Georgia
USA

Katarzyna J. Purzycka
Polish Academy of Sciences
Poland

Editorial Office
MDPI AG
St. Alban-Anlage 66
Basel, Switzerland

This edition is a reprint of the Special Issue published online in the open access journal *Viruses* (ISSN 1999-4915) in 2017 (available at: http://www.mdpi.com/journal/viruses/special_issues/retrotransposon).

For citation purposes, cite each article independently as indicated on the article page online and as indicated below:

Author 1; Author 2. Article title. *Journal Name* **Year**, *Article number*, page range.

First Edition 2017

ISBN 978-3-03842-540-3 (Pbk)
ISBN 978-3-03842-541-0 (PDF)

Table of Contents

About the Special Issue Editors

David J. Garfinkel Ph.D., is currently a Professor in the Department of Biochemistry & Molecular Biology at the University of Georgia, Athens Georgia USA. Dr. Garfinkel received a Ph.D. in Microbiology (1981) from the University of Washington, where he investigated the molecular genetics of Crown Gall tumorigenesis with Dr. Eugene W. Nester. He was a recipient of a Damon Runyon Postdoctoral Fellowship (1982–1984) to understand the mechanism of Ty1 transposition with Dr. Gerald R. Fink at the Whitehead Institute for Biomedical Research/Massachusetts Institute of Technology. Dr. Garfinkel was an independent investigator (1985–1990) and senior investigator (1990–2010) at the National Cancer Institute/NIH before joining the faculty at the University of Georgia (2010–present). The Garfinkel lab addresses how retrotransposon movement is controlled by element and host factors using the Ty1/yeast paradigm. Dr. Garfinkel and colleagues have published more than 80 peer reviewed articles, and his work has been supported by the National Cancer Institute/NIH, the National Institute of General Medical Sciences/NIH, the University of Georgia, and Bayer CropScience. In 2014, Dr. Garfinkel was elected a Fellow of the American Association for the Advancement of Science.

Katarzyna J. Purzycka earned a Bachelor's Degree in Chemistry from Poznan University in Poland in 2004. In 2009, she was awarded her Ph.D. in Biochemistry from the Institute of Bioorganic Chemistry, Polish Academy of Sciences, where she studied HIV-2 RNA and its nucleoprotein complexes as potential therapeutic targets. She continued her studies on retroviral RNAs during postdoctoral training at the National Cancer Institute, NIH, USA. In 2012, she was appointed Assistant Professor of Biochemistry at the Institute of Bioorganic Chemistry, Polish Academy of Sciences, where she led the Retroelements' Structure and Function Laboratory. Dr. Purzycka was a visiting professor at the University of Georgia, USA and she is a recipient of several awards, including a MNiSW fellowship for outstanding young scientists. Her major scientific interests are RNA structural elements and transitions on both 2D and 3D structural levels that specify RNA functions at different stages of retrovirus and endogenous retrovirus-like retrotransposon replication.

Preface to "Recent Progress in Understanding the Mechanism and Consequences of Retrotransposon Movement"

Retrotransposons move via an RNA intermediate and have had a profound impact on genome evolution and function in eukaryotes. In the Special Issue entitled "Recent Progress in Understanding the Mechanism and Consequences of Retrotransposon Movement" published in Viruses, three research articles and seven reviews explore different facets of the retrotransposon lifestyle. Here, we briefly summarize work presented in the Special Issue that should update the readership of Viruses on retroelements and endogenous retroviruses.

Arkhipova and coworkers provide new computational analyses suggesting that ORF3 from several LTR-retrotransposons present in Bdelloid rotifers can encode not only Envelope-like proteins but also a GDSL esterase/lipase or a DEDDy-like exonuclease. The results suggest that there is extensive gene sharing between different groups of retroelements. Work from the Curcio and Purzycka labs focuses on the structure and function of transcripts from the Ty1 retrotransposon of Saccharomyces cerevisiae. Curcio and coworkers propose a model for kissing loop interactions involved in dimerization of genomic Ty1 RNA. Purzycka and coworkers analyze how the structure of a recently discovered internal Ty1 (i) transcript impacts initiation of translation from two alternative start codons. Interestingly, Ty1i RNA encodes a restriction factor encompassing the C-terminal of Gag that is required for modulating retrotransposition.

The seven review articles cover a wide area of retroelement biology; from investigating specific steps in the process of retrotransposition to understanding the retroelement and pseudogene landscape in normal and malfunctioning cells. Le Grice and coauthors contribute a detailed review analyzing distinct topological differences between reverse transcription of the yeast retrotransposon Ty3 and its retroviral counterparts. Zaratiegui reviews and expands on recent work linking transposon integration with the universal process of DNA replication. Nefedova and Kim also review retroelement integration using retrotransposons from Drosophila as a guide. Their contribution highlights the roles of DNA sequence and structure, and protein contacts required for integration in different regions of the genome. Contributions from Magiorkinis and Tramontano and coauthors address the fascinating but incompletely understood roles that endogenous retroviruses play in mammalian cells. Hurst and Magiorkinis highlight the benefits of increased genetic variation and accelerated genome evolution mediated by endogenous retroviral insertions with detrimental effects that may result in disease. The Grandi and Tramontano review provides a detailed and critical analysis of endogenous retroviral insertions and their relationship to a variety of human diseases. Scott and Devine review a very active area of research, where somatic retrotransposition of LINE-1 has been implicated in human cancer. Importantly, Scott and Devine emphasize the challenge of determining whether LINE-1 insertions are "cancer drivers or passengers". Lastly, Kubiak and Makalowska review the interesting roles that retrocopies of protein coding genes play in evolving new genes or regulating expression of other genes.

In closing, we thank the authors for their thoughtful contributions, the external reviewers for their help evaluating and improving the manuscripts, and the staff of Viruses for expert editorial assistance. We also hope the Special Issue informs the readership of Viruses of the intimate relationships between transposons and viruses.

David J. Garfinkel and Katarzyna J. Purzycka

Special Issue Editors

viruses

MDPI

Article

LTR-Retrotransposons from Bdelloid Rotifers Capture Additional ORFs Shared between Highly Diverse Retroelement Types

Fernando Rodriguez [1], Aubrey W. Kenefick [1,2] and Irina R. Arkhipova [1,*]

[1] Josephine Bay Paul Center for Comparative Molecular Biology and Evolution, Marine Biological Laboratory, 7 MBL Street, Woods Hole, MA 02543, USA; frodriguez@mbl.edu (F.R.); awkenefick@ucdavis.edu (A.W.K.)

[2] Present address: UC Davis Genome Center-GBSF, University of California, Davis, CA 95616, USA

* Correspondence: iarkhipova@mbl.edu; Tel.: +1-508-289-7120

Academic Editors: David J. Garfinkel and Katarzyna J. Purzycka
Received: 31 January 2017; Accepted: 4 April 2017; Published: 11 April 2017

Abstract: Rotifers of the class Bdelloidea, microscopic freshwater invertebrates, possess a highly-diversified repertoire of transposon families, which, however, occupy less than 4% of genomic DNA in the sequenced representative *Adineta vaga*. We performed a comprehensive analysis of *A. vaga* retroelements, and found that bdelloid long terminal repeat (LTR)-retrotransposons, in addition to conserved open reading frame (ORF) 1 and ORF2 corresponding to *gag* and *pol* genes, code for an unusually high variety of ORF3 sequences. Retrovirus-like LTR families in *A. vaga* belong to four major lineages, three of which are rotifer-specific and encode a dUTPase domain. However only one lineage contains a canonical *env*-like fusion glycoprotein acquired from paramyxoviruses (non-segmented negative-strand RNA viruses), although smaller ORFs with transmembrane domains may perform similar roles. A different ORF3 type encodes a GDSL esterase/lipase, which was previously identified as ORF1 in several clades of non-LTR retrotransposons, and implicated in membrane targeting. Yet another ORF3 type appears in unrelated LTR-retrotransposon lineages, and displays strong homology to DEDDy-type exonucleases involved in 3′-end processing of RNA and single-stranded DNA. Unexpectedly, each of the enzymatic ORF3s is also associated with different subsets of *Penelope*-like *Athena* retroelement families. The unusual association of the same ORF types with retroelements from different classes reflects their modular structure with a high degree of flexibility, and points to gene sharing between different groups of retroelements.

Keywords: retrovirus-like transposable elements; envelope gene (ENV); DEDDy exonuclease; GDSL esterase; dUTPase

1. Introduction

Long terminal repeat (LTR) retrotransposons represent a major class of transposable elements (TEs), which move via reverse transcription of the full-length RNA intermediate by the element-encoded reverse transcriptase (RT) [1]. They are structurally similar to vertebrate retroviruses, and undergo the same steps of reverse transcription in their replication cycle [2]. Intracellular LTR retrotransposons typically encode only two genes, the *gag* gene which forms the nucleoprotein core, and the *pol* gene which combines protease, RT, RNase H, and integrase enzymatic activities. Retroviruses additionally code for an *env* (envelope) gene, which endows them with the capacity to interact with cellular membranes for viral entry and exit. The lack of an extracellular stage in the LTR retrotransposon life cycle can be occasionally overcome by capture of an *env* gene from DNA viruses (e.g., baculovirus, phlebovirus, or herpesvirus) [3]. The baculovirus-derived *env* gene in the *gypsy* retrotransposon of *Drosophila melanogaster* has been studied most extensively, revealing infectious and fusogenic properties [4–6]. Domestication of *env* genes from endogenous retroviruses has also occurred throughout evolution, giving rise to novel unanticipated host functions [7–9].

Bdelloid rotifers are microscopic freshwater invertebrates that reproduce asexually, are highly resistant to desiccation and ionizing radiation, and contain numerous genes of foreign origin in subtelomeric regions [10–12]. We previously showed that bdelloid genomes contain canonical LTR-retrotransposons, *Juno* and *Vesta*, forming a deep-branching clade [13], as well as telomere-associated, endonuclease-deficient *Penelope*-like retroelements named *Athena* [14]. Both *Juno* and *Vesta* contain an open reading frame (ORF) 3, which was assumed to code for *env* but revealed no clear-cut homologies to known viral envelope genes.

The genome of the first bdelloid representative, *Adineta vaga*, has been sequenced [15]. Over 8% of its gene content is made up of foreign genes originating from bacteria, fungi, plants, or protists. Known TE families make up to 4% of the 218-Mb assembly, with low copy numbers per family (on average, 1–2 full-length copies and 10 times as many fragments), and high family diversity (over 255 families). Of these, about one-half are represented by retrotransposons, including 24 families of LTR retrotransposons belonging to four clades (*Juno*, *Vesta*, *TelKA*, and *Mag*). Most LTR retrotransposons have transposed recently, as judged by very few or no differences between the two LTRs [15]. Here we focus in detail on their coding capacity, and report that they can code for a variety of extra ORFs of enzymatic origin, which are also found on giant telomeric retroelements called *Terminons* (Arkhipova et al., submitted). We also report that all bdelloid retrotransposon clades, except for *Mag*, carry a dUTPase domain found in certain retroviruses and in basidiomycete LTR-retrotransposons.

2. Materials and Methods

2.1. Bioinformatics

The annotated *A. vaga* scaffolds containing LTR retrotransposons were downloaded from the genome browser at http://www.genoscope.cns.fr/adineta. Each LTR retrotransposon was manually re-annotated to confirm the presence of intact full-length ORF1, ORF2, and ORF3. Sequences from *P. roseola* (accession numbers DQ985390, EU643489, EU643490) and a natural isolate *Adineta sp. 11* were also used in the analysis. Homology searches were performed with HHpred (Version HHSuite-2.0.16mod) [16] and visualized with Jalview (Version 2.10.1) [17]. Multiple sequence alignments were done by MUSCLE [18], followed by maximum-likelihood and neighbor-joining phylogenetic analysis, and the resulting trees were edited in MEGA (Version 7.0.18) [19]. Alignments are available from the corresponding author upon request. Coiled-coil motifs were predicted by COILS/PCOILS (Version 2.2) [16], and transmembrane domains with TMHMM (Version 2.0) [20].

For genome-wide analysis, LTR families were extracted from the initial annotation of known *A. vaga* TE families [21]. We estimated the numbers of fragmented copies (longer than 100 base pair (bp), including solo LTRs) and numbers of full-length copies by BLAT (Version 34) [22], using full-length sequences as queries. ORF annotations within each full-length copy were also identified by BLAT search, using family-specific ORF sequences as queries. Alignment of RNA-seq and small RNA reads (NCBI accession Nos. SRP020358 and SRP070765) to the reference genome was performed as in [21]. Aligned sequences were counted for each TE copy and each annotated ORF feature with htseq-count [23].

2.2. Nucleic Acid Manipulations

Clonal cultures of *A. vaga* were grown and collected for DNA extraction as described in [15]. We designed the exact-matching forward and reverse primers from the corresponding genomic scaffolds (Table S1) to amplify the full-length ORF3 from each desired element. Polymerase chain reaction (PCR) conditions were as follows: 0.5 U of Q5 High-Fidelity DNA Polymerase (New England Biolabs, Ipswich, MA, USA) in a 25 µL reaction, with 1 µM of each primer, 200 µM dNTPs, 1× Q5 Reaction Buffer and template DNA. Thermocycling parameters were set following the conditions

2

specified in the Q5 High-Fidelity DNA Polymerase manual, with Tm values adjusted for each primer pair. PCR products were electrophoresed in 1.5% agarose gels in 1 × TAE (Tris base, acetic acid, EDTA) buffer, and visualized under UV light. PCR amplicons of the expected size were purified using Wizard® SV Gel and PCR Clean-Up System (Promega, Madison, WI, USA). Prior to T/A cloning, addition of an untemplated dA was done with *Taq* DNA Polymerase (Promega). PCR products were cloned into pGEM-T vector (Promega) and transformed into JM109 (Promega) or DH5a (New England Biolabs) competent cells per the supplier's specifications. Clones were screened for inserts of the expected size by PCR amplification with the universal primers M13 Forward and M13 Reverse. Plasmid DNA was prepared from selected clones with Zyppy™ Plasmid Miniprep Kit (Zymo Research, Irvine, CA, USA). Templates were sequenced on an Applied Biosystems 3730XL DNA Analyzer at the W. M. Keck Ecological and Evolutionary Genetics Facility at the Marine Biological Laboratory. After inspection of the chromatogram files, the phred/cross_match pipeline [24] was applied to check for quality and to screen out vector sequences. Sequences obtained in this study were deposited in GenBank under accession numbers KY820831–KY820845. Consensus sequences of LTR retrotransposons were deposited in Repbase [25].

3. Results

3.1. An Overview of LTR Retrotransposon Structure in Bdelloids

Of all TE types, LTR-retrotransposons are arguably the easiest to detect and annotate in sequenced genomes due to their characteristic LTR structures. Our recent inventory of the LTR-retrotransposon families in *A. vaga* identified 12 *Vesta*-like families, five *Juno*-like families, six *TelKA*-like families, and one *Mag* family, which in total occupy ~580 kb of genomic DNA [15]. We supplemented this comprehensive dataset with additional LTR retrotransposons from sequenced fosmids from a genomic library of *Philodina roseola (Pr)*, a species from the bdelloid family Philodinidae, which separated from *A. vaga* tens of millions of years ago [26], and from a draft genome of a natural isolate *Adineta sp. 11 (As)*. Notably, not only each congeneric, but also each of the *P. roseola* LTR retrotransposons can be assigned to the corresponding *A. vaga* families (Figure 1A), indicating the early origin of the LTR families and/or extensive horizontal transfer between species.

All LTRs carry TG CA at the ends, vary in length between 159 and 551 bp, and display very few substitutions between two LTRs, which is indicative of recent transposition [15]. As expected, all families code for *pol* genes with a canonical set of enzymatic activities that includes protease (PR), RT, ribonuclease H (RNase H; RH), and integrase (IN), in that order (Figure 1A). Every *gag* gene, except for *Vesta1b*, codes for a typical $CX_2CX_4HX_4C$ Zn-knuckle; in addition, *Mag*, *Juno1*, *Juno2*, and *Vesta6c* code for an adjacent second Zn-knuckle. Curiously, *Vesta1b* not only lacks Zn-knuckles, but also lacks a *gag-pol* translational frameshift, a feature it shares with the *Mag* family. An extra Zn-knuckle upstream of RT in the *pol* gene of all *Juno1* and *Juno2* elements also represents a departure from the standard organization. The GPY/F motif at the integrase C-terminus [27] is present in *Juno1*–*Juno4* and *Vesta6*–*Vesta7*, is modified to GPC in *TelKA* and *Vesta6c*, reduced to a proline in *Vesta1*–*Vesta5*, and is missing from the *Mag* lineage altogether. No chromodomain was found C-terminally to the GPY/F module in any lineage.

Interestingly, all members of the *Juno*, *Vesta*, and *TelKA* families, but not the *Mag* family, contain a dUTPase (dut) domain between PR and RT (Figure 1A), followed by an extra Zn knuckle in *Juno*. The Dut domain often occurs in vertebrate retroviruses, where it can be variably positioned between gag and RT, between RT and IN, or after IN [28,29]. However, it is rarely found in retrotransposons, and has been reported only in basidiomycetes [30], where it is similarly placed between PR and RT.

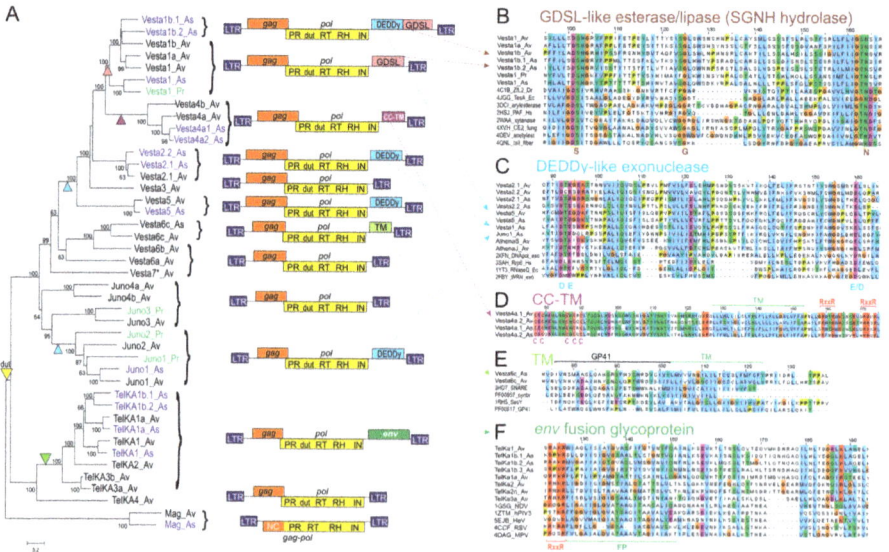

Figure 1. Structure, phylogeny and open reading frame 3 (ORF3) alignments of bdelloid long terminal repeat (LTR) retrotransposons. (**A**) Maximum likelihood phylogram of *pol* genes including protease (PR), dUTPase (dut), reverse transcriptase (RT), RNase H (RH), integrase (IN) domains and the associated ORF structure. Putative ORF3 acquisition/loss events are marked by triangles of matching color. Scale bar, amino acid substitutions per site; (**B–F**) Alignments of characteristic regions between retrotransposon ORF3s and selected GDSL esterases/lipases (**B**), DEDDy exonucleases (**C**), transmembrane (TM) proteins (**D–E**) and env fusion glycoproteins from paramyxoviruses (**F**). Also shown are catalytic S-G-N residues from SGNH block 3 (**B**), catalytic D/E residues from DEDDy block ExoI (**C**), Cys residues (**D**), TM domains (**D–E**), furin-like protease cleavage site (RXXR), and fusion peptide (FP) (**F**).

3.2. Types of Acquired Env-Like ORFs

An ORF3 downstream of the *pol* gene is usually assumed to code for an *env*-like protein, as in vertebrate retroviruses. Due to the low conservation of *env* sequences, such assignments often rely on computationally predicted features of broad applicability, such as TM domains, glycosylation sites, protease cleavage sites, or coiled-coil motifs, which are not restricted to *env* genes, but are commonly found in other proteins. Assignment of ORF3 to *env* genes can be unambiguous only when its origin can be traced to another virus [3].

In bdelloids, the *TelKA* clade contains a canonical *env*-like fusion glycoprotein about 600 aa in length, which is most similar to class IF proteins from paramyxoviruses—non-segmented negative-strand RNA viruses such as avian Newcastle disease virus (NDV), human parainfluenza (PIV), respiratory syncytial virus (RSV), metapneumovirus (MPV), and Hendra (HeV) [31] (pfam00523; HHpred alignment over the entire length with E-value = 1.9e^{-106}). Regions of high conservation (Figure 1F) include the furin-like protease cleavage site (RXXR), a hydrophobic region (FP, fusion peptide), a trimeric coiled-coil domain, a set of conserved cysteines for disulfide bridge formation between two protease cleavage products, and a C-terminal transmembrane (TM) anchor domain.

In the *Vesta4* clade, a much shorter (220–230 aa) ORF3 lacks detectable homology with known *env* genes, but nevertheless displays two hydrophobic transmembrane regions with a set of cysteine residues in between, followed by an RXXR motif and a coiled-coil domain (Figure 1E). Such structural organization is also suggestive of fusogenic properties, although other functions cannot be ruled out.

4

A possible *env*-like ORF3 is found in *Vesta6c* LTR retrotransposons from *A. vaga* and a congeneric natural isolate, *Adineta sp. 11* (Figure 1A). This ORF3 is characterized by the presence of TM domains, and an HHpred search reveals weak homology to the retroviral envelope glycoprotein gp41 (PF00517), which mediates fusion with the host cell (p-value = $9.2e^{-05}$) (Figure 1E). In *Adineta*, however, it does not represent a part of the larger gp120-like env precursor, and is instead coded by a small 210–230-aa ORF3. This ORF may be a remnant of an initially present full-length *env*-like ORF.

3.3. Unexpected Diversity of Non-Envelope ORF3 Functions

Functional assignment of an ORF3 is often far from straightforward, especially in the absence of a known viral source. For instance, several *copia*-like and *gypsy*-like LTR retrotransposons in plants have long been assumed to code for an envelope-like protein, although it is still unclear if they do [32–36]. Certain plant LTR retrotransposons and vertebrate retroviruses carry extra ORFs with no assignable function [29,37,38]. Surprisingly, we find that bdelloid LTR retrotransposons display a much higher degree of heterogeneity with respect to ORF3 than is typically observed in retroelements.

Use of sensitive HHpred searches allowed us to determine the origin of the remaining extra ORFs, which were previously classified as *env* due to the presence of computationally predicted motifs of broad specificity (TM domains, protease cleavage sites, N-glycosylation sites) [13]. We find that most members of the *Vesta* and *Juno* clades lack bona fide *env* genes, but instead have acquired different ORF3 coding for GDSL esterase/lipase and RNase D-like DEDDy-type exonuclease activities (Figure 1A–C). The DEDDy-type (or DnaQ-like) 3'-5' exonucleases perform 3'-end processing of various structured RNAs (RNase D, RNase T, exosome subunit Rrp6), but may also act on single-stranded DNAs (WRN, DnaQ, and proofreading subunits of A- and B-type DNA polymerases) [39]. GDSL esterases/lipases are hydrolytic enzymes with broad substrate specificity, named after a GDSL or similar sequence with the catalytic Ser in the first conserved block, and are also designated as SGNH hydrolases, named after the letters specifying the invariant catalytic S, G, N, and H residues in the four conserved blocks [40]. In each of these ORFs, the invariant residues are intact, indicating possible catalytic activity (Figure 1B). Motif DEDD is changed to DEED (Figure 1C). In *Adineta sp. 11*, both DEDDy and GDSL can occur within a single ORF3 (Figure 1A, top).

In the phylogram on Figure 1A, which depicts currently known families of bdelloid LTR retrotransposons, it may be seen that additional ORF3s, which are family-specific, are notably missing from the earliest branches (*TelKA4*; *Juno3–Juno4*; *Vesta6–Vesta7*). In the more recent branches, the DEDDy-like ORF has been independently acquired at least twice, by *Juno* and by *Vesta* (Figure 1A).

3.4. Different ORF3 Types Are Shared between Highly Diverse Retroelements

Interestingly, the diverse ORF3 types (GDSL, DEDDy, CC, TM) are not restricted to LTR retrotransposons. They can also be found in the highly unusual group of bdelloid retroelements which we recently described (Arkhipova et al., submitted). These retroelements, which we call *Terminons*, reveal an extraordinary degree of complexity, coding for multiple diverse ORFs and reaching 40 kb in length. As the principal polymerizing component, they contain *Athena*-like RTs belonging to the enigmatic class of *Penelope*-like retroelements (PLEs) [41]. *Terminons* also harbor a plethora of other ORFs of enzymatic and non-enzymatic nature, which in many families include DEDDy, GDSL, CC-, and TM-containing ORFs.

We performed phylogenetic analysis of DEDDy-like ORFs from bdelloid retrotransposons, and they are much more similar between the two retrotransposon groups than between TE-associated ORFs and their non-transposable cellular homologs, such as RNase D, mut-7, WRN, Rrp6, and DNA_pol_A exonucleases. Thus, these ORFs are less likely to have been captured from the host than they are likely to have been exchanged between different retroelement types (Figure 2A). This finding hints at the existence of a specialized DEDDy-like ORF pool utilized by diverse retroelements. To some extent, this is also applicable to GDSL-like ORFs: ORFs from *Vesta1* in *A. vaga* and *P. roseola* are apparently related to the GDSL domain in *Athena-I* (Figure 2B), which in turn reveals similarity to a stand-alone

GDSL-like ORF in the *A. vaga* host. Due to the absence of catalytic residues in GDSL derivatives from the *Athena-L* family, their origin is more difficult to determine, however, they are consistently grouped with a subfamily of SGNH hydrolases termed PC-esterases (Figure 2B), which are potentially involved in the modification of cell-surface glycoproteins [42]. The esterases found in selected non-LTR retrotransposon clades (L2, CR1, RTEX) [43,44] do not cluster with any of the above ORFs, indicating their independent capture (Figure 2B).

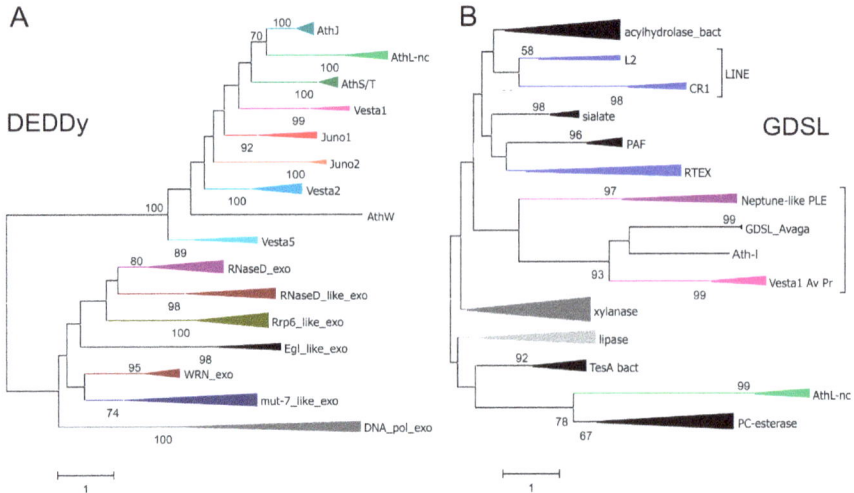

Figure 2. Diverse ORF3 functions in retrotransposons. (**A**) Amino acid sequence similarity between DEDDy-like ORFs from *Juno* and *Vesta* LTR retrotransposons, *Athena* retroelements, and different groups of cellular DEDDy exonucleases from cd09018 sequence cluster in the Conserved Domain Database (CDD); (**B**) GDSL-like ORFs in *Vesta1* LTR-retrotransposons, PLEs, non-LTR (or LINE-like) retrotransposons from esterase-containing clades CR1, L2, and RTEX, and representative groups from the cellular SGNH hydrolase superfamily (cd00229 cluster in CDD). All ORF3 sequences shown in Figure 1 were included and collapsed for better visualization. Branch support values exceeding 50% are shown. Scale bars, amino acid substitutions per site.

3.5. Transcription, Small RNA-Mediated Silencing, and Copy Numbers

In our earlier study investigating transcription and silencing of TE families in *A. vaga*, most LTR retrotransposons were found to be transcriptionally active [21]. However, their expression levels were determined from mapping to full-length TE annotations without subdivision into different ORFs, while ORF3s in LTR retrotransposons typically represent separate transcriptional units, and are expressed from spliced messages. To investigate whether the diverse ORF3s show transcriptional activity, we mapped *A. vaga* transcripts to each ORF individually. The results of RNA-seq profiling are shown in Figure 3A, which displays RPK (number of reads per kilobase) values for each ORF within the *A. vaga* LTR retrotransposon families. In most cases, LTR families display relatively low levels of transcription activity, although there are some notable exceptions. For instance, *Vesta4b* on the scaffold Av_1520 shows high transcript levels within each of the three ORFs (*gag*, *pol*, and CC), possibly reflecting recent arrival of an active element. Interestingly, this scaffold is circularly permuted, which may indicate that it was assembled from an extrachromosomal 1-LTR circle. High transcript levels are also observed for *Juno4b*, which lacks ORF3.

We also investigated whether each ORF type is subject to small RNA-mediated silencing. In *A. vaga*, pi-like small RNAs (sRNA) are preferentially mapped to annotated transposons, with most of the reads being in antisense orientation [21]. Mapping of sRNA read counts by ORF type (Figure 3B) demonstrates that the majority of sRNA reads (66.8%) are mapped to *pol* genes, which occupy most of the TE length. For LTR families with an annotated ORF3 (*env*, DEDDy, GDSL, CC, TM), 22% of sRNA reads are mapped to such ORFs, while 14% are mapped to *gag* and 64% to *pol* gene annotations. For LTR families without an ORF3, *gag* is covered by 29% and *pol* by 71% of the sRNA reads mapped, which is roughly equivalent in terms of read count per kilobase. Comparison of the RNA-seq and sRNA plots shows that transcriptional activity is typically accompanied by sRNA coverage, which involves every ORF type. However, the *env*-containing *TelKA1* and *TelKA1a* show higher levels of transcriptional activity and lower levels of sRNA coverage in comparison with other members of the TelKA clade, which may indicate that their recent arrival has not yet resulted in establishment of a robust piRNA silencing response.

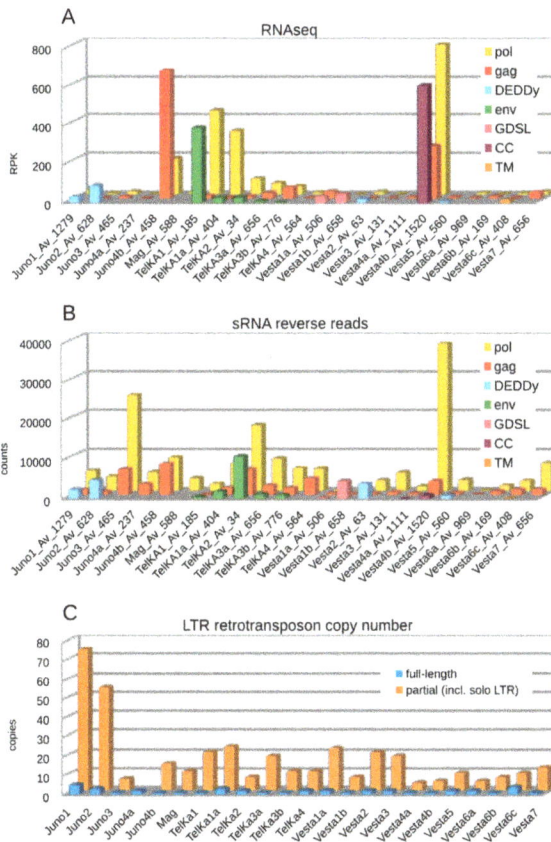

Figure 3. LTR retrotransposon copy numbers, and RNA profiles in *A. vaga*. Distribution of RNAseq reads with RPK (reads per kilobase) values (**A**), and small RNAs in reverse orientation (total counts) mapped to annotated ORFs (**B**) is shown for each family, along with the number of each reference scaffold. Numbers of full-length and fragmented copies (longer than 100 bp) estimated by BLAT, using full-length sequences as queries, are shown in (**C**). ORFs are color-coded as indicated.

We also attempted to reveal correlations between the presence of ORF3, transcriptional activity, and copy number of LTR retrotransposons. Figure 3C visualizes the number of full-length and partial copies of LTR retrotransposons in each family, with a large proportion of partial copies represented by solo LTRs. While full-length copies are indeed scarce, a DEDDy-like ORF in *Juno1* and *Juno2* might be correlated with a higher overall copy number, however, *Vesta2* and *Vesta5* with the same ORF are low-copy. It is possible that the latter represent earlier arrivals, as evidenced by their more basal position on the phylogenetic tree, and that most of these copies have undergone removal by LTR-LTR recombination, as described in [15].

3.6. Sequence Variation in Env-Like and GDSL-Like ORFs

To validate the correct assembly of ORF3 and to evaluate the level of its intraspecific nucleotide sequence variation, we chose the *env*-like ORFs from five *TelKA* families and GDSL-like ORFs from three *Vesta* families for PCR amplification and sequencing. Primers were designed for amplification of full-length ORF3s in each of the families, and the resulting amplicons were cloned and Sanger-sequenced. An additional *env*-like ORF from a non-autonomous family related to *TelKA2*, named *TelKA2n*, which is missing the C-terminal part of *gag* and most of the *pol* gene, was also amplified and sequenced. All families except *TelKA3a* yielded amplicons of the expected length.

The information on sequence polymorphisms is presented in Table 1. In 11 individual 1.7-kb long *env* clones, 16 out of 20 single nucleotide substitutions resulted in amino acid replacement, while in four 1.3-kb GDSL clones, 3 out of 11 substitutions changed the corresponding amino acid. Substitutions which were already present in one of the copies from the genome assembly were marked as "natural", and a few substitutions were marked as "unique" if they could not be found in the assembly contigs. Most of these mutations apparently reflect natural intragenomic variation, and most of the "unique" substitutions should represent de novo mutations which arose over the five-year period since the genome was sequenced, although a few may still correspond to PCR errors despite the use of the Q5 polymerase with high fidelity exceeding best polymerases by an order of magnitude [45]. If only "natural" variation is considered, we do not find evidence that the number of synonymous substitutions significantly exceeds that of non-synonymous substitutions or vice versa, indicating that intragenomic variation of ORF3s is mostly neutral, and that its level is approximately the same as found in *gag* and *pol* genes (not shown). Indeed, selective forces would be expected to operate during critical steps of the life cycle, such as inter-genomic transmission, while intragenomic *env* evolution is more likely to be neutral. Except for *TelKA1* and *Vesta1*, at least one cloned copy in each family was identical to the full-length reference copy, either at the nucleotide or at the amino acid sequence level. In future experiments, we plan to determine whether the *env*-like or GDSL-like ORFs can exhibit fusogenic or lipolytic properties, respectively.

Table 1. Nucleotide sequence variation in *env*-like and GDSL-like open reading frames (ORFs).

Clone	Reference Scaffold/Contig [1]	Substitutions, bp	Substitutions, aa	Natural aa Differences	Unique aa Differences
env1	1591/5150	4	4	R-Q, E-Q, T-I	I-V
env1a.1		3	2		I-T, V-A
env1a.4	1200/4393	0	0		
env1a.8		3	3	V-A, T-I	S-F
env2.1	34/303	0	0		
env2.2		2	1	I-T	
env2.3		4	3	V-I, I-T	M-I
env2n.1	680/3155	1	1		A-S
env2n.2		2	1		D-G
env3b.1	776/3459	0	0		
env3b.2		1	1	I-T	
ves1	494/2540	8	3	T-S, R-S, H-Q	
ves1a	506/2575	1	0	silent	
ves1b.1	658/3084	1	0	silent	
ves1b.3		1	0	silent	

[1] Scaffold numbering: http://www.genoscope.cns.fr/adineta (annotated), and Contig numbering: WGS shotgun assembly CAWI000000000.2 (unannotated); One reference scaffold/contig is listed for each family.

4. Discussion

Our studies uncover an unexpected diversity of additional ORFs in LTR-retrotransposons, which goes beyond their well-known ability to acquire *env* genes from other viruses to facilitate host entry and egress. Earlier studies of plant gypsy-like LTR retrotransposons and animal retroviruses, while revealing extra ORFs, failed to uncover homologies with known proteins, except for two ORFs randomly captured from the host [29,37,38]. In this study, we used sensitive profile-profile searches to detect remote homologs in the HMM profile databases, revealing enzymatic origin for two types of extra ORFs in LTR retrotransposons of microscopic freshwater invertebrates, bdelloid rotifers. It is still unclear whether these ORFs confer proliferative advantages to TEs harboring them, as our analysis of their transcriptional activity did not reveal unambiguous correlations with copy numbers, and their intragenomic evolution does not reveal significant departures from neutrality.

In principle, a DEDDy-like exonuclease might participate in the processing of the 3′-ends of retrotransposon-encoded RNAs, while a GDSL esterase/lipase might facilitate penetration through host membranes during entry and exit. However, the catalytic activity of these ORFs is yet to be demonstrated. A role in post-transcriptional silencing, such as that of *mut-7* in *Caenorhabditis elegans* [46], may also be entertained, although self-limiting TEs would not be expected to survive in the long term, as they would be out-competed [47]. It is also formally possible that enzymatic ORFs may still perform the *env*-like function, despite their diverse origins and the lack of similarities to viral *env* genes. Future experiments aimed at determining fusogenic and/or lipolytic properties of the extra ORFs might help to clarify this issue. However, unlike bona fide *env*-like ORFs in *TelKA*, the GDSL-like and DEDDy-like ORFs lack CC- or TM-domains, suggesting that they do not perform *env*-like functions, but could rather play auxiliary roles in the replication cycle. In selected non-LTR retrotransposons (CR1, RTEX, ZfL2), a catalytically active SGNH hydrolase/esterase, which occupies a *gag*-like position upstream of *pol* and can dimerize via its coiled-coil domain, is thought to play a role in ribonucleoprotein (RNP) assembly and in membrane-dependent transport or localization [43,44]. While DEDDy exonucleases have not yet been reported in retrotransposons, it is worth noting that the metazoan Maelstrom and EXD1 proteins involved in piRNA biogenesis represent catalytically inactive DEDD nuclease derivatives retaining the RNA binding function [48,49]. Maelstrom also contains a Cys-His-Cys motif involved in Zn^{2+} coordination, which can also be noted in *Vesta2* and *Vesta5* DEDDy ORFs.

It is even more perplexing that similar ORFs can be shared between retrotransposable elements of highly diverse nature, such as LTR-retrotransposons and PLEs. Even the esterases from distantly related canonical *Neptune*-like PLEs [50] from fish and mollusks exhibit some similarity, albeit with insufficient clade support (Figure 2B). A plausible explanation for ORF acquisition is the existence of a common step in their transposition cycles permitting RT-mediated template switches in intersecting cellular locations (e.g., sites of RNP assembly). While there is currently no information on the exact transposition mechanisms for complex retroelements, it may be thought that the shared ORF types may be used to confer advantages to different types of retroelements, regardless of specific details of their retrotransposition cycles.

The fact that the extra ORFs are largely detected in the more recent branches of LTR retrotransposons, while missing from the more basal branches, points at a relatively recent acquisition of these ORFs. Another interpretation is that the terminal branches represent recent arrivals and systematically lose extra ORFs, as they become adapted to the intragenomic mode of proliferation. It has been argued that loss of the *env* gene turns endogenous retroviruses into genomic "superspreaders" [51]; however, this is clearly not the case in bdelloids, as is evident from copy number comparisons between *env*-containing and *env*-less families (Figure 3C). LTR retrotransposons in bdelloids are frequently eliminated by LTR-LTR recombination, leading to accumulation of solo LTRs, and by microhomology-mediated deletions, resulting in the formation of partial copies [15]. Thus, acquisition of an *env* gene or its equivalent may be regarded as a path to effective escape, facilitating horizontal mobility. While the role of lipases or exonucleases in this process remains to be determined, it may substitute for the obvious function of envelope genes in unexpected ways, which could be uncovered in future experiments.

Supplementary Materials: The supplementary materials are available online at www.mdpi.com/1999-4915/9/4/78/s1. Table S1. Primers used for ORF3 amplification.

Acknowledgments: We thank Irina Yushenova for advice on PCR cloning and sequencing. This work was supported by the National Institutes of Health grant GM111917 to I.A.; A.K. was supported by the Research Experiences for Undergraduates supplement to the National Science Foundation grant MCB-1121334 to I.A.

Author Contributions: I.A. conceived the study; I.A. and F.R. designed the experiments; F.R. and A.K. performed the experiments; I.A. and F.R. analyzed the data and wrote the paper.

Conflicts of Interest: The authors declare no conflict of interest. The funding sponsors had no role in the design of the study; in the collection, analyses, or interpretation of data; in the writing of the manuscript, and in the decision to publish the results.

References

1. Craig, N.L.; Chandler, M.; Gellert, M.; lambowitz, A.M.; Rice, P.A.; Sandmeyer, S.B. *Mobile DNA III*; ASM Press: Washington, DC, USA, 2015.
2. Arkhipova, I.R.; Mazo, A.M.; Cherkasova, V.A.; Gorelova, T.V.; Schuppe, N.G.; Ilyin, Y.V. The steps of reverse transcription of Drosophila mobile genetic elements and U3-R-U5 structure of their LTRs. *Cell* **1986**, *44*, 555–563. [CrossRef]
3. Malik, H.S.; Henikoff, S.; Eickbush, T.H. Poised for contagion: Evolutionary origins of the infectious abilities of invertebrate retroviruses. *Genome Res.* **2000**, *10*, 1307–1318. [CrossRef] [PubMed]
4. Kim, A.; Terzian, C.; Santamaria, P.; Pélisson, A.; Prud'homme, N.; Bucheton, A. Retroviruses in invertebrates: The gypsy retrotransposon is apparently an infectious retrovirus of Drosophila melanogaster. *Proc. Natl. Acad. Sci. USA* **1994**, *91*, 1285–1289. [CrossRef] [PubMed]
5. Song, S.U.; Gerasimova, T.; Kurkulos, M.; Boeke, J.D.; Corces, V.G. An env-like protein encoded by a Drosophila retroelement: Evidence that gypsy is an infectious retrovirus. *Genes Dev.* **1994**, *8*, 2046–2057. [CrossRef] [PubMed]
6. Misseri, Y.; Cerutti, M.; Devauchelle, G.; Bucheton, A.; Terzian, C. Analysis of the Drosophila gypsy endogenous retrovirus envelope glycoprotein. *J. Gen. Virol.* **2004**, *85*, 3325–3331. [CrossRef] [PubMed]
7. Mi, S.; Lee, X.; Li, X.-p.; Veldman, G.M.; Finnerty, H.; Racie, L.; LaVallie, E.; Tang, X.-Y.; Edouard, P.; Howes, S.; et al. Syncytin is a captive retroviral envelope protein involved in human placental morphogenesis. *Nature* **2000**, *403*, 785–789. [PubMed]
8. Malik, H.S.; Henikoff, S. Positive selection of Iris, a retroviral envelope-derived host gene in Drosophila melanogaster. *PLoS Genet.* **2005**, *1*, e44. [CrossRef] [PubMed]
9. Malfavon-Borja, R.; Feschotte, C. Fighting fire with fire: Endogenous retrovirus envelopes as restriction factors. *J. Virol.* **2015**, *89*, 4047–4050. [CrossRef] [PubMed]
10. Gladyshev, E.A.; Meselson, M.; Arkhipova, I.R. Massive horizontal gene transfer in bdelloid rotifers. *Science* **2008**, *320*, 1210–1213. [CrossRef] [PubMed]
11. Gladyshev, E.; Meselson, M. Extreme resistance of bdelloid rotifers to ionizing radiation. *Proc. Natl. Acad. Sci. USA* **2008**, *105*, 5139–5144. [CrossRef] [PubMed]
12. Mark Welch, D.B.; Mark Welch, J.L.; Meselson, M. Evidence for degenerate tetraploidy in bdelloid rotifers. *Proc. Natl. Acad. Sci. USA* **2008**, *105*, 5145–5149. [CrossRef] [PubMed]
13. Gladyshev, E.A.; Meselson, M.; Arkhipova, I.R. A deep-branching clade of retrovirus-like retrotransposons in bdelloid rotifers. *Gene* **2007**, *390*, 136–145. [CrossRef] [PubMed]
14. Gladyshev, E.; Arkhipova, I.R. Telomere-associated endonuclease-deficient Penelope-like retroelements in diverse eukaryotes. *Proc. Natl. Acad. Sci. USA* **2007**, *104*, 9352–9357. [CrossRef] [PubMed]
15. Flot, J.F.; Hespeels, B.; Li, X.; Noel, B.; Arkhipova, I.; Danchin, E.G.; Hejnol, A.; Henrissat, B.; Koszul, R.; Aury, J.M.; et al. Genomic evidence for ameiotic evolution in the bdelloid rotifer Adineta vaga. *Nature* **2013**, *500*, 453–457. [CrossRef] [PubMed]
16. Alva, V.; Nam, S.-Z.; Söding, J.; Lupas, A.N. The MPI bioinformatics Toolkit as an integrative platform for advanced protein sequence and structure analysis. *Nucleic Acids Res.* **2016**, *44*, W410–W415. [CrossRef] [PubMed]
17. Waterhouse, A.M.; Procter, J.B.; Martin, D.M.A.; Clamp, M.; Barton, G.J. Jalview Version 2—A multiple sequence alignment editor and analysis workbench. *Bioinformatics* **2009**, *25*, 1189–1191. [CrossRef] [PubMed]

18. Edgar, R.C. MUSCLE: A multiple sequence alignment method with reduced time and space complexity. *BMC Bioinform.* **2004**, *5*, 113. [CrossRef] [PubMed]
19. Kumar, S.; Stecher, G.; Tamura, K. MEGA7: Molecular Evolutionary Genetics Analysis version 7.0 for bigger datasets. *Mol. Biol. Evol.* **2016**, *33*, 1870–1874. [CrossRef] [PubMed]
20. Krogh, A.; Larsson, B.; von Heijne, G.; Sonnhammer, E.L.L. Predicting transmembrane protein topology with a hidden markov model: Application to complete genomes. *J. Mol. Biol.* **2001**, *305*, 567–580. [CrossRef] [PubMed]
21. Rodriguez, F.; Arkhipova, I.R. Multitasking of the piRNA silencing machinery: Targeting transposable elements and foreign genes in the bdelloid rotifer Adineta vaga. *Genetics* **2016**, *203*, 255–268. [CrossRef] [PubMed]
22. Kent, W.J. BLAT—The BLAST-like alignment tool. *Genome Res.* **2002**, *12*, 656–664. [CrossRef] [PubMed]
23. Anders, S.; Pyl, P.T.; Huber, W. HTSeq—A Python framework to work with high-throughput sequencing data. *Bioinformatics* **2015**, *31*, 166–169. [CrossRef] [PubMed]
24. Gordon, D.; Green, P. Consed: A graphical editor for next-generation sequencing. *Bioinformatics* **2013**, *29*, 2936–2937. [CrossRef] [PubMed]
25. Bao, W.; Kojima, K.K.; Kohany, O. Repbase Update, a database of repetitive elements in eukaryotic genomes. *Mob. DNA* **2015**, *6*, 11. [CrossRef] [PubMed]
26. Hur, J.H.; Van Doninck, K.; Mandigo, M.L.; Meselson, M. Degenerate tetraploidy was established before bdelloid rotifer families diverged. *Mol. Biol. Evol.* **2009**, *26*, 375–383. [CrossRef] [PubMed]
27. Malik, H.S.; Eickbush, T.H. Modular evolution of the integrase domain in the Ty3/Gypsy class of LTR retrotransposons. *J. Virol.* **1999**, *73*, 5186–5190. [PubMed]
28. Hizi, A.; Herzig, E. dUTPase: The frequently overlooked enzyme encoded by many retroviruses. *Retrovirology* **2015**, *12*, 70. [CrossRef] [PubMed]
29. Chong, A.Y.; Kojima, K.K.; Jurka, J.; Ray, D.A.; Smit, A.F.A.; Isberg, S.R.; Gongora, J. Evolution and gene capture in ancient endogenous retroviruses-insights from the crocodilian genomes. *Retrovirology* **2014**, *11*, 71. [CrossRef] [PubMed]
30. Riccioni, C.; Rubini, A.; Belfiori, B.; Passeri, V.; Paolocci, F.; Arcioni, S. Tmt1: The first LTR-retrotransposon from a Tuber spp. *Curr. Genet.* **2008**, *53*, 23–34. [CrossRef] [PubMed]
31. Lamb, R.A.; Paterson, R.G.; Jardetzky, T.S. Paramyxovirus membrane fusion: Lessons from the F and HN atomic structures. *Virology* **2006**, *344*, 30–37. [CrossRef] [PubMed]
32. Laten, H.M.; Majumdar, A.; Gaucher, E.A. SIRE-1, a copia/Ty1-like retroelement from soybean, encodes a retroviral envelope-like protein. *Proc. Natl. Acad. Sci. USA* **1998**, *95*, 6897–6902. [CrossRef] [PubMed]
33. Du, J.; Tian, Z.; Hans, C.S.; Laten, H.M.; Cannon, S.B.; Jackson, S.A.; Shoemaker, R.C.; Ma, J. Evolutionary conservation, diversity and specificity of LTR-retrotransposons in flowering plants: Insights from genome-wide analysis and multi-specific comparison. *Plant J.* **2010**, *63*, 584–598. [CrossRef] [PubMed]
34. Wright, D.A.; Voytas, D.F. Athila4 of Arabidopsis and Calypso of Soybean Define a Lineage of Endogenous Plant Retroviruses. *Genome Res.* **2002**, *12*, 122–131. [CrossRef] [PubMed]
35. Peterson-Burch, B.D.; Wright, D.A.; Laten, H.M.; Voytas, D.F. Retroviruses in plants? *Trends Genet.* **2000**, *16*, 151–152. [CrossRef]
36. Vicient, C.M.; Kalendar, R.; Schulman, A.H. Envelope-class retrovirus-like elements are widespread, transcribed and spliced, and insertionally polymorphic in plants. *Genome Res.* **2001**, *11*, 2041–2049. [CrossRef] [PubMed]
37. Steinbauerová, V.; Neumann, P.; Novák, P.; Macas, J. A widespread occurrence of extra open reading frames in plant Ty3/gypsy retrotransposons. *Genetica* **2011**, *139*, 1543–1555. [CrossRef] [PubMed]
38. Aiewsakun, P.; Katzourakis, A. Marine origin of retroviruses in the early Palaeozoic Era. *Nat. Commun.* **2017**, *8*, 13954. [CrossRef] [PubMed]
39. Zuo, Y.; Deutcher, M.P. Exoribonuclease superfamilies: Structural analysis and phylogenetic distribution. *Nucleic Acids Res.* **2001**, *29*, 1017–1026. [CrossRef] [PubMed]
40. Akoh, C.C.; Lee, G.C.; Liaw, Y.C.; Huang, T.H.; Shaw, J.F. GDSL family of serine esterases/lipases. *Prog. Lipid Res.* **2004**, *43*, 534–552. [CrossRef] [PubMed]
41. Evgen'ev, M.B.; Arkhipova, I.R. Penelope-like elements—A new class of retroelements: Distribution, function and possible evolutionary significance. *Cytogenet. Genome Res.* **2005**, *110*, 510–521. [CrossRef] [PubMed]

42. Anantharaman, V.; Aravind, L. Novel eukaryotic enzymes modifying cell-surface biopolymers. *Biol. Direct* **2010**, *5*, 1. [CrossRef] [PubMed]

43. Kapitonov, V.V.; Jurka, J. The esterase and PHD domains in CR1-like non-LTR retrotransposons. *Mol. Biol. Evol.* **2003**, *20*, 38–46. [CrossRef] [PubMed]

44. Schneider, A.M.; Schmidt, S.; Jonas, S.; Vollmer, B.; Khazina, E.; Weichenrieder, O. Structure and properties of the esterase from non-LTR retrotransposons suggest a role for lipids in retrotransposition. *Nucleic Acids Res.* **2013**, *41*, 10563–10572. [CrossRef] [PubMed]

45. Hestand, M.S.; Houdt, J.V.; Cristofoli, F.; Vermeesch, J.R. Polymerase specific error rates and profiles identified by single molecule sequencing. *Mutat. Res./Fundam. Mol. Mech. Mutagen.* **2016**, *784–785*, 39–45. [CrossRef] [PubMed]

46. Ketting, R.F.; Haverkamp, T.H.; van Luenen, H.G.; Plasterk, R.H. Mut-7 of C. elegans, required for transposon silencing and RNA interference, is a homolog of Werner syndrome helicase and RNaseD. *Cell* **1999**, *99*, 133–141. [CrossRef]

47. Arkhipova, I.; Meselson, M. Deleterious transposable elements and the extinction of asexuals. *Bioessays* **2005**, *27*, 76–85. [CrossRef] [PubMed]

48. Chen, K.-M.; Campbell, E.; Pandey, R.R.; Yang, Z.; McCarthy, A.A.; Pillai, R.S. Metazoan Maelstrom is an RNA-binding protein that has evolved from an ancient nuclease active in protists. *RNA* **2015**, *21*, 833–839. [CrossRef] [PubMed]

49. Yang, Z.; Chen, K.-M.; Pandey, R.R.; Homolka, D.; Reuter, M.; Janeiro, B.K.R.; Sachidanandam, R.; Fauvarque, M.-O.; McCarthy, A.A.; Pillai, R.S. PIWI Slicing and EXD1 Drive Biogenesis of Nuclear piRNAs from Cytosolic Targets of the Mouse piRNA Pathway. *Mol. Cell* **2016**, *61*, 138–152. [CrossRef] [PubMed]

50. Arkhipova, I. Distribution and phylogeny of Penelope-like elements in eukaryotes. *Syst. Biol.* **2006**, *55*, 875–885. [CrossRef] [PubMed]

51. Magiorkinis, G.; Gifford, R.J.; Katzourakis, A.; De Ranter, J.; Belshaw, R. Env-less endogenous retroviruses are genomic superspreaders. *Proc. Natl. Acad. Sci. USA* **2012**, *109*, 7385–7390. [CrossRef] [PubMed]

viruses

MDPI

Article

Structure-Function Model for Kissing Loop Interactions That Initiate Dimerization of Ty1 RNA

Eric R. Gamache [1], Jung H. Doh [1], Justin Ritz [2], Alain Laederach [2], Stanislav Bellaousov [3], David H. Mathews [3] and M. Joan Curcio [1,4,*]

[1] Laboratory of Molecular Genetics, Wadsworth Center, New York State Department of Health, Albany, NY 12201, USA; egamache@iu.edu (E.R.G.); junghdoh@gmail.com (J.H.D.)
[2] Department of Biology, University of North Carolina, Chapel Hill, NC 27599, USA; justin.ritz1@gmail.com (J.R.); alain@unc.edu (A.L.)
[3] Department of Biochemistry and Biophysics and Center for RNA Biology, University of Rochester Medical Center, Rochester, NY 14642, USA; stasuofr@gmail.com (S.B.); David_Mathews@URMC.Rochester.edu (D.H.M.)
[4] Department of Biomedical Sciences, University at Albany-SUNY, Albany, NY 12201, USA
* Correspondence: joan.curcio@health.ny.gov; Tel.: +1-518-473-4213

Academic Editors: David J. Garfinkel and Katarzyna J. Purzycka
Received: 31 January 2017; Accepted: 21 April 2017; Published: 26 April 2017

Abstract: The genomic RNA of the retrotransposon Ty1 is packaged as a dimer into virus-like particles. The 5′ terminus of Ty1 RNA harbors *cis*-acting sequences required for translation initiation, packaging and initiation of reverse transcription (TIPIRT). To identify RNA motifs involved in dimerization and packaging, a structural model of the TIPIRT domain in vitro was developed from single-nucleotide resolution RNA structural data. In general agreement with previous models, the first 326 nucleotides of Ty1 RNA form a pseudoknot with a 7-bp stem (S1), a 1-nucleotide interhelical loop and an 8-bp stem (S2) that delineate two long, structured loops. Nucleotide substitutions that disrupt either pseudoknot stem greatly reduced helper-Ty1-mediated retrotransposition of a mini-Ty1, but only mutations in S2 destabilized mini-Ty1 RNA in *cis* and helper-Ty1 RNA in trans. Nested in different loops of the pseudoknot are two hairpins with complementary 7-nucleotide motifs at their apices. Nucleotide substitutions in either motif also reduced retrotransposition and destabilized mini- and helper-Ty1 RNA. Compensatory mutations that restore base-pairing in the S2 stem or between the hairpins rescued retrotransposition and RNA stability in *cis* and trans. These data inform a model whereby a Ty1 RNA kissing complex with two intermolecular kissing-loop interactions initiates dimerization and packaging.

Keywords: long terminal repeat-retrotransposon; Ty1; *Saccharomyces cerevisiae*; RNA secondary structure; RNA packaging; RNA kissing complex; pseudoknot; kissing loop; SHAPE analysis

1. Introduction

Long terminal repeat (LTR)-retrotransposons and related families of endogenous retroviruses are mobile genetic elements that are widespread in eukaryotic genomes. These elements encode the enzymatic machinery to reverse transcribe RNA and integrate the resulting cDNA into the host genome. They mobilize their own RNA, that of non-autonomous mobile elements, and, more rarely, "hitchhiker" transcripts including coding and non-coding RNAs. The genomic incorporation of cDNA derived from cellular RNAs results in the duplication or replacement of cellular genes and the formation of novel chimeric genes and regulatory non-coding genes, insertional mutations and chromosomal rearrangements [1–4]. In *Saccharomyces cerevisiae*, for example, it has been argued that most protein coding genes have been replaced with cDNA copies lacking introns through the activity of retrotransposons [5–7]. In addition, chimeric cDNAs are incorporated at telomere ends in the absence of telomerase, leading to

13

gross chromosomal rearrangements [8]. Because of the mutagenic and regulatory potential of cDNAs derived from cellular transcripts, the factors that govern the specificity of RNA selection for reverse transcription are of great interest, yet little is known about the principles that govern recognition of RNAs for packaging into virus-like particles (VLPs), the site of reverse transcription. This question is addressed here by investigating the determinants of Ty1 RNA packaging. Ty1 is the most active LTR-retrotransposon family in *S. cerevisiae* [9]. The positive-strand genomic Ty1 RNA initiates in the 5′ LTR and terminates in the 3′ LTR. Ty1 RNA is translated into p49-Gag and p199-Gag-Pol precursor proteins. These proteins assemble into an immature VLP, with p49-Gag binding to Ty1 RNA as a dimer to encapsidate the RNA genome [10]. Inside the VLP, p49-Gag is processed to form p45-Gag, resulting in VLP maturation, which in turn results in stabilization of the Ty1 RNA dimer. The p199-Gag-Pol precursor is processed into p45-Gag, protease (PR), integrase (IN) and reverse transcriptase (RT). Ty1 RNA functions as a template for synthesis of cDNA that is transported to the nucleus and integrated into the genome.

A domain of Ty1 RNA consisting of the 53-nucleotide 5′ UTR and 327 nucleotides of the *GAG* coding region are required in *cis* for translation initiation, packaging and the initiation of reverse transcription (TIPIRT domain; Figure 1) [11]. Mutational analysis has identified several RNA motifs within the TIPIRT domain that play a role in reverse transcription. These regions include the primer-binding site (PBS; nucleotides 95–104), which is complementary to the 3′ end of tRNA$_i$^{Met}. The tRNA$_i$^{Met} is selectively packaged into Ty1 VLPs and serves as the primer for initiation of reverse transcription [12,13]. Three adjacent 6- or 7-nucleotide regions of TIPIRT, known as Box 0 (nucleotides 110–116), Box 1 (nucleotides 144–149) and Box 2.1 (nucleotides 162–168) [14,15], are complementary to sequences within the T or D hairpins of tRNA$_i$^{Met}. Analyses of mutations in both Ty1 RNA and tRNA$_i$^{Met} have established a role for an extended interaction between tRNA$_i$^{Met} and the PBS, Box 0 and Box 1 regions of Ty1 RNA in the initiation of reverse transcription [15,16]. Overlapping Box 2.1 is a 14-nucleotide motif known as CYC5 (nucleotides 155–168), which is perfectly complementary to a sequence in the 3′ UTR known as CYC3. CYC5:CYC3 complementarity promotes efficient reverse transcription in vitro and retrotransposition in vivo [17,18]. In addition, intramolecular pairing of nucleotides 1–7 to nucleotides 264–270 promotes efficient reverse transcription [19,20].

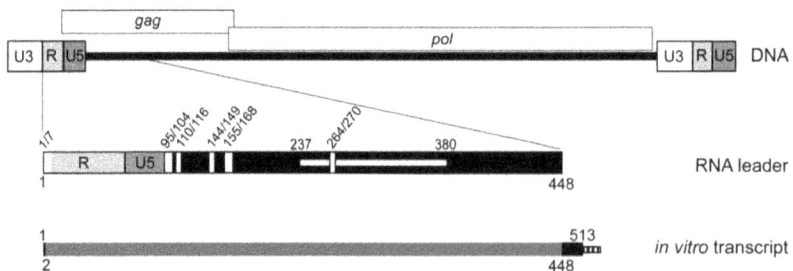

Figure 1. Schematic of the Ty1 element DNA, the Ty1 RNA 5′ TIPIRT domain and in vitro transcripts analyzed by SHAPE chemistry. Ty1 retrotransposon DNA consists of two 334 bp long terminal repeats (LTRs; represented by tripartite rectangles) composed of U3 (unique to the 3′ end of the RNA), R (repeated at the 5′ and 3′ ends of the RNA) and U5 (unique to the 5′ end of Ty1 RNA). LTRs flank a central coding region (black bar). The *GAG* and *POL* ORFs are denoted by rectangles above the element. Below the DNA, the 5′ leader of Ty1 RNA from nucleotide 1 (the beginning of "R") to nucleotide 448 (in the *GAG* ORF), which includes the TIPIRT domain (nucleotides 1–380), is represented below the Ty1 element DNA. Vertical white rectangles denote sequences that are essential for initiation of reverse transcription (1/7 and 264/270 pseudoknot S1 stem; 95/104-PBS; 110/116-Box 0; 144/149-Box 1; 155/168-CYC5, including Box 2.1). The horizontal white rectangle spanning nucleotide 237–380 denotes a region required for Ty1 RNA packaging. The schematic at the bottom represents the in vitro transcript (nucleotides 1–513, plus an FTL tag indicated by the striped box) that was analyzed by SHAPE. Grey shading (nucleotide 2–448), region for which SHAPE reactivities were obtained.

A secondary structure model of the 5′ terminus of Ty1 RNA within VLPs was derived from SHAPE (selective hydroxyl-acylation analyzed by primer extension) data [21]. In this model, nucleotides 1–325 form a long-range pseudoknot in virio. The pseudoknot core consists of two 7 bp stems with a 1-nucleotide interhelical connector, and long structured loops that bridge the stems. The model supports many aspects of earlier structural models that were based on secondary structure prediction and mutational analyses [16,19], including pairing of the tRNA$_i$^{Met} to the PBS, Box 0 and Box 1 regions of TIPIRT and circularization of Ty1 RNA via the CYC5:CYC3 interaction. Moreover, the functionally defined pairing of nucleotides 1–7 to nucleotides 264–270 forms the S1 stem of the pseudoknot. All of the RNA motifs that are known to be required for reverse transcription are in S1 or its multibranched loop (L1), suggesting that this domain may be functionally as well as structurally distinct from S2 and its loop (L3). Using nucleotide substitutions and compensatory mutations, it was shown that the S2 stem is required for retrotransposition, but, an S2 stem-destabilizing mutation, U260C, had no effect on reverse transcription [20].

In contrast with *cis*-acting sequences required for reverse transcription, Ty1 RNA sequences that are necessary for dimerization and packaging within VLPs have not been precisely defined [22]. An internally deleted mini-Ty1 RNA containing the 380-nucleotide TIPIRT domain and 357 nucleotides of the 3′ terminus of Ty1 RNA including the 3′ polypurine tract and 3′ LTR, was shown to be sufficient for retrotransposition when *GAG* and *POL* proteins were expressed in *trans* from a helper-Ty1 element [11]. Deletion of nucleotides 237–380 abolished retrotransposition and co-purification of mini-Ty1 RNA with VLPs, suggesting that *cis*-acting sequences required for Ty1 RNA packaging reside in this domain. This region includes one strand of the S1 stem as well as the S2 stem and its structured loop [21]. However, mutations that destabilize S1 pairing, or the U260C mutation in the S2 stem did not diminish Ty1 RNA packaging [20].

RNA elements required for encapsidation of retroviral RNA within virions, known as ψ (psi) sites, are at least 100 nucleotides long, contain multiple stem-loop structures and are in the 5′ UTR, sometimes extending into *GAG*. RNA elements that facilitate dimerization are located near those that promote RNA encapsidation, and dimerization and packaging are tightly coupled processes, both facilitated by the nucleocapsid activity of Gag [23,24]. Prior to recruitment into assembling virions, dimerization of retroviral genomes is initiated by an intermolecular "kissing loop" interaction between single-stranded loop sequences of stem-loops in the RNA. Subsequently the interaction extends into palindromic sequences in the stems to form stable dimers. Purzycka et al. [21] identified three palindromic sequences (PAL1–PAL3) in the 5′ terminus of Ty1 RNA that were less reactive in virio than ex virio. Based on analogy with retroviral dimerization sites, the authors proposed that PAL sequences are sites where the nucleic acid chaperone activity of Gag could promote a transition from intramolecular pairing to intermolecular pairing [25]. However, potential kissing loop sequences that initiate dimerization of Ty1 RNA have not been identified.

In this work, we present a SHAPE-directed structural model of the 5′ TIPIRT domain of Ty1 RNA in vitro. The model corroborates previously proposed models of the 5′ terminus of Ty1 RNA in virio and in vitro [20,21]. We overlay nucleotide conservation of *Saccharomyces* Ty1 and Ty2 element sequences onto the secondary structure model to identify conserved secondary structures with potential roles in packaging. The biological significance of structural elements was investigated by introducing mutations into mini-Ty1 RNA and measuring helper-Ty1-mediated retrotransposition in vivo. We confirmed the requirement for both stems of the pseudoknot in retrotransposition [20], and found that separating the stems by four nucleotides has no effect on retrotransposition, raising the possibility that the stems play roles in non-overlapping functions. In addition, complementary 7-nucleotide motifs at the apices of two stem-loops, SL1a and SL3a, were shown to be required for efficient retrotransposition. Unlike S1 stem mutations [20], nucleotides substitutions in the S2 stem subject Ty1 RNA to rapid degradation in *cis* and in *trans*. Also, SL1a and SL3a loop substitutions result in slow and fast degradation, respectively, in *cis* and in *trans*. Trans-complementation of the helper-Ty1 RNA instability by compensatory mutations in the mini-Ty1 RNA SL1a and SL3a apical motifs suggests that these motifs form intermolecular duplexes. Based on these data, we propose that intermolecular pairing between the apical motifs of these stem-loops, one of which has been implicated in dimerization [21], and the other of which may be dependent on the S2 stem for its stability, forms a Ty1 RNA kissing complex that initiates dimerization of Ty1 RNA for packaging into VLPs.

2. Materials and Methods

2.1. In Vitro Transcription and RNA Purification

The DNA template for in vitro transcription was generated by PCR with primers PJ502 (5'-CCTAATACGACTCACTATAGGGGAGGAGAACTTCTAGTATATTCTG-3') and PJ745 (5'-ATGAGCTCCCAGATTCGTCAGAATTATCAGTAAATGTATTACCTGACTCAGG-3') and plasmid pGTy1his3AI-[Δ1] [26] as a template. The reaction yielded a DNA fragment with the T7 promoter, 513 bp corresponding to nucleotide 1–513 of Ty1-H3 RNA and 27 bp complementary to the Cy5-FTL primer. The PCR product was gel-purified using a Gel Extraction kit (Qiagen, Germantown, MD, USA). In vitro transcription reactions were performed using ~150 ng of the purified DNA template in a 20 μL MEGAscript T7 transcription kit (Invitrogen, Carlsbad, CA, USA) reaction, which was incubated at 37 °C for 4 h. The RNA was purified using the MEGAclear kit (Invitrogen). RNA was stored at −80 °C.

2.2. Selective 2'-Hydroxyl Acylation Analyzed by Primer Extension

A 10 picomole sample of the in vitro transcribed RNA was brought to a total volume of 12 μL by addition of 0.5× TE. The RNA was heated at 95 °C for 2 min and cooled on ice for 2 min. Following addition of 6 μL 3.3× RNA Folding Buffer (1×: 100 mM HEPES [pH 8.0], 20 mM $MgCl_2$, 100 mM KCL), the reaction was split into two samples of equal volume. RNA was renatured by incubation at 37 °C for 20 min. To one sample, 1 μL N-methylisotoic anhydride (NMIA) in DMSO was added, and to the other, 1 μL DMSO was added (control). Reactions were incubated at 37 °C for 45 min. The NMIA-modified RNA and control RNA samples were ethanol-precipitated by adding 90 μL H_2O, NaCl to 44 mM, glycogen to 44 μg/μL and EDTA to 44 μM. After adding 3.5× volumes of ethanol, the RNA was precipitated at −80 °C for 30 min. The RNA was pelleted at 4 °C, and washed with 70% ethanol. Pellet was dried in a Savant SpeedVac concentrator and resuspended in 10 μL of 0.5× TE buffer (1×: 10 mM Tris-HCl [pH 8], 1 mM EDTA). The Cy5-FTL primer (5'-ATAATTCTGACGAATCTGGGAGCTCAT-3') was annealed to the 3' end of each RNA at 65 °C for 15 min, and then 35 °C for 15 min. RNA was reverse transcribed by first adding Superscript III First-Strand Synthesis buffer (Invitrogen), 5 mM DTT, 40U of RNAseOUT (Invitrogen), and 500 μM dNTPs to the reaction, followed by incubation at 52 °C for 1 min. Superscript III Reverse Transcriptase (200 units; Invitrogen) was added, and the reaction was incubated at 52 °C for 15 min. RNA was hydrolyzed by addition of NaOH to 180 mM, followed by heating to 95 °C for 5 min. Reactions were neutralized by addition of an amount of HCl equivalent to the NaOH added. The remaining nucleic acid was precipitated using sodium acetate at a final concentration of 75 mM, $MgCl_2$ at 25 mM, and 3.3× volumes of ethanol followed by cold centrifugation. The resulting pellet was washed with 70% ethanol and dried in a Savant SpeedVac concentrator. The pellet was resuspended in 40 μL Sample Loading Solution (Beckman Coulter, Indianapolis, IN, USA), and 1.1 μL of the 600-bp Beckman Coulter sequencing ladder was added. Sequencing ladder reactions were performed in the same way as the control reaction above, with addition of 2 μL of one 5 mM dideoxyNTP (ddNTP). Two different ddNTP reactions were run for each sample, using a different ddNTP for each. Primer extension products were resolved by capillary electrophoresis using a Beckman Coulter CEQ8000 Genetic Analysis System.

Experimental datasets from three technical replicates were individually corrected for signal variation, and peak intensities were integrated using MatLab (MathWorks, Inc., Natick, MA, USA) and ShapeFinder [27]. Reactivities were normalized by dividing the peak intensities by the average of the 10% most reactive peaks excluding outliers, which were determined by boxplot analysis as those peaks showing reactivity greater than 1.5× the interquartile range. The standard deviation

(SD) of the normalized reactivities of each nucleotide position was calculated across datasets, and reactivities with SD > 0.7 were also excluded. Normalized reactivities of overlapping nucleotides from the three RNA species were averaged to obtain a composite dataset spanning nucleotides 1 to 615 of the Ty1 RNA. Composite reactivity data was used to determine a pseudo-free energy change restraint added to the nearest-neighbor thermodynamic parameters [28,29]. Structure prediction was performed using ShapeKnots [30]. Collapsed diagrams were generated using XRNA (http://RNA. ucsc.edu/RNAcenter/xRNA/xRNA.html). Diagrams were edited using Adobe Illustrator. The raw SHAPE data is available in SNRNASM format as supplemental data in this manuscript (Table S1) [31].

2.3. Conservation of Sequences in the Ty1 RNA 5' Terminus

Clustal X [32] was used to align sequences corresponding to nucleotides 1 to 615 of the transcript of 31 genomic Ty1 elements and 15 Ty2 elements in *S. cerevisiae* strain S288C (www.yeastgenome.org), 35 Ty1 and 17 Ty2 sequences from other strains of *S. cerevisiae* (www.ncbi.nlm.nih.gov/genome? term=txid4932[orgn]), and four Ty1-like elements from other *Saccharomyces* species, including one element from *Saccharomyces weihenstephen* (accession number gb | ABPO01001678.1 |); one element from *Saccharomyces mikatae* (gb | AACH01000084.1 |); one element from *Saccharomyces paradoxus* (gb | AABY01000078.1 |); and one element from *Saccharomyces kluyveri* (gb | AF492702.1 |). Each nucleotide position was assigned to one of three categories based on whether the nucleotide was conserved in all 102 Ty elements and if not, whether the nucleotide was conserved in the 66 Ty1 elements of *S. cerevisiae*.

2.4. Plasmids

The helper-Ty1 plasmid, pEIB, was a kind gift of Leslie Derr and Jeffrey Strathern. It is a 2 μ-based, *TRP1*-marked plasmid harboring the *GAL1* promoter fused to nucleotides 241-5561 of Ty1-H3 DNA [33]. This region of Ty1-H3 includes the R and U5 regions of the 5' LTR and *GAG* and *POL* ORFs, but lacks the plus-strand polypurine tract (PPT1) and the 3' LTR. The Ty1-H3 sequence in pEIB also harbors mutations (T335C, T338A, A339T, G340C, C341A, C344A, T347C) that disrupt annealing of the tRNA$_i^{Met}$ primer but preserve the amino-acid sequence of *GAG*.

The mini-Ty1*his3AI* plasmid, pJC994, is a 2 μ-based, *URA3*-marked plasmid that was constructed by deleting the HpaI-SnaBI fragment of pGTy1*his3AI-[Δ1]* (nucleotides 818–5463 of Ty1-H3 DNA) [26]. Mutations were introduced into plasmid pJC994 using the QuikChange Lightning Site-Directed Mutagenesis kit (Agilent Technologies, Santa Clara, CA, USA) and the standard protocol. Plasmid DNA was purified and Ty1 sequences were confirmed by DNA sequencing. The sequence of primers used for site-directed mutagenesis is available upon request.

Plasmid pGAL1:GAG$_{NT}$:GFP is a CEN-based *LEU2*-marked plasmid consisting of vector pRS415 carrying an ApaI-EagI fragment containing the *GAL1* promoter, the 575 bp XhoI-HpaI fragment of Ty1-H3 (nucleotides 241–815), a 7-nucleotide linker including a *BamHI* site, and the *GFP(S65T)* ORF and *ADH1* terminator from plasmid pFA6-GFP(S65T)-HIS3MX [34]. Mutations in Ty1-H3 sequence were introduced into pGAL1:GAG$_{NT}$:GFP by PCR-amplification of Ty1 sequences from derivatives of pJC994 containing various mutations in mini-Ty1, digestion with *XhoI-BamH*1, and substitution of the resulting fragment for the XhoI-BamHI fragment of pGAL1:GAG$_{NT}$:GFP.

2.5. Quantitative Transposition Frequency Assay

Plasmids pEIB and pJC994 or its mutagenized derivatives were co-transformed into strain JC5839 (MATa *his3Δ1 ura3Δ0 leu2Δ0 met15Δ0 trp1::hisG spt3Δ::kanMX*), a derivative of strain BY4741 [35]. Single colony isolates of each strain grown in SC-Ura-Trp 2% glucose broth at 30 °C for 2 days were pelleted and resuspended in 5 volumes of SC-Ura-Trp 2% galactose, 2% raffinose broth. Each culture was divided into seven 1-mL cultures, which were grown at 20 °C. After 48 h, 1 mL of YEPD broth was added to each culture, which was incubated at 20 °C for 18 h. A 1 μL aliquot of each culture was removed, and dilutions were plated onto YEPD agar to determine the total number of colony forming units. Aliquots of the remaining culture were plated onto SC-His 2% glucose agar. The retrotransposition frequency for each culture is the number of His$^+$ prototrophs divided by the total number of colony forming units in the same volume of culture. The median retrotransposition frequency among the seven biological replicates was determined, and the 95% confidence interval was calculated.

2.6. Northern Analysis

Single colony isolates of each strain were grown in SC-Ura-Trp 2% glucose broth at 30 °C. Cells were pelleted, washed in water, and resuspended in SC-Ura-Trp 2% galactose 2% raffinose broth at an OD_{600} of 0.05 and grown for 20–24 h at 20 °C to an OD_{600} of 0.4. Cells were pelleted and washed in water, and cell pellets were frozen at −80 °C. RNA was extracted from cell pellets thawed on ice for 30 min using the MasterPure (Epicentre, Madison, WI, USA) kit. A 10 µg sample of total RNA and an equal volume of Ambion NorthernMax glyoxal loading buffer (Thermo Fisher Scientific, MA, USA) was incubated for 30 min at 50 °C. Samples were fractionated on a 1% agarose gel in 10 mM $NaPO_4$, pH 6.5 at 100 V for 2.5 h. RNA was transferred to a Hybond-XL (GE Healthcare, Troy, NY, USA) membrane using alkaline transfer conditions for 3 h, and then crosslinked to the membrane using a Spectrolinker (Spectronics Corporation, Westbury, NY, USA) set to "optimal". In vitro transcribed RNA probes were synthesized using SP6 or T7 polymerase in conjunction with ^{32}P-rCTP. Antisense Ty1 RNA (nucleotides 815–2173) transcribed from plasmid pGEM-TyA1 [36] was used to specifically detect helper-Ty1 RNA, sense-strand *HIS3* transcript from plasmid pGEM-HIS3 [36] detected mini-Ty1*his3AI* RNA, and an antisense 18S rRNA transcript from plasmid pBDG512 [37] detected 18S rRNA. Probes were incubated sequentially in NorthernMax hybridization buffer (Ambion). After washing, blots were exposed to phosphor screens and scanned using a Typhoon phosphorimager (GE Healthcare). Images were quantitated using ImageQuant software (GE Healthcare) by normalizing to the 18S rRNA signal. Blots were stripped in boiling 0.1% SDS, rinsed and stored in 5× SSC before reprobing.

2.7. GFP Activity

Plasmid pGAL1:GAG$_{NT}$:GFP, derivatives bearing nucleotide substitutions and vector pRS415 were transformed into the *spt3Δ::kanMX* derivative of strain BY4741 [35]. Two transformants of each plasmid were grown in SC-Leu 2% glucose overnight at 30 °C. Cells were spun down, and pellets resuspended in an equal volume of SC-Leu 2% raffinose 2% sucrose. A 1:20 dilution in SC-Leu 2% raffinose 2% sucrose broth was grown overnight at 20 °C. Cultures were diluted to an OD_{600} of 0.2 and grown for 3 h at 20 °C. Galactose (2% final) was added and cultures were incubated for 2.5 h at 20 °C. A 1 mL aliquot of each culture was spun down at $1000 \times g$ for 10 min. The medium was aspirated and cell pellets were resuspended in 500 µL sterile water. The geometric mean of the GFP activity in 10,000 cells was quantified by flow cytometry using a FACSCalibur (Becton, Dickinson and Company, Franklin Lakes, NJ, USA), and the average of the geometric mean of GFP activity in the two biological replicates of each strain was determined.

3. Results

3.1. Secondary Structure Model of Ty1 RNA TIPIRT Domain

The goal of this study was to identify RNA secondary structures and motifs within the Ty1 TIPIRT domain that are involved in RNA dimerization and packaging into VLPs. To begin, a secondary structure model of the Ty1 RNA leader sequences was developed using average SHAPE reactivities and the ShapeKnots algorithm [30]. SHAPE analysis involves treating a folded RNA with an electrophilic agent that forms 2'-O-ester adducts with reactive nucleotides in RNA. The SHAPE reactivity of each nucleotide is inversely correlated to the contribution of that nucleotide to base-pairing or tertiary interactions. Adduct formation on each nucleotide is measured as the degree of impediment to primer extension by reverse transcriptase. The ShapeKnots algorithm [30] combines a pseudoknot discovery algorithm with one that reconciles experimental SHAPE reactivities with traditional free energy rules to obtain a structure that is maximally compatible with the experimental data.

An in vitro transcript corresponding to nucleotides 1–513 of Ty1-H3 RNA, which encompasses the Ty1 TIPIRT domain, plus a 27-nucleotide tag was subject to SHAPE analysis (Figure 1). The transcript was folded in 100 mM KCl and 6.7 mM $MgCl_2$ and then treated with N-methylisotoic anhydride (NMIA), which forms $2'$-O-ester adducts with reactive nucleotides. The reaction was performed under conditions that promote the formation of a single adduct per RNA molecule. The reactivity of individual nucleotides was determined by reverse transcriptase-mediated primer extension analysis of the transcript that was treated with NMIA or, as a control, untreated. Extension reactions were performed using a fluorescently labeled primer hybridized to the 27-nucleotide tag at the $3'$ end of the transcript. The products of primer extension reactions were resolved by capillary electrophoresis. Nucleotides modified by $2'$-O-adducts were detected as stops to primer extension, resulting in a peak. The reactivity of each nucleotide was determined by integrating individual peaks from NMIA-treated samples. Three independent repetitions were performed and the average SHAPE reactivity at each nucleotide was determined. The average SHAPE reactivities were used to restrain computational predictions of secondary structure models by the ShapeKnots algorithm.

A model of the secondary structure of the Ty1 RNA TIPIRT domain annotated by the average SHAPE reactivity of each nucleotide position is shown in Figure 2. A prominent feature of the model is a pseudoknot formed by long-range interactions of sequences spanning the first 326 nucleotides of Ty1 RNA, which is within the functionally defined 380-nucleotide TIPIRT domain. This pseudoknot is similar to those predicted previously in the $5'$ terminus of in vitro transcribed Ty1 RNA and in Ty1 RNA isolated from VLPs, although earlier modeling did not make use of a pseudoknot discovery algorithm [20,21]. The pseudoknot core consists of the 7-bp S1 pairing (Figure 2, blue shading) and the 8-bp S2 pairing (Figure 2, green shading) connected by a 1-nucleotide interhelical loop (L2) (Figure 2, yellow shading). The S1 stem of the pseudoknot, formed by pairing of the seven 5'-terminal nucleotides of Ty1 RNA to nucleotides 264–270, has an established function during retrotransposition [19,20]. Nucleotides 255–262 interact with nucleotides 319–326 of Ty1 RNA to form the S2 pairing of the pseudoknot (Figure 2, green shading). The S2 stem contains an additional base-pair (C255–G326) that was not predicted in earlier models [20,21].

All but one of the nucleotides within the pseudoknot core had low reactivity with NMIA, including the unpaired L2 nucleotide, suggesting that the pseudoknot is a thermodynamically stable tertiary interaction within the Ty1 RNA. This conclusion is supported by the fact that other RNA structure prediction algorithms that do not employ SHAPE data, such as pknotsRG and IPknot [38,39], also predict a pseudoknot with identical S1 and S2 stems and L2 nucleotide in the $5'$ leader of Ty1 RNA (Figures S1 and S2).

The multibranched L1 loop (8/254) of the pseudoknot, formed by stem S1, contains three nested stem-loops (SL1a-SL1c). The first stem-loop (13/32; SL1a) has a single bulged nucleotide and short loop (Figure 2, pink shading). PAL1 and PAL2 sequences, which were proposed to interact intermolecularly in the dimeric RNA of VLPs [25], are contained in the SL1a hairpin. The second stem-loop (39/204; SL1b) is an extended domain containing two nested stem-loops. SL1b contains the sequences that pair with tRNA$_i$^{Met} and with $3'$ terminal sequences of Ty1 RNA (Figure 2, black outlines) in the model of Ty1 gRNA in virio [21]. The third (206/248; SL1c) is a stem-loop with a bulge loop and an internal loop. L1 sequences include the entire $5'$ UTR (1/53) of Ty1 RNA and the AUG codon of *GAG* (Figure 2, highlighted in grey).

The L3 loop of the pseudoknot (271/318) is formed by the S2 pairing and composed almost entirely of the low reactivity SL3a stem-loop (272/318) (Figure 2, purple shading), which has two small internal loops. S2 and L3 are within a region of Ty1 RNA that is necessary for packaging into VLPs (238/380) [11]. Beyond the pseudoknot, the $3'$ terminal region of the Ty1 in vitro transcript harbors three stem-loops, SL4, SL5 and SL6. SL6 contains PAL3 (423/428), a putative site of Ty1 RNA dimerization in VLPs [21].

19

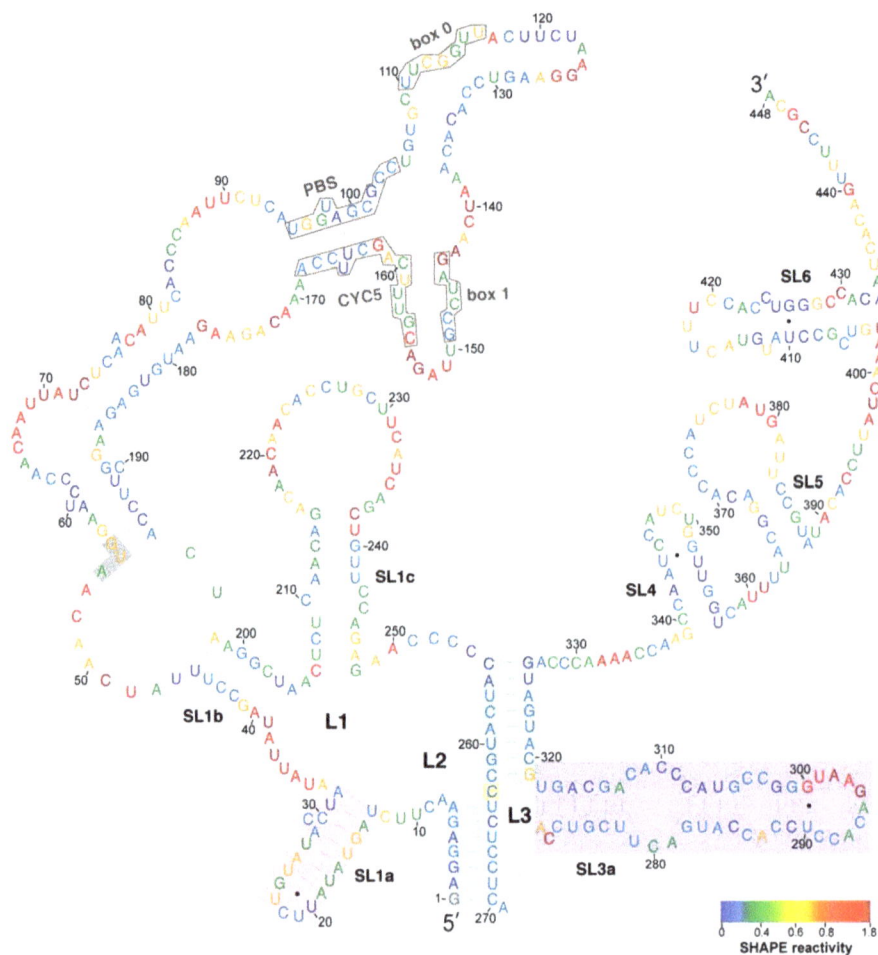

Figure 2. SHAPE reactivities and secondary structure model of the 5′ leader of Ty1 RNA. Nucleotides are colored according to their SHAPE reactivities, which are indicated on the color bar at the bottom left. Regions of low reactivity have a high probability of being constrained within secondary or tertiary structure. The AUG nucleotides shaded in grey comprise the start codon of *GAG*. The pseudoknot core contains stem S1 (blue shading), loop L2 (nucleotide 263, yellow shading), and stem S2 (green shading). Pseudoknot loops L1 (nucleotides 9–254) and L3 (nucleotides 271–318) are not shaded. The SL1a hairpin (pink shading) and SL3a hairpin (purple shading) are indicated.

Regions of the structural model that differ from previous SHAPE analysis-derived structural models of Ty1 RNA in virio [21] and the 5′ terminus of Ty1 RNA in vitro [20] include: (a) the presence of the 255C-326G base-pair in the S2 pseudoknot stem, as noted above; (b) extension of the SL1a stem by two base-pairs by inclusion of a 1-nucleotide bulge in our model; (c) the presence of a large loop at the apex of stem-loop SL1c in our model, compared to a bulge-stem-loop structure at the apex of SL1c in previous models; (d) extension of the SL3a stem-loop by two base-pairs by inclusion of a 1-nucleotide bulge in our model; and (e) the presence of SL4, which is not present in previous models. As expected, no evidence of interactions seen in virio between motifs in SL1b and tRNA$_i$^{Met} or between CYC5 and CYC3 was observed because neither tRNA$_i$^{Met} nor CYC3 are present in our system.

The location of hairpin SL3a within an essential packaging domain prompted us to look for features that could function in the formation of a Ty1 RNA kissing complex. We noticed that the ACAGAAU (293/299) sequence in the SL3a loop is perfectly complementary to an AUUCUGU (19/25) motif in the loop and two apical base-pairs of the SL1a stem (G-U and U-A). The tertiary structure of the pseudoknot might allow these complementary motifs to pair intramolecularly. However, 4 of the 7 nucleotides (296/299) in the SL3a loop are highly reactive in SHAPE analysis of RNA in vitro (Figure 2) [20]; therefore, it is unlikely that the SL1a and SL3a motifs are base-paired in vitro. The loop of SL3a is also highly reactive in virio [21], suggesting that the SL1a and SL3a motifs are also not base-paired in VLPs. Another intriguing possibility is that the complementary apical motifs of SL1a and SL3a base-pair intermolecularly to form a symmetrical Ty1 RNA kissing complex with two kissing loops (Figure 3). In vitro, where the TIPIRT domain RNA is monomeric in the absence of Gag [17,40], and in VLPs, where the Ty1 RNA is a mature dimer [10,21], the motif in SL3a is mostly reactive, arguing against base-pairing of the complementary SL1a-SL3a motifs in these RNA forms. Nonetheless, pairing between the SL1a and SL3a apical motifs on different Ty1 RNA molecules could form a transient symmetrical kissing complex that initiates packaging of Ty1 RNA into VLPs, and then is converted to a stable dimer linkage within the mature VLP.

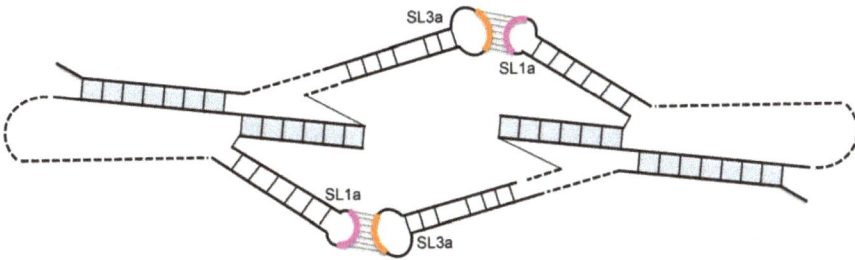

Figure 3. Model of a symmetrical Ty1 RNA kissing complex containing two Ty1 RNA pseudoknots interacting via two 7-base-pair intermolecular RNA duplexes formed between apical motifs in stem-loop SL1a (pink arc) and SL3a (orange arc). The pseudoknot stems are shaded in blue.

3.2. Conservation of Ty1 RNA TIPIRT Domain

We compared the conservation of nucleotides within the Ty1 RNA 5′ terminus to the secondary structure model to ascertain whether there are conserved structural features that could function in *cis* in retrotransposition. Because most *S. cerevisiae* Ty1 elements are mobile or recently mobile [41], and therefore have a high degree of sequence identity [42,43], we also compared their sequences to that of Ty2 elements, a closely related family of LTR-retrotransposons in *S. cerevisiae*. The 5′ terminal sequence of 66 Ty1 elements and 32 Ty2 elements from a variety of laboratory, industrial and natural *S. cerevisiae* strain genomes [44], as well as four Ty1 elements from other *Saccharomyces* species were aligned. Each nucleotide position was assigned to one of three categories based on the degree of conservation at that position: (1) conserved in all 102 *Saccharomyces* Ty1 and Ty2 elements (Figure 4, red coloring); (2) conserved in all 66 *S. cerevisiae* Ty1 elements (Figure 4, purple coloring); or (3) variable among the *S. cerevisiae* Ty1 elements analyzed (Figure 4, grey coloring).

The alignment indicates that nucleotides in the pseudoknot core are very highly conserved. S1 nucleotides are invariant in all *Saccharomyces* Ty1 and Ty2 elements. S2 nucleotides, including C255 and G326, whose pairing is predicted uniquely in the structural model presented here, are invariant, with the exception of three nucleotides at the base of S2. Two of these nucleotides (C262 and C320) are substituted in a few Ty2 elements, while the third nucleotide, G319, is a U nucleotide in four of the 66 emphS. cerevisiae Ty1 elements, but is otherwise conserved. Similarly, the L2 nucleotide C263 is substituted by an A nucleotide in three *S. cerevisiae* Ty1 elements. Thus, every residue of the pseudoknot core is invariant or has limited variation, in agreement with the conclusion of Huang et al. [20].

The entire 326-nucleotide pseudoknot domain has a high degree of conservation overall. Sequences that are very highly conserved among *S. cerevisiae* Ty1 elements include those that bind tRNA$_i^{Met}$ (PBS, Box 0 and Box 1; Figure 4, black outlines) and those within sequence regions that are predicted to be base-paired, including the SL1a stem, regions of the SL1b stem such as the pairing between nucleotides 39–45 and 198–204 and the SL1c stem. While most regions that are predicted to be single stranded have low nucleotide conservation, nucleotides 8–12, nucleotides 34–38, nucleotides 63–69, and the SL3a loop are conserved. The SL1a loop is conserved in *S. cerevisiae* Ty1 elements but not in Ty2 elements. Within the 53-nucleotide 5′ UTR, 34 nucleotides (64%) are invariant amongst all 102 *Saccharomyces* Ty1 and Ty2 elements analyzed, while 44 nucleotides (83%) are conserved among 66 *S. cerevisiae* Ty1 elements.

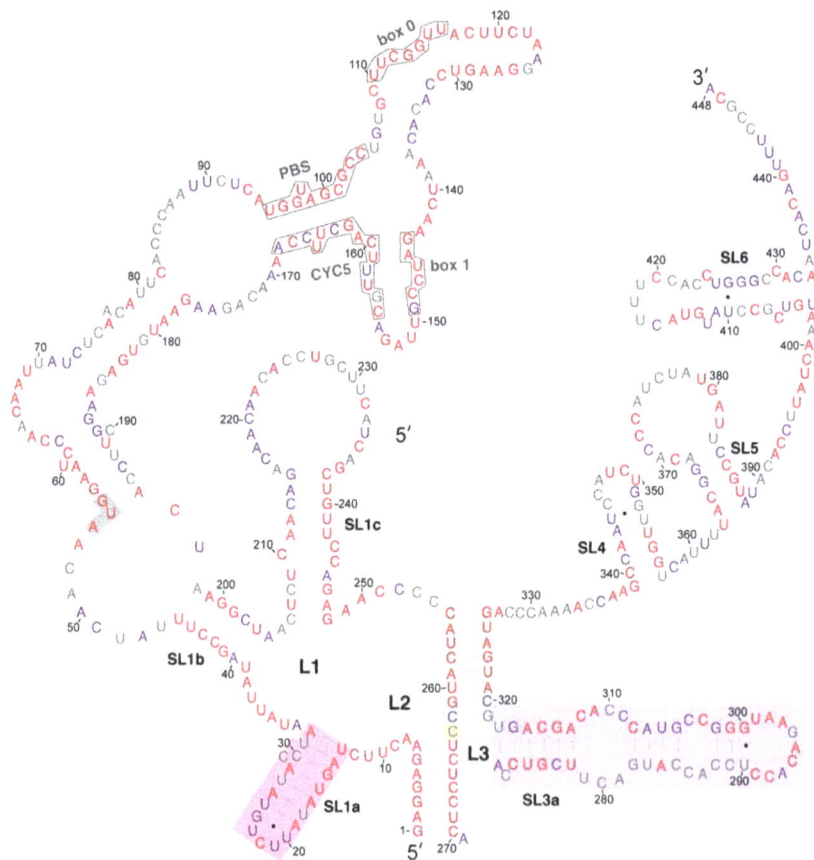

Figure 4. Relative evolutionary conservation of each nucleotide overlayed on the secondary structure model of the 5′ leader of Ty1 RNA. The color of each RNA base indicates its degree among conservation among 102 Ty1 and Ty2 elements from the genus *Saccharomyces*. Categories of conservation are as follows: red, 100% conserved among 102 Ty1 and Ty2 elements in the genus *Saccharomyces*; purple, 100% conserved in 66 *Saccharomyces cerevisiae* Ty1 elements; grey, not 100% conserved in either set.

3.3. Requirement for Pseudoknot Stems S1 and S2 in Retrotransposition

To identify the role of Ty1 RNA secondary structures in retrotransposition, we used an established helper-Ty1/mini-Ty1 assay in which two defective but complementing Ty1 elements are co-expressed, each from a plasmid-based *GAL1* promoter (Figure 5) [11]. The helper-Ty1 element encodes functional

Gag and Gag-Pol proteins, and its RNA is packaged in VLPs but cannot be used in reverse transcription because it harbors silent substitutions in the PBS and lacks the 3′ polypurine tract and LTR [11]. The mini-Ty1his3AI element has an internal deletion of most of the GAG ORF and the entire POL ORF; nonetheless, 5′ leader sequences corresponding to nucleotides 1–575 of Ty1 RNA as well as the last 357 nucleotides of Ty1, including the 3′ polypurine tract and LTR, are retained. Together, these regions are sufficient for mini-Ty1 RNA to be used as a template for retrotransposition when Ty1 proteins are supplied in trans. Mini-Ty1his3AI also carries the his3AI retrotransposition indicator gene, which allows cells harboring transposed reverse transcripts to be detected as His+ prototrophs [45]. The plasmids were expressed in an spt3Δ strain, which lacks expression of endogenous Ty1 RNA. The median retrotransposition frequency in the strain co-expressing the mini-Ty1his3AI with wild-type sequences and the helper-Ty1 was 1.82×10^{-6}. The frequency of His+ prototrophs in the absence of helper-Ty1 was 1.8% of that in its presence. This background of His+ prototrophs may be due to a low frequency of recombination events that introduces full-length genomic Ty1 sequences into the mini-Ty1his3AI plasmid.

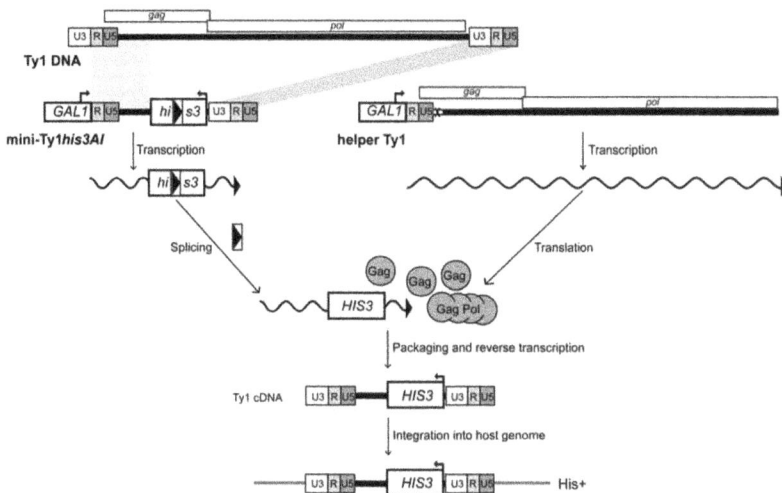

Figure 5. Assay for helper-mediated retrotransposition of mini-Ty1his3AI. A complete Ty1 element is shown at the top for reference. The mini-Ty1his3AI element and helper-Ty1 element are each expressed from the GAL1 promoter (labeled rectangle), which is fused to the transcription start site of Ty1-H3 at the first nucleotide of the R domain in the 5′ LTR. GAL1:mini-Ty1his3AI is carried on a URA3-based plasmid and GAL1:helper-Ty1 is contained on a TRP1-based plasmid (not illustrated). The elements are co-expressed in an spt3Δ strain lacking endogenous Ty1 element transcription. The internally deleted mini-Ty1his3AI element contains 5′ sequences corresponding to nucleotides 1–575 of Ty1 RNA, as well as the last 357 nucleotides of Ty1, including the 3′ polypurine tract (not illustrated) and 3′ LTR. The his3AI retrotransposition indicator gene, consisting of the HIS3 marker gene interrupted by an antisense intron (boxed arrowhead), is inserted in the mini-Ty1 between the 5′ leader and 3′ LTR. The direction of mini-Ty1his3AI transcription from the GAL1 promoter (denoted by the arrow atop the GAL1 rectangle) is opposite to the direction of his3AI transcription (denoted by an arrow atop the HIS3 rectangle), so the intron is only be spliced from the Ty1his3AI transcript. The helper-Ty1 element carries functional GAG and POL ORFs, but the polypurine tract and 3′ LTR are deleted. In addition, silent nucleotide substitutions in the PBS (denoted by a white rectangle marked with an "X") block the binding of tRNA$_i$Met. Splicing is illustrated by removal of the boxed arrowhead representing the intron from the rectangle that denotes the HIS3 gene. Gag and Gag-Pol proteins translated from the helper-Ty1 RNA form VLPs that package the spliced mini-Ty1HIS3 RNA, which is reverse transcribed to form Ty1HIS3 cDNA. Integration of the cDNA into the host genome allows the cell to be detected as a His+ prototroph.

Mutations were introduced into structural elements of the TfP1RT domain of the mini-Ty1*his3AI* plasmid. All mutations and compensatory mutations introduced into *GAG* maintained an open reading frame but not necessarily the amino acid sequence of the truncated Gag product. An UC264AG substitution that disrupts S1 complementarity in mini-Ty1*his3AI* RNA reduced helper-Ty1 mediated retrotransposition to 4% of that of the mini-Ty1*his3AI* with wild-type sequence (Figure 6, M1). A compensatory mutation that reestablishes S1 complementarity restored retrotransposition to levels equivalent to the wild-type mini-Ty1*his3AI* (Figure 6, CM1). Similar results were obtained with the identical substitutions in a previous study [20]; therefore, these findings validate the helper-Ty1/mini-Ty1 assay and confirm the role of the S1 pairing in retrotransposition [11,19].

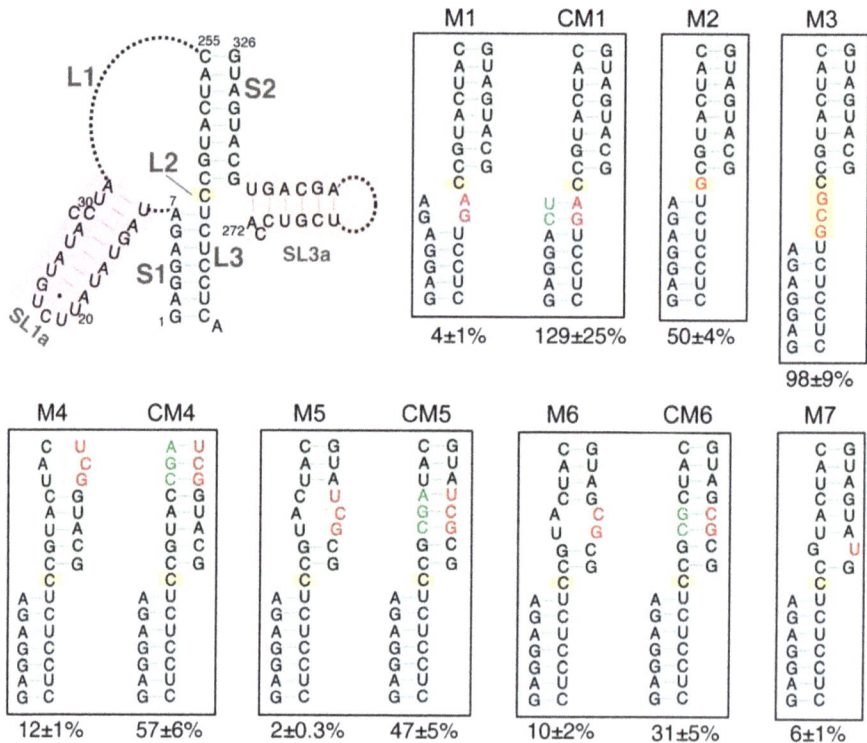

Figure 6. Retrotransposition of mini-Ty1*his3AI* elements with mutations in the Ty1 pseudoknot core. The schematic (top left) shows the secondary structure of the Ty1 pseudoknot core and portions of the L1 and L3 loops. Blue shading, stem S1; yellow shading, loop L2; green shading, stem S2; pink shading, SL1a hairpin, a segment of the L1 loop; orange shading, SL3a hairpin, a portion of the L3 loop. Dotted lines represent bases in loops L1 and L3 that are not shown. Labeled, boxed schematics show the nucleotide substitutions or additions in each mutant mini-Ty1*his3AI* element analyzed. Black letters represent wild-type nucleotides; red letters represent nucleotide substitutions or additions; and green letters represent compensatory substitutions that restore base-pairing with nucleotide substitutions. The percentage below each box is the median frequency of helper-mediated retrotransposition of the mini-Ty1*his3AI* bearing the indicated mutation divided by the median helper-mediated retrotransposition frequency of the mini-Ty1*his3AI* element with wild-type Ty1-H3 sequence, +/− the 95% confidence interval.

We analyzed the requirement for pseudoknot stem S2 by introducing double and triple nucleotide substitutions that disrupt S2 complementarity. These mutations reduced retrotransposition to 2–12% of that of the wild-type mini-Ty1*his3AI* (Figure 6, M4, M5 and M6). Even the single C320U substitution, which is predicted to change a G-C base-pair to a G-U base-pair, reduced retrotransposition to 6% of wild-type activity (Figure 6, M7). Reestablishing S2 complementarity in the mutants harboring double and triple nucleotide substitutions by introduction of compensatory mutations restored retrotransposition up to 31–57% of the wild-type mini-Ty1*his3AI* (Figure 6, CM4, CM5 and CM6). Compensatory mutations may not fully reconstitute the activity of the wild-type mini-Ty1*his3AI* because the base composition of S2 or ensemble folding of mini-Ty1*his3AI* RNA is altered. Together, these data suggest that the S2 stem of the pseudoknot is as critical for retrotransposition as the S1 stem.

Many pseudoknots have 0 to 1-nucleotide interhelical loops that promote a stable pseudoknot conformation in which individual stems stack coaxially [46]. It has been suggested that S1 and S2 of the TIPIRT domain pseudoknot stack coaxially [20,21], even though the unreactive L2 nucleotide can be substituted without major effects on pseudoknot structure or function [20]. To determine the consequences of disrupting the potential for coaxial stacking of the pseudoknot stems, we increased the length of L2 from one to four nucleotides by addition of a GCG triplet (Figure 6, M3). This mutation had no effect on retrotransposition of mini-Ty1*his3AI*. We also confirmed that the C236G substitution of the L2 nucleotide reduced retrotransposition only modestly (50%) (Figure 6, M2). In summary, our data demonstrate that neither the length nor composition of L2 is a major determinant of pseudoknot conformation; therefore, coaxial stacking of S1 and S2 is not likely to be necessary for pseudoknot function.

3.4. Requirement for Complementary Motifs in SL1a and SL3a Hairpins in Retrotransposition

The SL3a hairpin (272/318) is in a region of the TIPIRT domain that contains essential Ty1 RNA packaging sequences [11]. The ACAGAAU (293/299) motif in the loop of SL3a is complementary to the AUUCUGU motif (19/25) encompassing the 3-nucleotide loop and first two base-pairs of the SL1a stem (Figure 7). Except for 1 nucleotide (G296) in SL3a, both sequences are invariant in *S. cerevisiae* Ty1 elements. Therefore, we hypothesized that intermolecular "kissing loop" interactions between the complementary sequences in SL1a and SL3a (Figure 3) could initiate dimerization of Ty1 RNA. To determine whether these complementary motifs are individually required for retrotransposition, we substituted U̲C̲UCUAA for ACAGAAU (293/299) in the SL3a loop, which reduced helper-Ty1-mediated mini-Ty1*his3AI* retrotransposition to 7% of wild-type activity (Figure 7, M13). Substitution of UU̲A̲GAGA for AUUCUGU (19/25) in SL1a reduced retrotransposition to 8% (Figure 7, M9). Both the AUUCUGU19UUAGAGA mutant and wild-type RNA have an A-U and G-U base-pair at the apex of the SL1a stem; thus, the retrotransposition defect of the AUUCUGU19UUAGAGA mutant is probably not due to disruption of the SL1a stem. Instead our findings indicate that complementary motifs in SL1a and SL3a are required in *cis* in Ty1 retrotransposition.

To determine whether reestablishing complementarity between apical sequences of the SL1a and SL3a hairpins restores retrotransposition, both AUUCUGU19UUAGAGA and ACAGAAU293UCUCUAA were introduced into a single mini-Ty1*his3AI* element. This double mutant transposed at 15% of the frequency of the wild-type mini-Ty1*his3AI* and about 2-fold more often than either single mutant (Figure 7, CM9/13). Partial restoration of retrotransposition rather than an additive decrease in retrotransposition in the SL1a-SL3a double mutant suggests that base-pairing between complementary apical sequences of SL1a and SL3a promotes retrotransposition. Restoration of retrotransposition is not as strong as that seen with other compensatory mutations in stem S1 or S2 of the pseudoknot, but such a difference is expected if the SL1a-SL3a interaction is intermolecular, as opposed to the intramolecular interactions that form stem S1 and S2. This is because a mini-Ty1 RNA bearing both SL1a and SL3a mutations would only be able to form a kissing complex with another mutant mini-Ty1 RNA and not with the wild-type helper-Ty1 RNA, and therefore the pool of kissing complexes that could be packaged into VLPs would be reduced. However, these data alone cannot differentiate between an intramolecular or intermolecular interaction between of the SL3a loop and complementary sequences in the SL1a stem-loop.

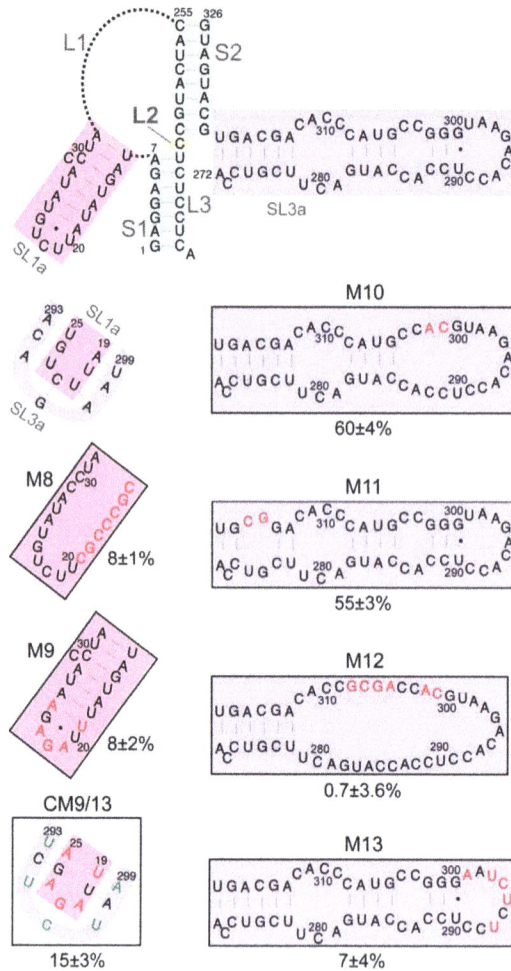

Figure 7. Retrotransposition of mini-Ty1*his3AI* elements with mutations in stem-loops SL1a and SL3a. The schematic (top) shows the secondary structure of the Ty1 pseudoknot core and loops L1, L2 and L3, with the SL1a hairpin (pink shading) and SL3a hairpin (purple shading) highlighted. A second schematic (second from top, left) shows the proposed kissing loop interaction between the seven apical nucleotides of hairpin SL1a (pink shading) and seven apical sequences of the SL3a hairpin. Labeled, boxed schematics show the nucleotide substitutions or additions in each mutant mini-Ty1*his3AI* element analyzed. Black letters indicate wild-type sequence; red letters indicated nucleotide substitutions or additions; and green letters indicate compensatory substitutions that are predicted to restore base-pairing with nucleotide substitutions. The percentage below each box is the median frequency of helper-mediated retrotransposition of each mini-Ty1*his3AI* bearing the indicated mutation divided by the median helper-mediated retrotransposition frequency of the mini-Ty1*his3AI* element with wild-type Ty1-H3 sequence, +/− the 95% confidence interval.

To examine the role of the SL3a bulged stem in retrotransposition, we introduced double mutations near the base and the loop of the SL3a stem. Nucleotides C324 and A325, and the bases with which they are predicted to pair (275/276) are invariant among Ty1 and Ty2 elements; however, disruption of this pairing caused only a minor decrease in retrotransposition (Figure 7, M11). Similarly, a two-nucleotide substitution of CA for GG (301/302) near the SL3a loop also resulted in a minor retrotransposition defect (Figure 7, M10). In contrast, substitution of six nucleotides within the bulged stem of SL3a strongly decreased retrotransposition (Figure 7, M12).

Sequences that comprise the SL1a stem-loop are mostly conserved, particularly in *S. cerevisiae* Ty1 elements, despite the fact that this region is non-coding. A 7-nucleotide substitution that completely disrupts pairing in the S1 stem strongly reduced retrotransposition (Figure 7, M8). Mini-Ty1*his3AI* RNA with a two-nucleotide substitution in the SL1a stem could not be co-transformed with helper-Ty1 into the same yeast strain, even though several transformation strategies were attempted. In summary, major nucleotide substitutions in the stems of SL1a and SL3a hairpins strongly decreased retrotransposition, but it remains to be determined whether the secondary structure of the stems is the critical feature required.

3.5. Role for the S2 Stem and SL1a-SL3a Kissing Loops in Ty1 RNA Stability

Because the S2 stem and SL3a hairpin overlap with a region required for Ty1 RNA packaging, mutations in the S2 stem and SL3a loop, as well as apical mutations in the SL1a hairpin hypothesized to interact with SL3a, might inhibit retrotransposition by blocking packaging of Ty1 RNA. To explore this possibility, we first determined whether mutations in stem S2 and hairpins SL1a and SL3a affect RNA stability. The level of transcript from wild-type and mutant pGAL1:mini-Ty1*his3AI* elements was monitored by northern analysis using a probe specific to *his3AI*. Helper-Ty1 RNA was also quantitated using a probe in the Ty1 *POL* region; a discrete band of ~5.5 kb was detected despite the absence of the termination signal in the 3′ LTR. Strains were induced by growth in galactose for 24 h at 20 °C to mimic the conditions used in the retrotransposition assay. Levels of mini-Ty1*his3AI* RNA in the presence and absence of helper-Ty1 RNA were equivalent (Figure 8A, compare WT lanes plus (+) and minus (−) helper-Ty1), demonstrating that packaging of mini-Ty1*his3AI* RNA is not required for stability. The level of mini-Ty1*his3AI* RNA with a UC264AG mutation in pseudoknot stem S1 was decreased about 2-fold (Figure 8A, M1). This result is consistent with previous analyses of this and other stem S1 mutations in a full-length pGAL1:Ty1*his3AI* element in the absence of helper-Ty1 [20]. Thus, disruption of stem S1 minimally affects Ty1 RNA stability. In contrast, mini-Ty1*his3AI* RNA bearing the AUG321GCU mutation in stem S2 was undetectable (Figure 8A, M5). Surprisingly, helper-Ty1 RNA was also absent, indicating that expressing mini-Ty1*his3AI* RNA with the AUG321GCU mutation destabilizes helper-Ty1 RNA in *trans*. Mini-Ty1*his3AI* RNA with double compensatory mutations AUG321GCU/CAU258AGC was also present at very low levels, but the level of helper-Ty1 RNA in this strain was completely restored (Figure 8A, CM5). These findings support the idea that base-pairing of stem S2 is necessary for mini-Ty1 RNA and helper-Ty1 RNA stability. Instability of the AUG321GCU/CAU258AGC mini-Ty1 RNA was unexpected, because this compensatory mutant transposes at 47% of the frequency of the wild-type mini-Ty1*his3AI*. A possible explanation for this inconsistency is that two temporally or structurally distinct pools of the AUG321GCU/CAU258AGC mutant exist, one that is successfully packaged into VLPs and is used in retrotransposition, and another that is degraded.

To explore this possibility, we used a second, more sensitive approach to measure mini-Ty1 RNA levels, this time in the absence of helper-Ty1. The Ty1 sequences from each pGAL1:mini-Ty1*his3AI* plasmid was subcloned into an expression plasmid, creating an in-frame fusion of the 5′ UTR and first 522 nucleotides of *GAG* to the *GFP* ORF (Gag$_{NT}$:GFP). The pGAL1:mini-Ty1(Gag$_{NT}$:GFP) plasmids were introduced into the *spt3Δ* strain, and expression was induced for 2.5 h in galactose at 20 °C. The mean GFP activity in 10,000 cells bearing a plasmid with wild-type or mutant Ty1 sequences was measured by flow cytometry to monitor the presence of Ty1 RNA after a brief galactose-induction (Figure 8B). The GFP activities in isolates with plasmid pGAL1:mini-Ty1(Gag$_{NT}$:GFP) containing the UC264AG mutation or the UC264AG/GA6UC compensatory mutation in stem S1 were comparable to that of the plasmid with wild-type Ty1 sequence (Figure 8B, compare M1 and CM1 to WT), supporting the idea that mutations in pseudoknot stem S1 minimally destabilize Ty1 RNA [20]. A single nucleotide substitution at the base of stem S2, which changes a GC pair to a GU pair and decreases retrotransposition to 6% of wild-type also had no significant effect on GFP levels

(Figure 8B, M7). However, two triple mutations that disrupt pseudoknot stem S2, AUG321GCU and AUG324GCU, yielded GFP activities that were not detectable above the background fluorescence in a strain without *GFP* (Figure 8B, compare M4 and M5 to empty vector). These results mirror those seen for the AUG321GCU mutant (M5) in northern analysis and imply that disrupting stem S2 substantially destabilizes Ty1 RNA. In contrast, Gag$_{NT}$:GFP levels were restored to 100% or more of wild-type levels in strains carrying the double compensatory mutants, AUG321GCU/CAU258AGC or AUG324GCU/CAU255AGC in stem S2 (Figure 8B, CM4 and CM5). The AUG324GCU/CAU255AGC mutant RNA may be unstable when assayed by northern analysis (Figure 8A, CM5), but able to express Gag$_{NT}$:GFP because of a temporal lag between synthesis and degradation of the RNA, which is sufficient to allow AUG324GCU/CAU255AGC mutant RNA to be packaged and used for retrotransposition. Alternatively, it is possible that co-expression of helper-Ty1 is necessary for instability of the AUG321GCU/CAU258AGC mutant. Overall, these data suggest that disruption of the S2 stem results in rapid degradation of the mini-Ty1 RNA and promotes degradation of helper-Ty1 RNA in *trans*.

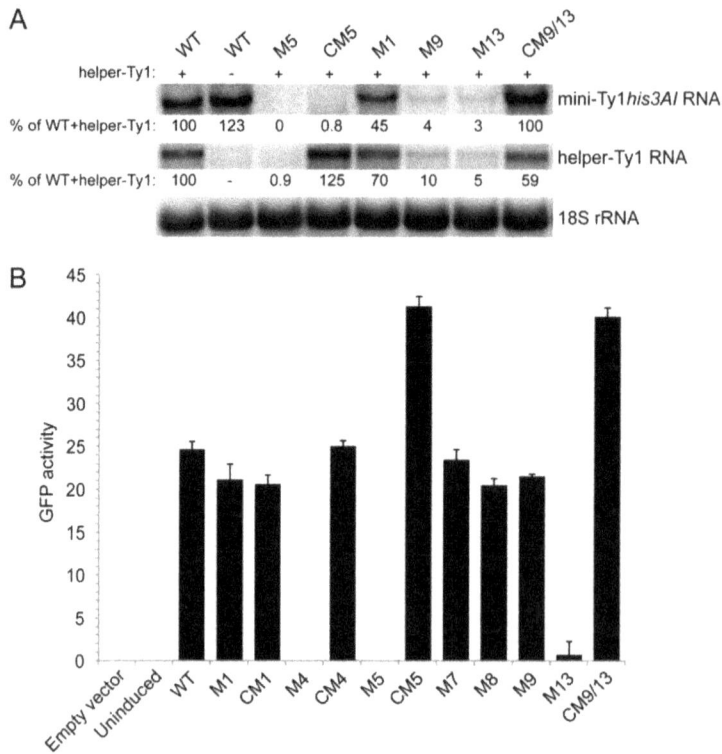

Figure 8. Levels of mini-Ty1*his3AI* RNA bearing different mutations and helper-Ty1 RNA. (**A**) Northern blot analysis of strains carrying the p*GAL1*:mini-Ty1*his3AI* plasmid harboring wild-type Ty1 sequences (WT) or mutant Ty1 sequences and the p*GAL1*:helper-Ty1 induced for 24 hours in galactose-containing medium. The presence or absence of the p*GAL1*:helper-Ty1 plasmid is indicated by + and – symbols, respectively, above the blot. Labels for mutations correspond to those in Figures 6 and 7; (**B**) Measurement of the median GFP activity in 10,000 cells of two different transformants of each p*GAL1:GAG$_{NT}$:GFP* plasmid containing wild type Ty1 TIPIRT domain sequences or derivatives with mutations named as in Figures 6 and 7. Strains were induced in galactose-containing medium for 2.5 h. Error bars are the standard deviation of the median GFP activity in each of two transformants.

Northern blot analysis also revealed that levels of the mini-Ty1*his3AI* RNA and the helper-Ty1 RNA were reduced ten-fold or more in mutants carrying the AUUCUGU19UUAGAGA substitutions at the apex of hairpin SL1a or the ACAGAAU293UCUCUAA substitutions in the SL3a loop of mini-Ty1 RNA (Figure 8A, M9 and M13). Moreover, levels of both the mini-Ty1*his3AI* and helper-Ty1 RNA were rescued in the compensatory mutant with restored SL1a/SL3a complementarity (Figure 8A, CM9/13). The ACAGAAU293UCUCUAA substitutions in SL3a also resulted in very low GFP activity in the Gag$_{NT}$:GFP assay; however, the AUUCUGU19UUAGAGA mutation in SL1a resulted in nearly wild-type levels of GFP activity (Figure 8B, M9 and M13). Interestingly, a 7-nucleotide substitution that disrupts the stem of hairpin SL1a also yielded Gag$_{NT}$:GFP activity that was similar to that of the wild-type plasmid (Figure 8B, M8). The Gag$_{NT}$:GFP activity of the AUUCUGU19UUAGAGA/ACAGAAU293UCUCUAA compensatory mutant is also similar to that of wild-type, suggesting that instability of the ACAGAAU293UCUCUAA mutation in SL3a is rescued by the compensatory mutation in SL1a (Figure 8B, CM9/13). Together, these findings suggest that the apices of SL1a and SL3a hairpins interact via 7 nucleotides of complementarity, and that lack of complementarity destabilizes Ty1 RNA in *cis* and in *trans*. Comparison of the northern and GFP assay results suggest that RNA with mutations in SL1a may be degraded more slowly than those in the SL3a loop or only degraded in the presence of the helper-Ty1. Overall, these data suggest that the S2 stem and kissing loop interactions between SL1a and SL3a may promote an intermolecular interaction between Ty1 RNAs, and that a symmetrical kissing complex with two SL1a-SL3a duplexes may be optimal for Ty1 RNA stability, particularly in the presence of Gag protein.

4. Discussion

This study reveals the conservation of sequence motifs and structural elements within the long-range pseudoknot in the T1PIRT domain of Ty1 RNA and describes novel functions for elements within the pseudoknot. We show that the pseudoknot stems can be separated by four nucleotides with no effect on retrotransposition and that mutations that disrupt pseudoknot stem S2 give rise to RNA instability phenotypes that are distinct from phenotypes that result from S1 mutations [19,20]. A major new finding of this work is that mutations that disrupt the S2 stem of the RNA pseudoknot or complementarity between apical sequences of a hairpin in pseudoknot loop L1 (SL1a) and a hairpin that comprises most of pseudoknot loop L3 (SL3a) not only inhibit retrotransposition but also destabilize mini-Ty1 RNA in *cis* and helper-Ty1 RNA in *trans*. Moreover, compensatory mutations that restore pairing in stem S2 or complementarity between SL1a and SL3a apices alleviate Ty1 RNA degradation in *cis* and in *trans* and suppress the retrotransposition defect of single mutants. Based on these findings, we propose a model in which two intermolecular interactions between complementary apical sequences in SL1a and SL3a form a symmetrical kissing complex (Figure 3), and that this kissing complex initiates Ty1 RNA dimerization and packaging. Furthermore, we propose that formation of only a single intermolecular SL1a-SL3a kissing loop targets both interacting RNAs for degradation. This model explains the phenotypes of apical SL1a and SL3a hairpin mutants and mutants with substitutions in the pseudoknot S1 and S2 stems as follows. When the mini-Ty1 with wild-type sequences is expressed, both homogeneous kissing complexes containing two mini-Ty1 or two helper-Ty1 RNAs and heterogenous kissing complexes with one mini-Ty1 RNA and one helper-Ty1 RNA are expected to form, since helper-Ty1 RNA has wild-type SL1a and SL3a sequences and can be packaged into VLPs [11]. We propose that mini-Ty1 mutants with nucleotide substitutions in complementary sequences of either SL1a or SL3a would not be able to form homogeneous mini-Ty1 RNA kissing complexes, and heterogeneous mini-Ty1/helper-Ty1 RNA kissing complexes would have only a single kissing loop, thereby targeting both RNAs for degradation. In the mini-Ty1 RNA with restored complementarity between SL1a and SL3a hairpins, both types of homogeneous kissing complexes could form, but heterogeneous mini-Ty1/helper-Ty1 RNA complexes could not form, even with a single kissing loop, and therefore we propose that neither mini-Ty1 nor helper-Ty1 RNA would be targeted for degradation. The fact that only homogeneous mini-Ty1 RNA kissing complexes would result in retrotransposition events could explain why the compensatory SL1a-SL3a mutant retrotransposes at a much lower frequency than the wild-type mini-Ty1, which can form both homogeneous mini-Ty1 RNA and heterogeneous mini-Ty1/helper-Ty1 RNA kissing complexes that lead to retrotransposition. In the case of the S2 stem, nucleotide substitutions that disrupt base-pairing may block formation of the SL3a stem-loop, as SL3a encompasses all but one nucleotide of the L3 loop between S1 and S2. Indeed, the SL3a hairpin is not present in a 1482 nt in vitro Ty1 transcript that lacks a pseudoknot [21]. One possible interpretation of these data is that the S2 stem is required for SL3a to form. In contrast, the SL1a stem-loop is predicted to form in the absence of a pseudoknot [21]. Therefore, it is possible that mini-Ty1 RNA with mutations that disrupt S2 do not form homogenous mini-Ty1 kissing complexes but instead form heterogeneous mini-Ty1/helper-Ty1 RNA complexes with one kissing loop, targeting both RNAs for degradation. Compensatory mutations that restore complementarity in the S2 stem would allow both SL1a and SL3a hairpins to form, allowing both heterogeneous and homogeneous complexes with two kissing loops to form. Although steady-state levels of RNA from a compensatory mutant in S2 are low, the RNA is stable long enough to express wild-type GFP levels in the Gag$_{NT}$:GFP assay, and, more importantly, the corresponding element is transpositionally active, indicating that at least some mini-Ty1 RNA survives packaging and functions as a template for retrotransposition. Finally, mutations in the S1 stem would not cause degradation of mini-Ty1 RNA despite the fact that the pseudoknot cannot form because neither the SL1a hairpin nor the SL3a hairpin depends on S1 stem formation [21]. Thus, S1 stem mutants could interact heterogeneously and homogeneously with two kissing loops, but retrotransposition would be blocked by a failure of reverse transcription to occur [20,21].

Notably, the complementary 7-nucleotide motifs in SL1a and SL3a are completely conserved within *S. cerevisiae* Ty1 except for one nucleotide (G296) in the SL3a loop; however, nucleotides 19–25 in SL1a are not conserved in Ty2 elements. Thus, if our model for the initial dimerization of Ty1 RNA is correct, the divergence between Ty1 and Ty2 RNA sequences in SL1a and SL3a could impede the packaging of Ty1 and Ty2 RNAs together in the same VLP where template switching during reverse transcription could create chimeric elements. Therefore, failure to form Ty1/Ty2 RNA dimers could explain how these elements are maintained as distinct families.

It is important to note that the data presented do not include physical evidence that the SL1a and SL3a hairpins interact intermolecularly. Nonetheless, we have shown that substitutions in the SL1a or SL3a apical motifs of mini-Ty1 destabilize helper-Ty1 in *trans*, and importantly, introduction of the corresponding co-varying substitutions in the mini-Ty1 RNA SL3a or SL1a motifs, respectively, complement the RNA instability defect of helper-Ty1 RNA in *trans*. Trans-complementation of the helper-Ty1 defect provides direct genetic evidence of an intermolecular interaction that has not been observed in monomeric Ty1 RNA or in dimeric packaged Ty1 RNA, suggesting that this essential interaction could occur within the transient Ty1 RNA kissing complex. Formally, it is also possible that intramolecular pairing between complementary SL1a and SL3a motifs enhances kissing complex formation, perhaps by promoting an RNA tertiary structure that is necessary for an intermolecular interaction between unidentified regions of the Ty1 TIPIRT domain. Although beyond the scope of this study, many aspects of the model we have proposed might be tested using the in vitro RNA dimerization assay of Cristofari et al. [17], as the RNAs bearing SL1a and SL3a mutations may be stable in vitro.

Retroviral RNAs typically form dimers that are packaged into nascent virions via one or two kissing loop interactions; the resulting kissing complex is converted to a stable dimer during proteolytic maturation of the viral particle [47]. Consistent with retroviral RNAs, Ty1 elements bearing a mutation that blocks proteolytic processing of Gag form dimers, but they are less stable than those formed in wild-type VLPs [10]. These findings suggest that the Ty1 RNA dimer also exists in two forms: an initial kissing complex that is recognized for packaging by the immature Gag protein and a mature dimer that is stabilized during proteolytic maturation of the VLP. Based on these findings, we suggest that two RNA duplexes formed between the complementary 7-nucleotide motifs in SL1a and SL3a result in formation of the initial kissing complex that undergoes a structural transition to the mature form of the Ty1 RNA dimer, which may no longer contain SL1a–SL3a duplexes. Purzyka et al. [25] have proposed that the dimer within VLPs contains interactions between the self-complementary PAL1 and PAL2 sequences within the SL1a stem, as well as a second interaction between PAL3 sequences, which are downstream of the pseudoknot in an area not strictly required for packaging. A possible mechanism that might explain the structural transition between SL1a-SL3a duplexes in the kissing complex and PAL1 and PAL2 duplexes in the mature dimer is that the melting of the first two base-pairs of the SL1a stem by SL1a-SL3a duplex formation could destabilize pairing in the rest of the SL1a stem. Melting of the SL1a stem would expose four of the six PAL1 and PAL2 nucleotides on each strand for duplex formation, and these partial PAL1 and PAL2 duplexes could then be extended by melting the remaining two base-pairs that are interacting with SL3a sequences.

Is Gag involved in the formation of the Ty1 kissing complex in vivo? Our data suggest that packaging of Ty1 RNA is not required for its stability, since the truncated Gag protein encoded by mini-Ty1 cannot form VLPs [48], yet mini-Ty1 RNA expressed in the absence of Gag from endogenous or helper-Ty1 elements is as stable as in its presence. This conclusion contrasts with that of Checkley et al. [49], who showed that Gag supplied in *trans* enhances the stability of a Ty1 RNA containing a premature stop codon adjacent to the start codon, rendering it untranslatable. It seems likely that our differing conclusions stem from the use of different Ty1 RNAs (mini-Ty1 versus untranslatable Ty1 RNA). In our system, it is possible that mini-Ty1 RNA molecules interact intermolecularly via two SL1a-SL3a duplexes in the absence of Gag, and this could stabilize the RNA. This would explain why mutations in the SL3a loop and S2 stem are unstable in the absence of Gag (Figure 8B). Notably, retroviral dimer initiation sites interact in vitro in the absence of Gag, and it has been argued that kissing interactions of retroviral RNA precede packaging [47,50]. However, dimerization of mini-Ty1 RNA in vitro is not detected in the absence of Gag or a C-terminal fragment of Gag harboring the nucleocapsid domain [17,40]. These findings are consistent with an alternative model in which kissing complex formation in vivo requires Gag binding. In this model, mini-Ty1 RNA would be stable either when kissing complexes do not form in the absence of Gag or when symmetrical kissing complexes form in the presence of Gag, but not when asymmetrical kissing complexes with one SL1a-SL3a duplex form. Notably, many of the mutants analyzed in the pseudoknot core and SL1a stem-loop are in Ty1 RNA sequences that are bound by Gag or the nucleocapsid domain [21,40], suggesting that altered binding of Gag to asymmetrical kissing complexes could be a contributing factor in the degradation of Ty1 RNA in *cis* and in *trans*.

Several lines of evidence confirm the conclusion that the Ty1 pseudoknot forms both in vitro in truncated Ty1 RNA leader sequences and in vivo in mini- and full-length Ty1 RNA and is biologically relevant [20,21]. First, the pseudoknot is predicted by several RNA structure prediction algorithms, even in the absence of constraints imposed by SHAPE reactivities, suggesting it is thermodynamically stable. Second, the core of the pseudoknot is almost completely unreactive, which suggests that both stems of the pseudoknot are base-paired within the same molecule of RNA. Third, both pseudoknot stems are required for efficient retrotransposition of Ty1 RNA in vivo [20]. While the findings suggest that the pseudoknot forms in vivo, they do not rule out the possibility that the individual stems form at different times and act at different steps in retrotransposition. For example, the L1 loop and S1 stem of the pseudoknot include all the 5′ sequences known to be required for initiation of reverse transcription [19,20], while the S2 stem and L3 loop coincide with an essential packaging region [11]. While the SL1a stem within the L1 loop has also been proposed to play a role in packaging, this stem-loop likely forms in the absence of the SL1 stem or pseudoknot [21]. In contrast, stem-loop SL3a does not form in the absence of the pseudoknot [21], and our data clearly suggest that the S2 pairing, like the SL3a kissing motif, is required for Ty1 RNA stability (Figure 8). A role for the individual pseudoknot stems in demarcating and stabilizing two separate structural domains is appealing because of the overlap between structurally and functionally defined domains that has been revealed in this and previous studies [11,19–21]. Formation of the pseudoknot versus formation of only the S1 stem or the S2 stem are not mutually exclusive possibilities, and there may be switching between one conformation that is stabilized by the pseudoknot, and others that contain only the S1 stem and the L1 loop, or only the S2 stem and the L3 loop. The idea that formation of the TIPIRT domain pseudoknot is regulated at different points in retrotransposition is attractive because the length and base composition of the 1-nucleotide interhelical L2 loop can be altered without substantial effects on retrotransposition. One interpretation of this finding is that the L2 nucleotide allows for a flexible pseudoknot conformation in vivo, and therefore that the tertiary architecture of the TIPIRT domain could change at different stages in the retrotransposition cycle. The ability of the TIPIRT domain to adopt multiple conformations is likely to be important, given the breadth of functions that the TIPIRT domain plays in retrotransposition.

The secondary structure model of the TIPIRT domain predicts that much of the 53-nucleotide 5' UTR of Ty1 RNA is sequestered by base-pairing, including the pseudoknot S1 stem. The SHAPE-directed structural model described here as well as earlier models revealed significant secondary structure within the 5' UTR that is potentially inhibitory to ribosomal scanning, including the base-pairing of nucleotides 1 to 7, stem-loop SL1a, base-pairing of nucleotides 39 to 45, and sequestration of the AUG codon in an helix of seven base-pairs and a 1×1 internal loop. Moreover, the 5' UTR and sequences that base-pair to portions of it are very highly conserved in Ty1 elements, especially in regions with secondary structure. The predicted thermodynamic stability of the Ty1 RNA pseudoknot suggest that its formation results in folding of the 5' terminus into a compact tertiary structure that would render it inaccessible for translation initiation and perhaps even 5'–3' degradation. The presence of significant secondary structure is unusual in 5' UTRs of S. cerevisiae genes [51]. Hence, translation of Ty1 RNA, a requisite step in retrotransposition, is not likely to be favored by formation of the pseudoknot. Regulation of the TIPIRT domain structure may play some role in several peculiarities of Ty1 RNA metabolism and function, including the unusually long half-life [52] and the sensitivity of Ty1 RNA translation to loss of translation initiation factor eIF4G1 and 40S rRNA subunit proteins [53–55]. Pseudoknots frequently play regulatory roles in gene expression; thus, regulation of the formation of the TIPIRT domain pseudoknot may be a critical factor governing the partitioning of Ty1 RNA between its different functions in translation, packaging and reverse transcription.

Supplementary Materials: The following are available online at www.mdpi.com/1999-4915/9/5/93/s1, Figure S1: Secondary structure of Ty1 RNA 5' leader by pknotsRG, Figure S2: Secondary structure of Ty1 RNA 5' leader by IPknot, Table S1: raw SHAPE data in SNRNASM format.

Acknowledgments: We thank Christine E. Hajdin and Kevin M. Weeks, University of North Carolina, Chapel Hill, for sharing the ShapeKnots prediction software before publication, Chetna Gopinath and Lauren Neulander-Davis for their assistance with collecting SHAPE data, and Sheila Lutz for plasmid construction and for helpful comments on the manuscript. DNA sequencing was performed by the Wadsworth Center Applied Genomics Technology Core and flow cytometry was performed in the Wadsworth Center Immunology Core. The work was supported by funds from NIH grant R01-GM52072 to M.J.C.; NIH grants R21-MH087336, R01-HL111527, R01-HG008133 and R01-GM101237 to A.L.; and NIH R01-GM076485 to D.H.M.

Author Contributions: E.R.G, A.L. and M.J.C. conceived and designed the experiments; E.R.G. and J.H.D. performed the experiments; E.R.G., J.H.D., J.R., A.L., S.B., D.H.M. and M.J.C. analyzed the data; J.R., A.L., S.B. and D.H.M. contributed analysis tools; E.R.G. and M.J.C. wrote the paper.

Conflicts of Interest: The authors declare no conflict of interest. The founding sponsors had no role in the design of the study; in the collection, analyses, or interpretation of data; in the writing of the manuscript, and in the decision to publish the results.

References

1. Tan, S.; Cardoso-Moreira, M.; Shi, W.; Zhang, D.; Huang, J.; Mao, Y.; Jia, H.; Zhang, Y.; Chen, C.; Shao, Y.; et al. LTR-mediated retroposition as a mechanism of RNA-based duplication in metazoans. *Genome Res.* **2016**, *26*, 1663–1675. [CrossRef] [PubMed]

2. Kalyana-Sundaram, S.; Kumar-Sinha, C.; Shankar, S.; Robinson, D.R.; Wu, Y.-M.; Cao, X.; Asangani, I.A.; Kothari, V.; Prensner, J.R.; Lonigro, R.J.; et al. Expressed pseudogenes in the transcriptional landscape of human cancers. *Cell* **2012**, *149*, 1622–1634. [CrossRef] [PubMed]

3. Khurana, E.; Lam, H.Y.K.; Cheng, C.; Carriero, N.; Cayting, P.; Gerstein, M.B. Segmental duplications in the human genome reveal details of pseudogene formation. *Nucleic Acids Res.* **2010**, *38*, 6997–7007. [CrossRef] [PubMed]

4. Fu, B.; Chen, M.; Zou, M.; Long, M.; He, S. The rapid generation of chimerical genes expanding protein diversity in zebrafish. *BMC Genom.* **2010**, *11*, 657. [CrossRef] [PubMed]

5. Fink, G.R. Pseudogenes in yeast? *Cell* **1987**, *49*, 5–6. [CrossRef]

6. Maxwell, P.H.; Curcio, M.J. Retrosequence formation restructures the yeast genome. *Genes Dev.* **2007**, *21*, 3308–3318. [CrossRef] [PubMed]

7. Aravind, L.; Watanabe, H.; Lipman, D.J.; Koonin, E.V. Lineage-specific loss and divergence of functionally linked genes in eukaryotes. *Proc. Natl. Acad. Sci. USA* **2000**, *97*, 11319–11324. [CrossRef] [PubMed]

8. Maxwell, P.H.; Curcio, M.J. Incorporation of y′-Ty1 cDNA destabilizes telomeres in *Saccharomyces cerevisiae* telomerase mutants. *Genetics* **2008**, *179*, 2313–2317. [CrossRef] [PubMed]

9. Curcio, M.J.; Lutz, S.; Lesage, P. The Ty1 LTR-retrotransposon of budding yeast, *Saccharomyces cerevisiae*. *Microbiol. Spectrum.* **2014**, *3*, 1–35. [CrossRef] [PubMed]

10. Feng, Y.X.; Moore, S.P.; Garfinkel, D.J.; Rein, A. The genomic RNA in Ty1 virus-like particles is dimeric. *J. Virol.* **2000**, *74*, 10819–10821. [CrossRef] [PubMed]

11. Xu, H.; Boeke, J.D. Localization of sequences required in cis for yeast Ty1 element transposition near the long terminal repeats: Analysis of mini-Ty1 elements. *Mol. Cell. Biol.* **1990**, *10*, 2695–2702. [CrossRef] [PubMed]

12. Chapman, K.B.; Bystrom, A.S.; Boeke, J.D. Initiator methionine tRNA is essential for Ty1 transposition in yeast. *Proc. Natl. Acad. Sci. USA* **1992**, *89*, 3236–3240. [CrossRef] [PubMed]

13. Keeney, J.B.; Chapman, K.B.; Lauermann, V.; Voytas, D.F.; Astrom, S.U.; von Pawel-Rammingen, U.; Bystrom, A.; Boeke, J.D. Multiple molecular determinants for retrotransposition in a primer tRNA. *Mol. Cell. Biol.* **1995**, *15*, 217–226. [CrossRef] [PubMed]

14. Wilhelm, M.; Wilhelm, F.X.; Keith, G.; Agoutin, B.; Heyman, T. Yeast Ty1 retrotransposon: The minus-strand primer binding site and a cis-acting domain of the Ty1 RNA are both important for packaging of primer tRNA inside virus-like particles. *Nucleic Acids Res.* **1994**, *22*, 4560–4565. [CrossRef] [PubMed]

15. Friant, S.; Heyman, T.; Wilhelm, M.L.; Wilhelm, F.X. Extended interactions between the primer tRNAi(met) and genomic RNA of the yeast Ty1 retrotransposon. *Nucleic Acids Res.* **1996**, *24*, 441–449. [CrossRef] [PubMed]

16. Friant, S.; Heyman, T.; Bystrom, A.S.; Wilhelm, M.; Wilhelm, F.X. Interactions between Ty1 retrotransposon RNA and the T and D regions of the tRNA(imet) primer are required for initiation of reverse transcription in vivo. *Mol. Cell. Biol.* **1998**, *18*, 799–806. [CrossRef] [PubMed]

17. Cristofari, G.; Ficheux, D.; Darlix, J.L. The gag-like protein of the yeast Ty1 retrotransposon contains a nucleic acid chaperone domain analogous to retroviral nucleocapsid proteins. *J. Biol. Chem.* **2000**, *275*, 19210–19217. [CrossRef] [PubMed]

18. Cristofari, G.; Bampi, C.; Wilhelm, M.; Wilhelm, F.X.; Darlix, J.L. A 5′-3′ long-range interaction in Ty1 RNA controls its reverse transcription and retrotransposition. *EMBO J.* **2002**, *21*, 4368–4379. [CrossRef] [PubMed]

19. Bolton, E.C.; Coombes, C.; Eby, Y.; Cardell, M.; Boeke, J.D. Identification and characterization of critical cis-acting sequences within the yeast Ty1 retrotransposon. *RNA* **2005**, *11*, 308–322. [CrossRef] [PubMed]

20. Huang, Q.; Purzycka, K.J.; Lusvarghi, S.; Li, D.; Legrice, S.F.; Boeke, J.D. Retrotransposon Ty1 RNA contains a 5′-terminal long-range pseudoknot required for efficient reverse transcription. *RNA* **2013**, *19*, 320–332. [CrossRef] [PubMed]

21. Purzycka, K.J.; Legiewicz, M.; Matsuda, E.; Eizentstat, L.D.; Lusvarghi, S.; Saha, A.; Grice, S.F.; Garfinkel, D.J. Exploring Ty1 retrotransposon RNA structure within virus-like particles. *Nucleic Acids Res.* **2013**, *41*, 463–473. [CrossRef] [PubMed]

22. Pachulska-Wieczorek, K.; Le Grice, S.F.; Purzycka, K.J. Determinants of genomic RNA encapsidation in the *Saccharomyces cerevisiae* long terminal repeat retrotransposons Ty1 and ty3. *Viruses* **2016**, *8*. [CrossRef] [PubMed]

23. Lu, K.; Heng, X.; Summers, M.F. Structural determinants and mechanism of HIV-1 genome packaging. *J. Mol. Biol.* **2011**, *410*, 609–633. [CrossRef] [PubMed]

24. Rein, A.; Datta, S.A.; Jones, C.P.; Musier-Forsyth, K. Diverse interactions of retroviral gag proteins with RNAs. *Trends Biochem. Sci.* **2011**, *36*, 373–380. [CrossRef] [PubMed]

25. Purzycka, K.J.; Garfinkel, D.J.; Boeke, J.D.; Le Grice, S.F. Influence of RNA structural elements on Ty1 retrotransposition. *Mob. Genet. Elements* **2013**, *3*. [CrossRef] [PubMed]

26. Scholes, D.T.; Banerjee, M.; Bowen, B.; Curcio, M.J. Multiple regulators of Ty1 transposition in *Saccharomyces cerevisiae* have conserved roles in genome maintenance. *Genetics* **2001**, *159*, 1449–1465. [PubMed]

27. Vasa, S.M.; Guex, N.; Wilkinson, K.A.; Weeks, K.M.; Giddings, M.C. Shapefinder: A software system for high-throughput quantitative analysis of nucleic acid reactivity information resolved by capillary electrophoresis. *RNA* **2008**, *14*, 1979–1990. [CrossRef] [PubMed]

28. Deigan, K.E.; Li, T.W.; Mathews, D.H.; Weeks, K.M. Accurate shape-directed RNA structure determination. *Proc. Natl. Acad. Sci. USA* **2009**, *106*, 97–102. [CrossRef] [PubMed]

29. Mathews, D.H.; Disney, M.D.; Childs, J.L.; Schroeder, S.J.; Zuker, M.; Turner, D.H. Incorporating chemical modification constraints into a dynamic programming algorithm for prediction of RNA secondary structure. *Proc. Natl. Acad. Sci. USA* **2004**, *101*, 7287–7292. [CrossRef] [PubMed]

30. Hajdin, C.E.; Bellaousov, S.; Huggins, W.; Leonard, C.W.; Mathews, D.H.; Weeks, K.M. Accurate shape-directed RNA secondary structure modeling, including pseudoknots. *Proc. Natl. Acad. Sci. USA* **2013**, *110*, 5498–5503. [CrossRef] [PubMed]

31. Rocca-Serra, P.; Bellaousov, S.; Birmingham, A.; Chen, C.; Cordero, P.; Das, R.; Davis-Neulander, L.; Duncan, C.D.S.; Halvorsen, M.; Knight, R.; et al. Sharing and archiving nucleic acid structure mapping data. *RNA* **2011**, *17*, 1204–1212. [CrossRef] [PubMed]

32. Larkin, M.A.; Blackshields, G.; Brown, N.P.; Chenna, R.; McGettigan, P.A.; McWilliam, H.; Valentin, F.; Wallace, I.M.; Wilm, A.; Lopez, R.; et al. Clustal W and Clustal X version 2.0. *Bioinformatics* **2007**, *23*, 2947–2948. [CrossRef] [PubMed]

33. Boeke, J.D.; Eichinger, D.; Castrillon, D.; Fink, G.R. The *Saccharomyces cerevisiae* genome contains functional and nonfunctional copies of transposon Ty1. *Mol. Cell. Biol.* **1988**, *8*, 1432–1442. [CrossRef] [PubMed]

34. Wach, A.; Brachat, A.; Alberti-Segui, C.; Rebischung, C.; Philippsen, P. Heterologous HIS3 marker and GFP reporter modules for PCR-targeting in *Saccharomyces cerevisiae*. *Yeast* **1997**, *13*, 1065–1075. [CrossRef]

35. Brachmann, C.B.; Davies, A.; Cost, G.J.; Caputo, E.; Li, J.; Hieter, P.; Boeke, J.D. Designer deletion strains derived from *Saccharomyces cerevisiae* S288c: A useful set of strains and plasmids for PCR-mediated gene disruption and other applications. *Yeast* **1998**, *14*, 115–132. [CrossRef]

36. Curcio, M.J.; Hedge, A.M.; Boeke, J.D.; Garfinkel, D.J. Ty RNA levels determine the spectrum of retrotransposition events that activate gene expression in *Saccharomyces cerevisiae*. *Mol. Gen. Genet.* **1990**, *220*, 213–221. [CrossRef] [PubMed]

37. Lee, B.S.; Lichtenstein, C.P.; Faiola, B.; Rinckel, L.A.; Wysock, W.; Curcio, M.J.; Garfinkel, D.J. Posttranslational inhibition of Ty1 retrotransposition by nucleotide excision repair/transcription factor TFIIH subunits Ssl2p and Rad3p. *Genetics* **1998**, *148*, 1743–1761. [PubMed]

38. Reeder, J.; Giegerich, R. Design, implementation and evaluation of a practical pseudoknot folding algorithm based on thermodynamics. *BMC Bioinform.* **2004**, *5*. [CrossRef] [PubMed]

39. Sato, K.; Kato, Y.; Hamada, M.; Akutsu, T.; Asai, K. Ipknot: Fast and accurate prediction of RNA secondary structures with pseudoknots using integer programming. *Bioinformatics* **2011**, *27*, i85–i93. [CrossRef] [PubMed]

40. Nishida, Y.; Pachulska-Wieczorek, K.; Blaszczyk, L.; Saha, A.; Gumna, J.; Garfinkel, D.J.; Purzycka, K.J. Ty1 retrovirus-like element gag contains overlapping restriction factor and nucleic acid chaperone functions. *Nucleic Acids Res.* **2015**, *43*, 7414–7431. [CrossRef] [PubMed]

41. Curcio, M.J.; Garfinkel, D.J. Heterogeneous functional Ty1 elements are abundant in the *Saccharomyces cerevisiae* genome. *Genetics* **1994**, *136*, 1245–1259. [PubMed]

42. Kim, J.M.; Vanguri, S.; Boeke, J.D.; Gabriel, A.; Voytas, D.F. Transposable elements and genome organization: A comprehensive survey of retrotransposons revealed by the complete *Saccharomyces cerevisiae* genome sequence. *Genome Res.* **1998**, *8*, 464–478. [PubMed]

43. Jordan, I.K.; McDonald, J.F. Evidence for the role of recombination in the regulatory evolution of *Saccharomyces cerevisiae* Ty elements. *J. Mol. Evol.* **1998**, *47*, 14–20. [CrossRef] [PubMed]

44. Bleykasten-Grosshans, C.; Friedrich, A.; Schacherer, J. Genome-wide analysis of intraspecific transposon diversity in yeast. *BMC Genom.* **2013**, *14*. [CrossRef] [PubMed]

45. Curcio, M.J.; Garfinkel, D.J. Single-step selection for Ty1 element retrotransposition. *Proc. Natl. Acad. Sci. USA* **1991**, *88*, 936–940. [CrossRef] [PubMed]

46. Brierley, I.; Pennell, S.; Gilbert, R.J. Viral RNA pseudoknots: Versatile motifs in gene expression and replication. *Nat. Rev. Microbiol.* **2007**, *5*, 598–610. [CrossRef] [PubMed]

47. Paillart, J.C.; Shehu-Xhilaga, M.; Marquet, R.; Mak, J. Dimerization of retroviral RNA genomes: An inseparable pair. *Nat. Rev. Microbiol.* **2004**, *2*, 461–472. [CrossRef] [PubMed]

48. Roth, J.F. The yeast Ty virus-like particles. *Yeast* **2000**, *16*, 785–795. [CrossRef]

49. Checkley, M.A.; Mitchell, J.A.; Eizenstat, L.D.; Lockett, S.J.; Garfinkel, D.J. Ty1 gag enhances the stability and nuclear export of Ty1 mRNA. *Traffic* **2013**, *14*, 57–69. [CrossRef] [PubMed]

50. Johnson, S.F.; Telesnitsky, A. Retroviral RNA dimerization and packaging: The what, how, when, where, and why. *PLoS Pathog.* **2010**, *6*, e1001007. [CrossRef] [PubMed]

51. Ringner, M.; Krogh, M. Folding free energies of 5′-UTRs impact post-transcriptional regulation on a genomic scale in yeast. *PLoS Comput. Biol.* **2005**, *1*, e72. [CrossRef] [PubMed]

52. Munchel, S.E.; Shultzaberger, R.K.; Takizawa, N.; Weis, K. Dynamic profiling of mRNA turnover reveals gene-specific and system-wide regulation of mRNA decay. *Mol. Biol. Cell.* **2011**, *22*, 2787–2795. [CrossRef] [PubMed]

53. Malagon, F.; Jensen, T.H. The T body, a new cytoplasmic RNA granule in *Saccharomyces cerevisiae*. *Mol. Cell. Biol.* **2008**, *28*, 6022–6032. [CrossRef] [PubMed]

54. Suresh, S.; Ahn, H.W.; Joshi, K.; Dakshinamurthy, A.; Kananganat, A.; Garfinkel, D.J.; Farabaugh, P.J. Ribosomal protein and biogenesis factors affect multiple steps during movement of the *Saccharomyces cerevisiae* Ty1 retrotransposon. *Mob. DNA* **2015**, *6*. [CrossRef] [PubMed]

55. Palumbo, R.J.; Fuchs, G.; Lutz, S.; Curcio, M.J. Paralog-specific functions of Rpl7a and Rpl7b mediated by ribosomal protein or snoRNA dosage in *Saccharomyces cerevisiae*. *G3* **2017**, *7*, 591–606. [CrossRef] [PubMed]

viruses

Article

Structure of Ty1 Internally Initiated RNA Influences Restriction Factor Expression

Leszek Błaszczyk [1], Marcin Biesiada [1], Agniva Saha [2], David J. Garfinkel [2] and Katarzyna J. Purzycka [1,*]

[1] Institute of Bioorganic Chemistry, Polish Academy of Sciences, Poznan 61-704, Poland; blaszcz@ibch.poznan.pl (L.B.); biesiada@ibch.poznan.pl (M.B.)
[2] Department of Biochemistry & Molecular Biology, University of Georgia, Athens, GA 30602, USA; agniva.saha@gmail.com (A.S.); djgarf@uga.edu (D.J.G.)
*   Correspondence: purzycka@ibch.poznan.pl; Tel.: +48-618-528-503

Academic Editor: Eric O. Freed
Received: 1 February 2017; Accepted: 3 April 2017; Published: 10 April 2017

Abstract: The long-terminal repeat retrotransposon Ty1 is the most abundant mobile genetic element in many *Saccharomyces cerevisiae* isolates. Ty1 retrotransposons contribute to the genetic diversity of host cells, but they can also act as an insertional mutagen and cause genetic instability. Interestingly, retrotransposition occurs at a low level despite a high level of Ty1 RNA, even though *S. cerevisiae* lacks the intrinsic defense mechanisms that other eukaryotes use to prevent transposon movement. p22 is a recently discovered Ty1 protein that inhibits retrotransposition in a dose-dependent manner. p22 is a truncated form of Gag encoded by internally initiated Ty1i RNA that contains two closely-spaced AUG codons. Mutations of either AUG codon compromise p22 translation. We found that both AUG codons were utilized and that translation efficiency depended on the Ty1i RNA structure. Structural features that stimulated p22 translation were context dependent and present only in Ty1i RNA. Destabilization of the 5' untranslated region (5' UTR) of Ty1i RNA decreased the p22 level, both in vitro and in vivo. Our data suggest that protein factors such as Gag could contribute to the stability and translational activity of Ty1i RNA through specific interactions with structural motifs in the RNA.

Keywords: RNA structure; Ty1 retrotransposon; Gag; translation regulation

---

1. Introduction

Ty1 is a long-terminal repeat (LTR) retrotransposon in the *Pseudoviridae* family and the most abundant mobile genetic element in the *Saccharomyces cerevisiae* reference strain [1]. Ty1 contains *GAG* and *POL* genes bracketed by LTRs and proliferates in the yeast genome by integrating new copies through an RNA-mediated mechanism [2]. Dimeric Ty1 RNA is present in virus-like particles (VLPs) [3] that are comprised of the capsid protein Gag and Gag-Pol; the latter being synthesized by a programmed +1 frameshift event that occurs at overlapping leucine codons in *GAG* and *POL* [4]. *POL* encodes protease (PR), reverse transcriptase (RT) and integrase (IN), which are required for protein maturation, reverse transcription and integration, respectively. Gag is a VLP structural component and is expressed as a 441-amino acid precursor (p49) that undergoes a C-terminal cleavage by PR to produce the mature 401-residue protein (p45). Ty1 Gag binds RNA in vitro [5,6] and serves as a multifunctional regulator that orchestrates retrotransposon replication [7].

Ty1 contributes to the genetic diversity of *S. cerevisiae* and closely related species, however, these elements can also act as insertional mutagens and cause genetic instability by recombination-mediated gene rearrangements. Overloading the genome with retrotransposon insertions is another scenario that could be lethal to the cell. Paradoxically, Ty1 retrotransposition occurs at low rate, despite a high level of Ty1 RNA [2]. *S. cerevisiae* also lack the intrinsic defense mechanisms to prevent retrotransposition that are typically active in other eukaryotes, including DNA methylation [8,9], and the expression of several host proteins, such as apolipoprotein B mRNA-editing enzyme catalytic polypeptide-like 3 (APOBEC3) family members [10] or RNAi components [11,12]. Early on, a region of Ty1 required for copy number control (CNC) was identified but the mechanism underlying CNC remained puzzling [13]. Recent genetic analysis of the CNC region identified mutations abrogating CNC that map within *GAG* downstream of two internal AUG codons [14,15]. The separation of function phenotype displayed by one of the *GAG* mutations suggests that Ty1 encodes a protein that restricts its movement. Indeed, the recently discovered protein p22 inhibits retrotransposition in a dose-dependent manner and mediates CNC. p22 is encoded by the C-terminal half of Ty1 *GAG*, and similar to Gag-p49, undergoes maturation by Ty1 protease to form p18. However, p22 is encoded by internally initiated Ty1i RNA that contains two closely spaced AUG codons. Ribosomal profiling analyses show preferential usage of AUG1, but mutational analysis of Ty1i RNA initiation codons AUG1 and AUG2 suggests that both have the potential to be utilized for p22 translation. p18 expressed from either AUG1 or AUG2 confers strong inhibition of Ty1 mobility that correlates with their level of expression. Also, p22/p18 target Gag and inhibit several steps in the process of retrotransposition prior to reverse transcription [14–16].

Like programmed Ty1 frameshifting, employing multiple start codons to initiate the synthesis of p22 is reminiscent of the non-canonical translation strategies that viruses use to maximize their coding potential [17]. Canonical 5′-end-dependent translation initiation generally permits only one protein to be synthesized from a particular mRNA. However, the leaky scanning mechanism allows the production of functionally distinct proteins from a single transcript containing multiple initiation codons. In these cases, a suboptimal sequence surrounding the first AUG codon limits its recognition, which allows ribosomal scanning and translation from downstream initiation codons [17]. This strategy is commonly employed by RNA viruses, including retroviruses [18].

We have shown that p22 translation is a cap-dependent event, however, our results suggest that the structure of 5′ UTR of Ty1i mRNA may contribute to the efficiency of translation [14]. Secondary and tertiary structures of 5′ UTRs play important roles in the regulation of translation by affecting the recruitment, positioning and movement of ribosomes [19]. Folding of the 5′ UTR into an ensemble of secondary structures may influence the initiation of translation either positively or negatively. The nature of this effect is attributed, at least in part, to the thermodynamic stability of the structural elements formed in the 5′ UTR, their guanine-cytosine (GC) content, and positioning in relation to the 5′ cap and AUG initiation codon. Hairpin structures of even moderate thermodynamic stability located close to the 5′-end of the mRNA prevent cap-dependent formation of the preinitiation complexes and can lead to translation inhibition [20–23]. On the other hand, secondary structures present in the coding region may stimulate translation if placed at particular distances downstream of the initiation codon [24,25]. This stimulatory effect may be caused by a hairpin structure that pauses migration of the preinitiation complexes. Hairpin structures can be important for mRNAs containing AUG codons located in suboptimal sequence contexts, and thus undergo translation via leaky scanning. Structure-dependent pausing of the preinitiation complexes provides more time for the recognition of AUG codons in an unfavorable context. Whether this is a general mechanism remains to be determined, however, analysis of the predicted secondary structures downstream of initiation codons suggests that this may be the case [26]. The structural context of the AUG codon can modulate translation efficiency [27]. Coding sequences can also participate in the folding of the 5′ untranslated regions that modulate RNA stability [28,29]. However, coding sequence contributions to translation initiation remain understudied since functional and structural characterization is usually conducted on isolated 5′ UTR sequences.

We set out to characterize how p22 translation is initiated. Our work suggests that both AUG codons can be utilized but AUG1 is used preferentially and translation efficiency strongly depends on the Ty1i RNA structure. Features stimulating p22 translation are context dependent as revealed by specific structures in Ty1i mRNA that are absent in full length genomic mRNA. The 5′ UTR of p22 mRNA interacts with the coding region and destabilization at the secondary or 3D structural levels results in a decrease in p22 translation. Also, our data supports the idea that protein factors such as Gag interact with a structural motif in Ty1i RNA to modulate its stability and translation.

2. Materials and Methods

2.1. Preparation of the RNA Constructs for Structure Probing Experiments and In Vitro Translation Assays

All DNA templates for secondary structure probing experiments and in vitro translation were amplified from plasmid pBDG433, which contains transcribed sequences of Ty1-H3 subcloned into the riboprobe vector pSP64 (Promega, Madison, WI, USA). Forward and reverse primers are listed in Table S1. Each construct was confirmed by DNA sequencing. In vitro transcription reactions were performed using MEGAscript or MEGAshortscript T7 transcription kits (ThermoFisher, Waltham, MA, USA), as recommended by the manufacturer. RNA transcripts were purified using Direct-zol RNA MiniPrep Kit (Zymo Research, Irvine, CA, USA) and their integrity was monitored by formaldehyde agarose gel electrophoresis. Capped transcripts were synthesised in the presence of the ARCA Cap Analog (ThermoFisher). RNA used for native gel electrophoresis was [^{32}P]-labelled at their 3′-ends with T4 RNA ligase (ThermoFisher) according to standard procedures.

2.2. Selective Acylation Analysed by Primer Extension (SHAPE)

The reaction mixture (100 µL) containing 20 pmol of RNA in SHAPE renaturation buffer (10 mM Tris-HCl pH 8.0, 100 mM KCl, 0.1 mM ethylenediaminetetraacetic acid (EDTA), pH 8.0) was heated at 95 °C for 3 min and placed on ice for 5 min. Fifty microliters of 3× SHAPE folding buffer (120 mM Tris-HCl pH 8.0, 600 mM KCl, 1.5 mM EDTA pH 8.0, 15 mM MgCl$_2$) was added and samples were incubated for 30 min at 37 °C. Folded RNA was separated equally into two reactions and mixed with the 20 mM N-methylisatoic anhydride (NMIA) in dimethyl sulfoxide (DMSO) (2 mM final concentration of NMIA) or DMSO alone. Both reactions were incubated for 45 min at 37 °C followed by purification of RNA using Direct-zol RNA MiniPrep Kit.

2.3. DMS Modification

RNA (20 pmol in 50 µL) was refolded using the same conditions as those employed in the SHAPE experiments, then divided equally into two 24 µL reactions. Refolded RNA samples were mixed with 1 µL of dimethyl sulphate (DMS) in ethanol (0.5% final concentration) or ethanol alone. Both reactions were incubated 1 min at room temperature and mixed with 475 µL of stop solution (200 mM sodium acetate, 4.8 M β-mercaptoethanol). RNA was purified using Direct-zol RNA MiniPrep Kit immediately after stopping the reaction.

2.4. Hydroxyl Radical Probing

RNA samples (10 pmol) were refolded by heating at 95 °C for 2 min in water followed by incubation at 25 °C for 5 min. Next, 3× SHAPE folding buffer was added and the reaction was incubated for 25 min at 37 °C, then diluted 20× with 20 mM Tris-HCl pH 8.0. To initiate the production of hydroxyl radicals, 1.5 µL of 2.5 mM (NH$_4$)Fe(SO$_4$)$_2$, 50 mM sodium ascorbate, 1.5% H$_2$O$_2$ and 2.75 mM EDTA were applied separately to the wall of the tube followed by centrifugation. Six microliters of water were added to the control reaction. Reactions were incubated for 10 s at room temperature, then quenched by the addition of thiourea and EDTA to final concentrations of 20 mM and 40 mM, respectively. RNA was recovered using Direct-zol RNA MiniPrep Kit.

2.5. Reverse Transcription and Data Processing

A reaction containing 2–5 pmol RNA, 10 pmol of fluorescently labelled primer PR5 or PR6 (Table S1) (Cy5 (+reagent) or Cy5.5 (control reaction)) and 0.1 mM EDTA pH 8.0 was incubated at 95 °C for 3 min, 37 °C for 10 min and 55 °C for 2 min, and then reverse transcribed at 50 °C for 45 min using Superscript III Reverse Transcriptase (ThermoFisher) as described previously [30]. Sequencing reactions were carried out using primers fluorescently labelled with LicorIR-800 (ddT) or WellRed D2 (ddA) and a Thermo Sequenase Cycle Sequencing Kit, according to the manufacturer's protocol (Affymetrix, Santa Clara, CA, USA). Reverse transcription reactions and sequencing ladders were purified using ZR DNA Sequencing Clean-up Kit (ZymoResearch). cDNA samples were analysed on a GenomeLab GeXP Analysis System (Beckman–Coulter, Brea, CA, USA). Raw data were processed as described [31]. At least four repetitions were obtained for each reaction.

2.6. In Vitro Translation

In vitro translation experiments were carried out using wheat germ extract (WGE) as recommended by the manufacturer (Promega). The reaction mixture containing 12.5 µL of WGE lysate, 80 µM amino acid mixture minus methionine, 1.25 µL of [^{35}S]-labelled methionine (1000 Ci/mmol) (Hartmann Analytic, Braunschweig, Germany), 79 mM potassium acetate, 20 units of ribonuclease inhibitor (ThermoFisher) and 1 pmol of refolded capped or uncapped RNA in the final volume of 25 µL was incubated for 1 hour at 25 °C. Translation products were resolved on sodium dodecyl sulphate (SDS)-polyacrylamide gels followed by radioisotope imaging using a FLA 5100 image analyser (Fuji, Minato, Tokyo, Japan). Bands intensities were analysed using MultiGauge software (Fuji). At least three repetitions were obtained for each in vitro translation reaction.

2.7. Native Gel Electrophoresis

[^{32}P]-labelled RNA was refolded in SHAPE renaturation buffer by heating at 95 °C for 5 min and 4 °C for 5 min. SHAPE folding buffer contained increasing MgCl$_2$ concentrations ranging from 0.1 to 10 mM. The reaction mixture (15 µL) was incubated at 37 °C for 25 min following the addition of 1.5 µL of 25% ficoll. Samples were analysed by native polyacrylamide gel electrophoresis using 12% gels in 0.5× TB at 4 °C. Electrophoresis was carried out at a gel temperature of 4 °C (DNApointer, Biovectis, Warsaw, Poland) [32]. Gels were dried, exposed to a phosphorimager screen, and scanned using FLA 5100 image analyser.

2.8. Ty1 Gag Expression and Purification

A Ty1 Gag-p45-GST fusion protein was expressed in *Escherichia coli* (*E. coli*) strain BL21(DE3)pLysS (Invitrogen, Carlsbad, CA, USA). Six liters of cells were grown in Luria-Bertani (LB) medium containing 50 µg/mL ampicillin and 34 µg/mL chloramphenicol at 28 °C to an OD$_{600}$ of 0.7. Prior to isopropyl β-D-1-thiogalactopyranoside (IPTG) induction, cells were incubated for 30 min at 18 °C. Following the addition of IPTG (0.8 mM), the culture was induced at 18 °C overnight. Cells were pelleted by centrifugation at 4000 g for 10 min at 4 °C and resuspended in lysis buffer (50 mM Tris-HCl pH 8.0, 1 M NaCl, 10 mM β-mercaptoethanol, 2.5 mM DTT, 0.1 mM ZnCl$_2$, 0.5 mg/mL lysozyme, and protease inhibitor (Roche, Basel, Switzerland)). The cell suspension was sonicated 40 × 2 s on ice with a 30 s pause after each pulse. Debris was removed by centrifugation at 20,000 g for 20 min at 4 °C. Nucleic acids were precipitated using 0.45% polyethyleneimine and pelleted by centrifugation at 30,000 g for 30 min at 4 °C. The supernatant was mixed with 1.5–2 mL of Glutathione Sepharose 4B (GE Healthcare, Little Chalfont, UK) and incubated for 1 h at 4 °C with gentle agitation followed by centrifugation at 700 g for 5 min. The Glutathione Sepharose beads were loaded onto a column and washed with 10 column volumes (10 mL/wash) of wash buffer (50 mM Tris-HCl pH 8.0, 1 M NaCl, 10 mM β-mercaptoethanol, 2.5 mM DTT, 0.1 mM ZnCl$_2$). The glutathione S-transferase (GST) tag was removed by thrombin cleavage (GE Healthcare) at 4 °C for 12 h with gentle agitation. Ty1 Gag p45 was eluted using wash buffer, concentrated with centrifugal filtration (Millipore, Billerica, MA, USA), aliquoted and stored at −80 °C.

2.9. Filter Binding Assay

Reactions were performed in binding buffer (50 mM Tris-HCl pH 7.5, 40 mM KCl, 2 mM $MgCl_2$, 0.01% Triton X-100) containing different concentrations of NaCl (50, 100, 150, 200, 250, 500 mM). [^{32}P]-labeled domain I of Ty1i RNA (0.2 nM) was incubated for 4 min at 95 °C without magnesium ions and Triton X-100, and slowly cooled to 37 °C. $MgCl_2$ and Triton X-100 were added following incubation for 10 min at 37 °C. Ty1 Gag protein solutions were prepared by sequential two-fold dilution of Gag in binding buffer. The binding reaction was initiated by mixing equal volumes of RNA and Gag protein in a microplate (final concentration of RNA was 0.1 nM). The reactions were incubated for 15 min at 24 °C, filtered and washed with 2 × 200 µL binding buffer containing 50 mM NaCl. A 96-well dot-blot (Minifold, Whatman, Maidstone, UK) was used with nitrocellulose (Protran, Whatman, Maidstone, UK) on top and charged nylon (Hybond N+, GE Healthcare) membranes on the bottom. Prior to use, both membranes were soaked in binding buffer containing 50 mM NaCl. After filtration, membranes were dried and exposed to a phosphoimager screen. Data were fitted to the Hill equation using Origin 8.5 software (OriginLab, Northampton, MA, USA).

2.10. H1Δ Plasmid and Yeast Strains

The H1Δ deletion (T1015 - A1035) was generated by overlap PCR using flanking oligonucleotides Ty335F (5′-TGGTAGCGCCTGTGCTTCGGTTAC-3′) and RP1 (5′-ATAGTCAATAG CACTAGACC-3′), and overlapping oligonucleotides B (5′-GAAAGAATTTTCATGATAGGATGTCT TTGACCCAGGTAGGTAG-3′) and C (5′-GGTCAAAGACATCCTATCATGAAAATTCTTTCCAAAAG TATTGAAAAAA-3′). Wild-type pGPOLΔ (pBDG1130) [14] was used as the template for PCR. Nucleotide sequences correspond to the reference Ty1-H3 element (GenBank M10876.1). The H1Δ PCR product was cloned into pGPOLΔ using XhoI and BglII. The resulting plasmid pBAS47 is denoted as H1Δ. The H1Δ insert in pBAS47 was verified by DNA sequencing. Plasmids pBDG1130 and pBAS47 were transformed into the following strains: DG2196 (1 Ty1) [13] to generate DG2374 and YAS89, and DG3582 (0 Ty1) [14] to generate YAS85 and YAS87, respectively.

2.11. Northern and Western Blotting

Yeast cultures for total cellular RNA and protein extraction were grown in SC-Ura + 2% glucose medium at 22 °C for 24 h. RNA was extracted using the MasterPure Yeast RNA purification kit (Epicenter Biotechnologies, Madison, WI, USA) [14]. For each strain, 8 µg total RNA was separated on a 1.2% formaldehyde-agarose gel and subjected to Northern blot analysis using [^{32}P]-labeled riboprobes corresponding to Ty1 nucleotides 1266–1601 and ACT1, followed by phosphorimaging using a STORM 840 phosphorimager and ImageQuant software (GE Healthcare) [13]. Protein isolation and Western blot analysis to detect p22 was performed as described previously [14]. A rabbit polyclonal antisera against Pgk1 (kindly provided by Jeremy Thorner) was used at a 1:100,000 dilution. Immune complexes were detected with enhanced chemiluminescence (ECL) reagent (GE Healthcare). The amount of p22 relative to Pgk1 was estimated by densitometry using Quantity One software (Bio-Rad). Northern and Western analyses using the 0 Ty1 and 1 Ty1 strains containing pGPOLΔ or pH1Δ were repeated twice and representative results are presented. Also, independent Western analyses using the 0 Ty1 strain containing pGPOLΔ or pH1Δ were repeated three more times.

Ty1his3-AI mobility frequencies were determined as described previously [13,33]. Briefly, a single colony was resuspended in 1 mL water and four; 1 mL SC-Ura cultures were inoculated with 5 µL of cell suspension. Quadruplicate cultures for each strain were grown at 22 °C for three days. Cells were pelleted, resuspended in 1 mL water, and dilutions spread on SC-Ura and SC-Ura-His plates were incubated at 30 °C for 4 days. The frequency of Ty1his3-AI mobility was calculated by the number of His$^+$ Ura$^+$ colonies/the number of Ura$^+$ colonies per mL of culture.

2.12. RNA 3D Structure Prediction

Structure prediction experiments were performed by RNAComposer [34] webserver [35]. The AUG1AUG2 RNA domain I sequence: GGGUCAAAGACAUCCUAUCCGUUGAUUAUACGGAUA UCAUGAAAAUUCUUUCCAAAAGUAUUGAAAAAAUGCAAUCUGAUACCC and secondary structure topology in dot bracket notation: ((((...(((((((..((((((((........))))))).............(((.((........)).))).)))...)))...)))) were used as input data. The 3-way junction of domain I of AUG1AUG2 RNA was generated by RNAComposer, therefore, it was substituted by the elements introduced by the user. This element was chosen from RNA structures deposited in Research Collaboratory for Structural Bioinformatics (RCSB) Protein Data Bank (PDB) database following the criteria of the highest homology of secondary structure topology and sequence. More than 10 batches with different three-way junction structures were run. Ten models were generated for every batch. The resulting models were clustered based on the agreement with the hydroxyl radical cleavage data and the energy. Hydroxyl radical cleavage reactivity indexes from experiments were compared with indexes denoting atomic crowding around phosphorus at the corresponding nucleotide residue. The models with correct energy [36] and the best similarity were accepted.

3. Results and Discussion

3.1. Both AUG Codons in Ty1i RNA Can Be Recognized for Translation Initiation

Our previous results demonstrated that p22 translation can be initiated from AUG1 and AUG2 codons and is strictly cap-dependent. Also, either AUG1 or AUG2 can function to initiate translation when the other is mutated [14]. However, a number of questions remain unanswered: (i) Are both AUGs active for translation when present in the same RNA? (ii) Or is one codon translated preferentially? (iii) Does leaky scanning account for p22 synthesis from AUG2? Moreover, deleting the 5' UTR or mutating AUG1 or AUG2 decreases the level of p22 in vivo. For AUG1 and AUG2 codon mutants, the decrease in the p22 level is significantly larger than expected considering that one AUG codon is still present. These results suggest that the structure of the 5' terminal part of Ty1i RNA may influence p22 translation.

Translational activity of both AUG codons could be beneficial and contribute to the evolutionary diversification of p22. To gain insights into translation from AUG1 and AUG2 in Ty1i RNA, we performed in vitro translation assays using three derivatives of AUG1AUG2 RNA [14]. AUG1AUG2 RNA started at nt 1000 of Ty1, comprised the 5' UTR and p22 open reading frame (ORF), and ended with a natural stop codon (Figure 1). The difference between p22 proteins translated from AUG1 and AUG2 is only 10 amino acid residues (30 nt). Such a small size difference makes the two proteins difficult to separate by gel electrophoresis and obscures simultaneous analysis of the translation levels from both AUGs. To overcome this difficulty, we synthesized AUG1AUG2* RNA in which AUG2 (including its Kozak context) is 30 nucleotides downstream of the original AUG2, and introduced a GCG alanine codon in place of AUG2 (Figure 2). This modification increased the distance between AUG1 and AUG2* to 60 nt (20 amino acids), which allowed separation of the two translation products. A frameshift mutation (insertion of AU between U1050 and C1051) was introduced in AUG1[frs]AUG2 RNA (Figure 2). In this case, translation from AUG1 occurred out of frame in relation to AUG2 and resulted in the synthesis of a 49-amino acid peptide. Translation of the AUG1AUG2* and AUG1[frs]AUG2 RNAs allowed us to determine if both AUGs were recognized for translation. The third RNA, AUG1[stop]AUG2, contained an insertion of a single U between U1060 and U1061, which introduced a premature stop codon following translation from AUG1 (Figure 2). This RNA mutation was designed to help determine the level of p22 translated from AUG2. Each construct was also designed to avoid the introduction of rare codons that could obscure translation.

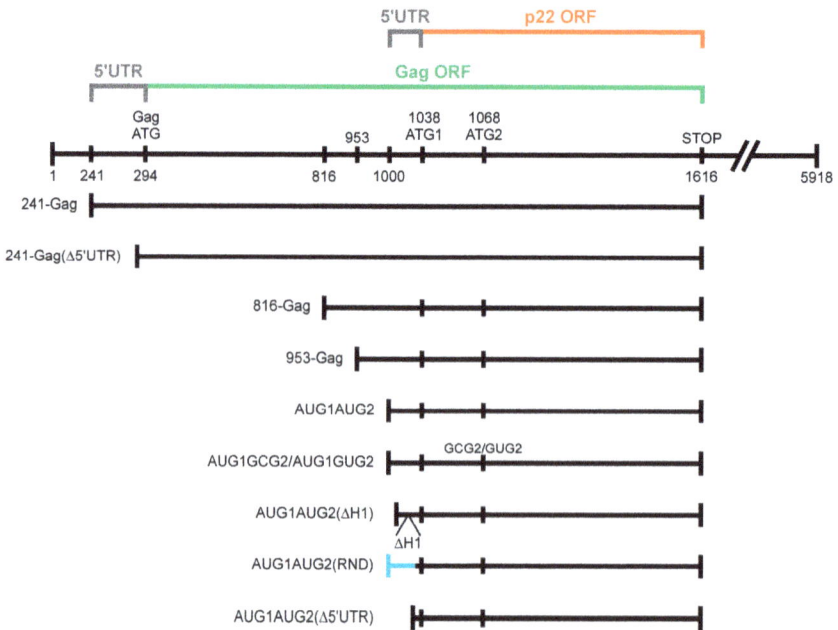

Figure 1. RNA constructs used in this study. Nucleotide positions correspond to the Ty1H3 DNA sequence (GenBank accession M18706.1) [15]. 5' UTR: 5' untranslated region, ORF: open reading frame.

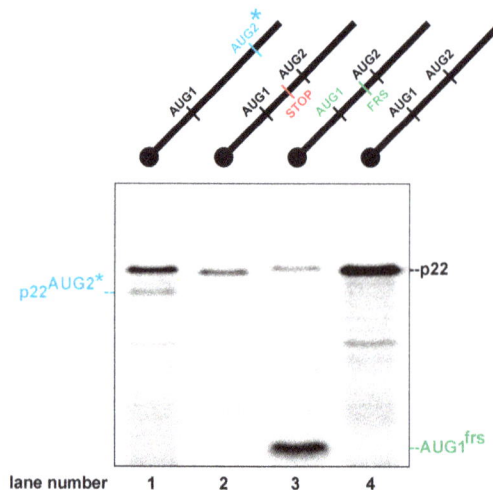

Figure 2. In vitro translation of Ty1i RNA and its derivatives in wheat germ extract. In vitro transcribed, capped RNA AUG1AUG2*, AUG1^{stop}AUG2, AUG1^{frs}AUG2 and AUG1AUG2 were translated in the presence of ^{35}S-methionine followed by electrophoresis and autoradiography. Schematic representation of RNA molecules is shown above the gel (see text for details).

AUG1AUG2* RNA was translated into two products: p22^{AUG1} synthesized from the natural AUG1 and the shorter protein p22^{AUG2*} (Figure 2, lane 1). p22^{AUG1} / p22^{AUG2*} were synthesized in a ratio of 5:1, which indicates that AUG1 is the main site of p22 translation initiation in AUG1AUG2*

RNA. However, the translational activity of AUG1AUG2* RNA decreased 75% when compared with wild-type AUG1AUG2 RNA. Two proteins were also translated from the AUG1^{frs}AUG2 RNA: a faster migrating out of frame AUG1^{frs} peptide, and p22^{AUG2}, which originated from the natural AUG2 triplet (Figure 2, lane 3). AUG1^{frs}/p22^{AUG2} were synthesized in a ratio of 6:1, which is similar to AUG1AUG2*, and confirms that AUG1 is utilized preferentially for p22 initiation in these two RNAs. As expected, p22^{AUG2} that initiated from AUG2 was detected with AUG1^{STOP}AUG2 RNA (Figure 2, lane 2). The level of AUG2-initiated p22 was low but comparable between different constructs.

Taken together, the results of in vitro translation show that both AUG codons present in Ty1i RNA can be actively translated and AUG1 is preferentially utilized to initiate p22 synthesis. Our results also suggest that leaky scanning is the most likely mechanism for p22 translation from AUG2. Experimental support for leaky scanning is illustrated by the decrease of AUG2 translation levels from AUG1AUG2* and AUG1^{frs}AUG2 RNAs (having both p22 AUG codons) in comparison to GCG1AUG2 RNA mutant where only AUG2 is present [14]. Moreover, the translational activity of AUG1AUG2* and AUG1^{frs}AUG2 RNAs was significantly lower when compared to wild-type AUG1AUG2 RNA. These results raise the possibility that AUG1AUG2* and AUG1^{frs}AUG2 RNAs affect the structure of the 5' UTR of Ty1i RNA, leading to translation inhibition, and that the 5' UTR may also regulate the production of p22.

3.2. The 5' UTR of mRNA Interacts with the p22 Coding Region

Significant loss of translational activity from AUG1 in AUG1GCG2 [14] (Figure 1), AUG1AUG2* and AUG1^{frs}AUG2 RNAs suggests that the structure of the region containing AUG1 and AUG2 is important for p22 translation. Therefore, we performed selective 2'-hydroxyl acylation analyzed by primer extension (SHAPE) [37] on the 5' terminal region of Ty1i RNA to examine its secondary structure. N-methylisatoic anhydride (NMIA) preferentially modifies 2'OH groups of single-stranded and flexible nucleotides in RNA. Primer extension of fluorescently labeled primers by reverse transcriptase is blocked at modified positions in RNA, and these truncated DNA products can be identified using capillary electrophoresis. Secondary RNA structures were obtained by computational analysis of the reverse transcription products. Secondary structure probing experiments were carried out on AUG1AUG2 RNA that was used in the in vitro translation studies. This ~630 nt long RNA contained the 5' UTR of Ty1i RNA (37 nt) and coding sequence of p22 (Figure 1).

Figure 3 shows a secondary structure model of the 5' terminal part of the Ty1i RNA [15] predicted using the *RNAstructure* software [38,39] which incorporates experimental constraints from SHAPE mapping.

Our results suggest that Ty1i RNA folds into two major domains. The smaller domain I (G1000–1083) and larger domain II (A1096–U1501) were connected by a 12nt-long single-stranded region (A1084–G1095).

Interestingly, domain I included the Ty1i 5' UTR and p22 coding sequence, and contained both p22 initiation codons (Figure 3). This structure is organized by the interaction of the proximal part of the 5' UTR (G1000–U1012) with a stretch of coding sequence (A1068–C1083; stems S1–S3). Also, two hairpin structures were present. Hairpin H1 (U1015–A1035) was composed of residues from the 5' UTR while hairpin H2 (U1048–A1066) contained nucleotides from the coding sequence. A three-way junction connected hairpins H1, H2 and stem S1.

The data from SHAPE probing support the predicted structure of domain I. Nucleotides within single-stranded regions were reactive towards the SHAPE reagent, including apical loops of both hairpins, internal loops, bulges and mismatches. The presented structure was also supported by dimethyl sulfate (DMS) probing. DMS methylates N1 of adenosines and N3 of cytidines that have an accessible Watson–Crick edge of the base rings [40]. In our structure, almost every A and C residue predicted to be single-stranded was susceptible to DMS methylation. However, some nucleotides in the hairpin H2 stem were methylated moderately by DMS but remained unreactive towards NMIA.

These results support the idea that the C1052 and A1064-A1066 hairpin region is constrained by non-standard base pairing.

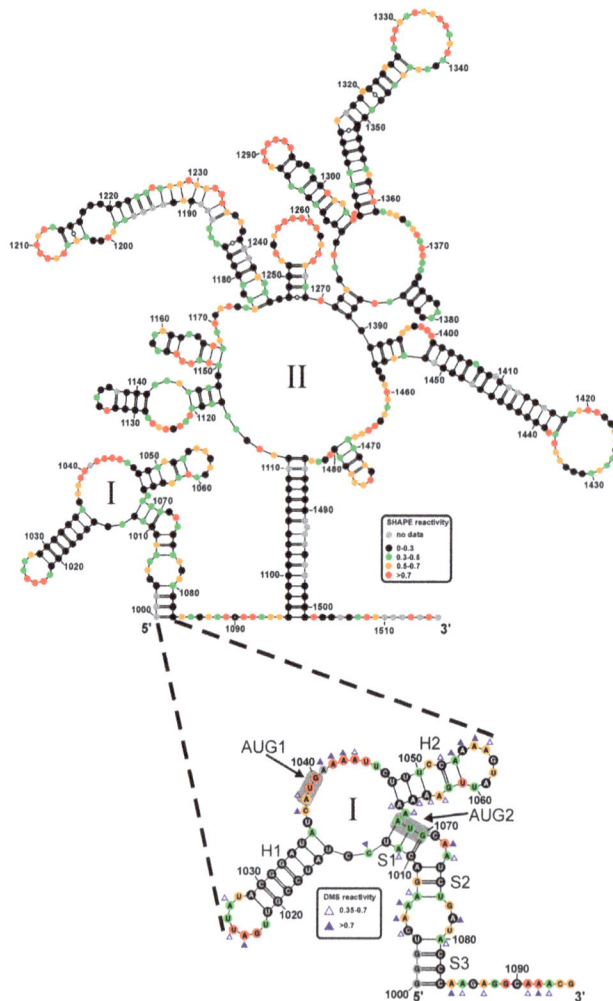

Figure 3. Secondary structure model of the 5′ terminal segment of Ty1i RNA (upper panel) and a detailed view of domain I (bottom panel) predicted by the *RNAstructure* software with experimental constraints [38]. Nucleotides are coloured according to their selective 2′-hydroxyl acylation analyzed by primer extension (SHAPE) reactivity (black, green, orange, red). The blue triangles (filled and open) represent dimethyl sulfate (DMS) modifications.

Interestingly, domain I contained both p22 initiation codons localized in different structural contexts (Figure 3). AUG1 constituted part of the 12nt-long single-stranded region U1036–C1047 while AUG2 was embedded in the double-stranded S1 stem that was formed by interactions of nts 1068–1070 with the residues of the 5′ UTR (C1010–U1012). The S1 stem may be thermodynamically unstable since the AUG2 triplet was somewhat reactive against NMIA.

Domain II folded into a large multibranched structure (Figure 3) organized by extensive pairing between A1096–C1111 and G1485–U1501. As a result, a 16 bp duplex region was formed. Domain II contained a complex junction that connected six simple hairpin structures and one branched region in a three-way junction motif. The majority of the single stranded regions were well mapped by NMIA. Importantly, the NMIA modification pattern of nucleotides spanning domain II in AUG1AUG2 RNA was very similar to the same region mapped inside VLPs using in virio SHAPE [30] (please note that the numbering herein corresponds to the complete Ty1H3 element while the numbering in reference [30] corresponds to Ty1 genomic RNA [30]). This result suggests that our in vitro folding conditions recapitulate the native structure of Ty1 RNA.

3.3. The 3D Structural Integrity of Domain I Affects p22 Translation

We reported that the combined level of p22 synthesized from AUG1GCG2 and GCG1AUG2 RNA constitutes only 30% of that obtained from wild-type AUG1AUG2 RNA [14]. Secondary structure probing of AUG1AUG2 RNA revealed that both p22 initiation codons were located within the same domain. Thus, mutation of AUG1 or AUG2 could cause structural perturbations that inhibit p22 translation. Since the in vitro translation results (Figure 2) identified AUG1 as a main translation initiation site for p22 synthesis, we hypothesized that mutating AUG2 to GCG strongly inhibited translation from AUG1 due to changing the structural context of AUG1 in domain I. The AUG2 to GCG mutation also introduced a U–G wobble pair as well as A–C mismatch that could affect the double-stranded character of the S1 stem.

To determine if the GCG mutation altered the structure of domain I, we performed secondary structure probing of AUG1GCG2 RNA using SHAPE. Although the overall reactivity pattern of the AUG1GCG2 RNA was preserved (Figure 4A), the region of domain I containing the GCG mutation (A1066–A1071) became highly reactive. This alteration suggests that the mutant RNA residues in the S1 stem are single-stranded or this region is highly unstable. Additionally, several nucleotides in hairpin H2 displayed a different pattern of reactivity: G1057–A1059, U1061 and G1062 exhibited higher reactivity while A1055 had decreased reactivity. Surprisingly, the structural motifs in the neighborhood of AUG1 remained essentially the same in wild type and mutant AUG1AUG2 RNA. Moreover, the GCG mutation did not change the secondary structure of domain II (data not shown). Overall, our data suggests that the GCG mutation disrupts the three-dimensional structure of domain I, which in turn inhibits the translation of p22 from AUG1.

Our model suggests that a three-way junction element (Figure 3) governs the special organization of domain I. By disrupting the S1 stem, the GCG mutation might change the topology and relative positioning of the H1 and H2 hairpins. Changes in the three-dimensional structure of RNA molecules can be monitored by native polyacrylamide gel electrophoresis [41]. Therefore, we subjected the isolated domain I (nts G1000–C1083) containing the GCG mutation (domain I^{GCG2}) along with the wild-type domain I to native gel electrophoresis (Figure 4B). We observed a slower mobility of domain I^{GCG2} RNA, which may reflect a change in the three-dimensional structure of domain I when compared with wild type. Migration of both wild-type and GCG mutated domain I remained unchanged at a higher concentration of Mg^{2+} ions, suggesting that this part of Ty1i RNA undergoes unimolecular folding [42].

The results obtained by native gel electrophoresis suggest that the double-stranded character of the S1 stem is an important factor stabilizing the three-dimensional structure of domain I. To help preserve the double-stranded character of stem S1, we mutated AUG2 to a GUG valine codon that changed only the first U–A pair to a U–G wobble pair (Figure 1). Secondary structure probing of AUG1GUG2 mutant RNA indicated that the S1 stem was slightly destabilized (Figure S1). Moreover, two residues directly upstream of the S1 stem (A1066 and A1067) were more reactive, suggesting an enhancement of local flexibility. A1066 and A1067 were also strongly modified in AUG1GCG2 mutant RNA. Some of the nucleotides in the H2 hairpin that changed their reactivity in AUG1GCG2 RNA behaved in a similar manner in AUG1GUG2 RNA. Higher reactivity of U1058 and A1062 as well as lack of reactivity of A1055 was detected. A1063 was also less reactive in AUG1GUG2 RNA when compared to wild type AUG1AUG2. Importantly, the structural context of AUG1 was preserved, which is similar to the AUG1GCG2 and AUG1GUG2 mutants. Taken together, our data suggest that the GUG2 mutation destabilized the S1 stem much less than the GCG2 mutation, and the structural integrity of the S1 stem and hairpin H2 are important determinants for the proper three-dimensional structure of domain I.

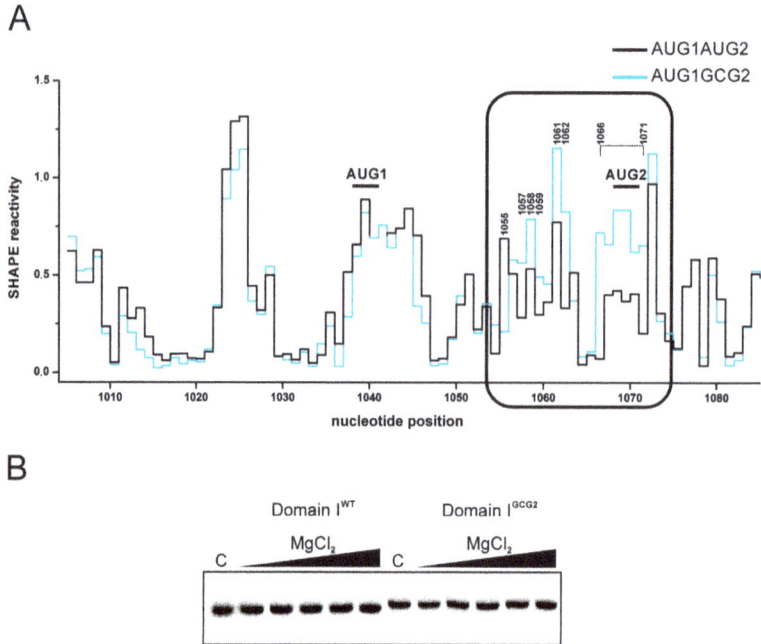

Figure 4. (A) SHAPE reactivity profile of the AUG1AUG2 (black) and AUG1GCG2 (blue) domain I as a function of nucleotide position. Nucleotides that changed their reactivity in domain I^{GCG2} are indicated. **(B)** Native gel electrophoresis of the [^{32}P]-labeled wild-type and mutated domain I of Ty1i RNA at increasing concentrations of MgCl$_2$. C: control reaction without MgCl$_2$. WT: wild type.

Mutation of AUG2 to GCG2 markedly inhibits p22 translation (Figure 2) [14]. Since we determined that the GUG2 mutation had a less profound effect on the domain I secondary structure, we analyzed the translational activity of capped and uncapped AUG1GUG2 RNA along with AUG1GCG2 and AUG1AUG2 RNA in vitro (Figure 5A). In agreement with our previous study [14], p22 translation from AUG1GCG2 RNA was inhibited to ~15% of the initial value calculated for AUG1AUG2 RNA. Interestingly, the translation of p22 from AUG1GUG2 RNA was also inhibited to ~20% when compared with wild type RNA. These results further extend our finding that the structural integrity of the domain I of Ty1i RNA contributes significantly to the efficient translation of the p22 from AUG1, and even small structural changes impair translation in vitro.

Placement of the initiation codon in thermodynamically stable secondary structures can decrease its translational activity [43]. However, the calculated thermodynamic stability [44] of domain I in wild-type Ty1i RNA was only −25.2 kcal/mol, and AUG1 was predicted to reside in a long single-stranded region (Figure 3). To assess the thermodynamic stability of the 5′ terminal segment of Ty1i RNA, we determined the reactivity profile of AUG1AUG2 RNA by SHAPE mapping at different temperatures (Figure 5B). SHAPE analysis at 37 °C and 60 °C identified residues within domain I that changed their reactivity at 60 °C. Interestingly, the most pronounced effects were observed in the regions prone to destabilization in RNA mutants AUG1GCG2 and AUG1GUG2 (Figure 4 and Supplementary Figure S1). At 60 °C, the nucleotide stretch A1067–G1077 (including AUG2) as well as the opposite strand A1005–C1013 became highly reactive, suggesting that the strands dissociate. Also, several residues located in the hairpin H2 stem (U1049–C1052) and in the apical loop (A1059–U1061) were altered, suggesting that the region containing AUG2 and hairpin H2 is less stable than other parts of domain I.

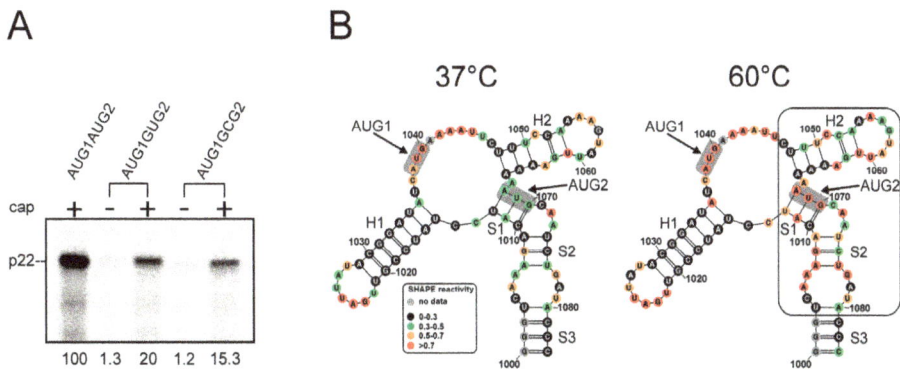

Figure 5. In vitro translation of AUG2 mutational variants of Ty1i RNA and melting profile of AUG1AUG2 RNA. (**A**) In vitro transcribed capped or uncapped transcripts were translated using wheat germ extract in the presence of ^{35}S-methionine. Calculated translation activity (in relation to the capped AUG1AUG2 RNA) is shown below the gel. (**B**) Melting of AUG1AUG2 RNA followed by SHAPE at 37 °C and 60 °C. Nucleotides are coloured according to their reactivity (black, green, orange, red). The segment of domain I with the strongest changes at 60 °C is boxed.

3.4. Structure of Domain I Specific for Ty1i RNA Stimulates p22 Translation

In vitro translation and secondary structure probing of the 5′ terminal part of wild-type and mutant Ty1i transcripts suggest that domain I plays an important role in the efficient translation of p22 from AUG1. Previous results show that p22 is not translated from the full-length genomic RNA [15]. These findings motivated us to ask whether the structure of domain I was stable in the context of a larger RNA that more closely resembles Ty1 genomic RNA. To this end, we analyzed a ~1400 nt RNA (nts 241–999 using the coordinates of the complete Ty1H3 element), termed 241-Gag RNA, that began from the first nucleotide of the genomic Ty1 RNA, and included the structured 5′ UTR [30,45] and Gag coding sequence (Figure 1). Comparison of SHAPE reactivity profiles of 241-Gag and AUG1AUG2 RNAs revealed different modification patterns of domain I (Figure 6A).

The reactivity of the region encompassing AUG2 (A1067–A1072) increased in 241-Gag RNA while the proximal part of the single-stranded region connecting domains I and II (A1084–G1089) lost accessibility to NMIA modification. The observed alterations suggest that domain I and the neighboring regions fold differently when the 5′-terminal sequence of genomic RNA is present in the transcript.

The secondary structure of the full-length Ty1 RNA has been determined inside virus-like particles (VLPs) by in virio SHAPE analysis [30]. In the proposed structure for Ty1 genomic RNA, the sequence encompassing domain I is folded differently than in Ty1i RNA (Figure 6B). Interactions between C979–U983 and A1085–G1089 extended domain I in the full-length transcript. Moreover, the structural context of the p22 initiation codons differed significantly. Unlike their context in Ty1i RNA, AUG1 was fully paired with the C1010–U1012 in full-length Ty1 RNA. Interestingly, the C1010–U1012 region was also paired but with the AUG2 codon forming the S1 stem in Ty1i RNA (Figure 3). AUG2 was localized in the stem of a predicted unstable hairpin G1057–C1071. The only common structural element within the region encompassing domain I in the full-length Ty1 and Ty1i RNAs was hairpin H1, suggesting that hairpin H1 folds independently of the structural elements present in its vicinity.

Importantly, comparing the reactivity profiles of 241-Gag and full-length Ty1 RNA [30] revealed that domain I folding was similar (Figure 6B). The main difference was AUG1 reactivity, which was high in 241-Gag RNA and low in full-length Ty1 RNA. This difference suggests that the cellular environment in this region, such as the presence of the Gag chaperone, folds the RNA into a more stable structure.

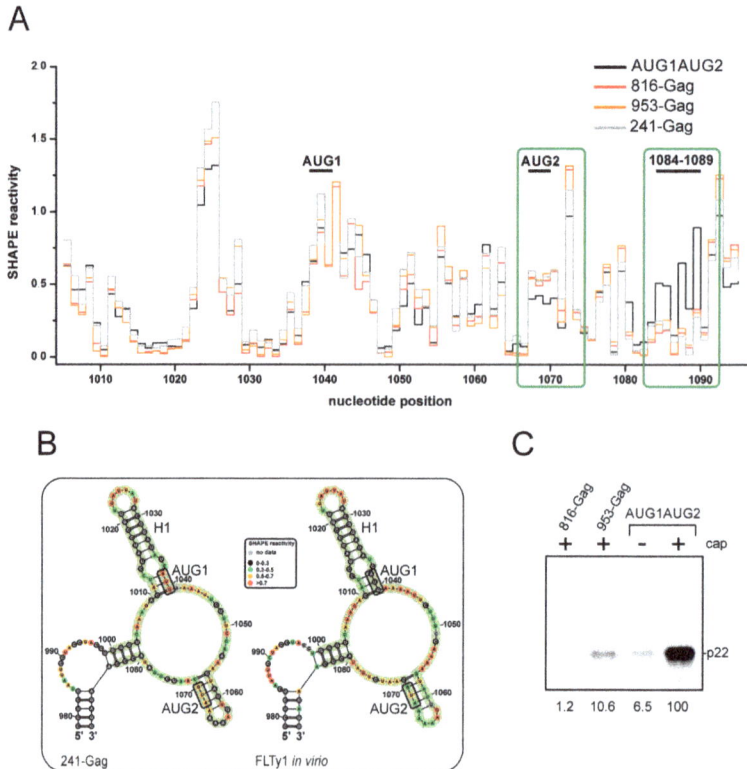

Figure 6. Secondary structure probing and the in vitro translation of 241-Gag RNA and its derivatives. (A) Reactivity plot of nucleotides spanning domain I in AUG1AUG2 RNA (black), 816-Gag RNA (red), 953-Gag RNA (orange) and 241-Gag RNA (grey). Regions showing consistent differences in reactivity are boxed (green). (B) Comparison of the secondary structure models of domain I obtained in vitro for 241-Gag RNA (left) and full-length genomic Ty1 RNA within virus-like particles (VLPs) (in virio conditions; right). Nucleotides that cover domain I in Ty1i RNA are marked (green background). p22 initiation codons and the H1 hairpin are also highlighted. (C) In vitro translation of sequential variants of Ty1 genomic RNA. Capped or uncapped transcripts were translated in wheat germ extract in the presence of the ^{35}S-methionine. Quantitation of the translation products is shown below the gel.

The distinct structure of the region encompassing domain I in the full-length Ty1 RNA raised a question concerning how domain I might influence p22 translation. The initiation of p22 synthesis from the 241-Gag RNA is unlikely to occur, which raises the possibility that p22 synthesis requires a specific structure of domain I in Ty1i RNA [14]. The presence of the Gag AUG initiation codon as well as seven internal in-frame AUG codons before encountering AUG1 would preclude migration of the preinitiation complexes downstream of the AUG1 and AUG2 initiation codons. Additionally, the 5′ UTR of Ty1i RNA in the 241-Gag RNA would be extended to over 700 nucleotides, which could greatly affect the scanning mechanism. To address whether a specific structure of the domain I of Ty1i RNA is necessary for the efficient translation of p22, we synthesized 816-Gag and 953-Gag RNAs (Figure 1). Both RNA molecules were designed to possess full-length folding of domain I, which is supported by their similar reactivity profile when compared to 241-Gag RNA (Figure 6A). The 816-Gag and 953-Gag RNAs were translated in vitro in wheat germ extract (Figure 6C). We observed that p22 protein was poorly translated from both RNA molecules and could be detected only when capped transcripts were used. Low levels of translation from extended Ty1 transcripts with the full-length-like folding of the region 1000–1083 suggests that the structure of the domain I observed in Ty1i RNA specifically stimulates p22 translation from AUG1.

3.5. The Ty1i RNA 5′ UTR Stimulates p22 Translation

To further understand the role of the Ty1i 5′ UTR in p22 translation, we analyzed in vitro several mutant RNA constructs (Figure 1). In AUG1AUG2(Δ5′ UTR), 32 of 37 nucleotides of the 5′ UTR have been deleted while in AUG1AUG2(RND) the same sequence was replaced by 32 random nucleotides. In AUG1AUG2(ΔH1), the common structural element of full-length Ty1 and Ty1i RNA (hairpin H1) was deleted (nts 1015–1031). Also, all transcripts maintained an intact Kozak context adjacent to the AUG1 initiation codon.

We observed significant inhibition of p22 translation from all three RNA constructs (Figure 7A). Deleting the 5′ UTR inhibited p22 translation by 40% when compared to wild-type AUG1AUG2 RNA. These results suggest that the Ty1i 5′ UTR is required for efficient p22 synthesis. Since shortening the 5′ UTR to only six nucleotides could interfere with ribosome scanning [46–48], we analyzed 241-Gag(Δ5′ UTR) RNA possessing 5′ UTR that was also reduced to six nucleotides. However, the translation of Gag was unaffected (Figure 7B). This result suggests that the inhibitory effect observed for AUG1AUG2(Δ5′ UTR) may impair the structure of domain I. The important role of the 5′ UTR in p22 translation was also supported by the translation of AUG1AUG2(RND) and AUG1AUG2(ΔH1) RNAs. Despite having a 5′ UTR of the same length as wild-type, AUG1AUG2(RND) RNA displayed >70% inhibition in p22 translation. A 55% inhibition of p22 synthesis was also observed with AUG1AUG2(ΔH1) RNA. Taken together, our data suggest a stimulatory role for the Ty1i 5′ UTR in the translation of p22 due to its involvement in the folding of domain I.

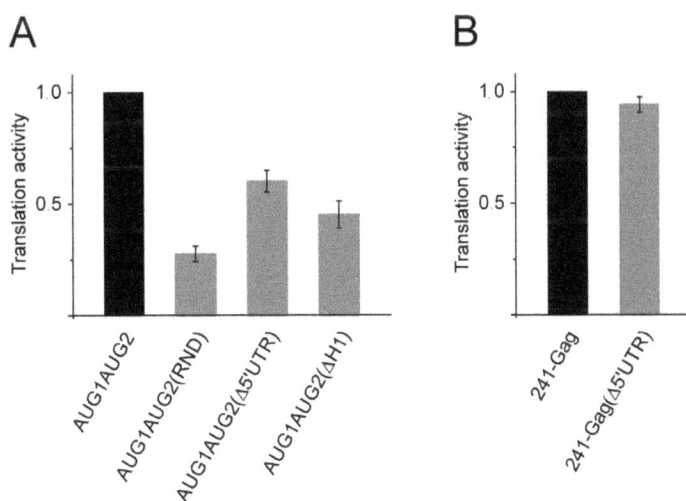

Figure 7. In vitro translation of the capped variants of the AUG1AUG2 RNA and 241-Gag RNA. Translational efficiency was normalized to the amount of the protein product synthesized from AUG1AUG2 RNA (**A**) or 241-Gag RNA (**B**).

3.6. Gag Interacts Specifically with Ty1i Domain I In Vitro

Translation initiation can be regulated not only by RNA structure but also by protein factors that interact with structural elements in mRNAs [19]. Since the amount of Gag and p22 determines the level of inhibition of Ty1 mobility [49], perhaps Gag modulates the efficiency and/or timing of p22 translation. Potential Gag binding sites in the 5′ terminal part of Ty1i RNA were detected by hydroxyl radical footprinting of AUG1AUG2 RNA complexed with recombinant Gag-p45 (Figure 8A). The protected sequences were identified by comparing the reactivity profiles of AUG1AUG2 RNA in the presence and absence of Gag. Only regions in domain I displayed decreased susceptibility to hydroxyl radical cleavage in the presence of Gag, including residues A1011–C1019 that comprise part of the S1 stem and the hairpin H1 stem. Another potential Gag binding site was localized in the p22 coding region (nts A1084–G1095) connecting domains I and II. In particular, C1081–C1090 was protected from the cleavage in the presence of Gag (Figure 8A,B).

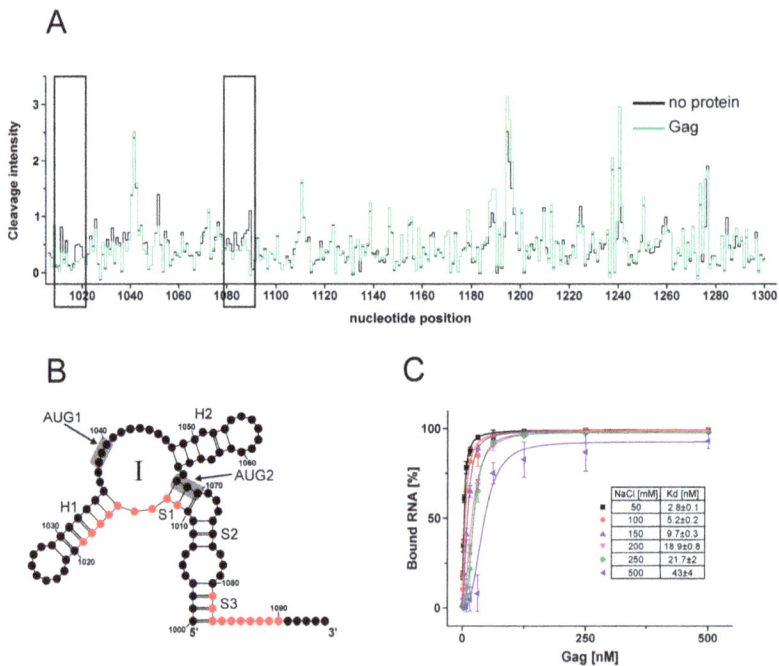

Figure 8. RNA binding properties of recombinant Ty1 Gag-p45. (**A**) Hydroxyl radical reactivity plots of protein free AUG1AUG2 RNA (black) in comparison with RNA probed in the presence of Gag (green). Regions showing consistent decreased reactivity over several nucleotides in the presence of Gag are boxed. (**B**) 2D structure model of Ty1i domain I with the positions protected from hydroxyl radical cleavage in the presence of the Ty1 Gag are indicated (red). (**C**) Filter-binding assay performed with Ty1i domain I RNA and Gag at different concentrations of NaCl (50–500 mM). The lines correspond to the best fit of the data. The error bars represent standard deviations. Kd: dissociation constant.

To further investigate the interaction between Gag and domain I, we calculated dissociation constants of RNA/protein complex formation using a double filter binding assay (Figure 8C). We used isolated domain I that was extended by the single-stranded stretch connecting domain I and II (RNA $I^{1000-1095}$) to encompass both Gag binding sites. The calculated dissociation constant (Kd ~3 nM) suggests that there is a high affinity binding site for Gag in domain I. To examine whether Gag binding is specific, we determined the Kd with increasing concentrations of NaCl, which is often used to compete out non-specific RNA/protein interactions [31]. The Gag/domain I interaction was slightly affected in the 100–250 mM NaCl range and persisted even at 500 mM NaCl (Kd ~43 nM). Taken together, the results from chemical footprinting and filter binding suggest that the interaction between Gag and domain I is strong and highly specific.

3.7. Deleting the Hairpin H1 Sequence Decreases Stability of Ty1i RNA In Vivo

To investigate the effects of the H1 hairpin on Ty1i RNA and p22 expression in vivo as well as on Ty1 transposition, a mutated pGPOLΔ plasmid was constructed (pBAS47, termed H1Δ) that expresses Ty1i RNA lacking the H1 sequence (U1015–A1035) from the 5' UTR (Figure 9). Wild type pGPOLΔ is a multicopy expression plasmid containing most of the Ty1 5'LTR and *GAG* that is driven by the *GAL1* promoter [15]. When yeast cells containing pGPOLΔ are grown in glucose media, *GAL1* promoted transcription of Ty1 is repressed. However, Ty1i RNA and p22 are still expressed from pGPOLΔ under glucose repression since Ty1i RNA is transcribed from internal initiation sites.

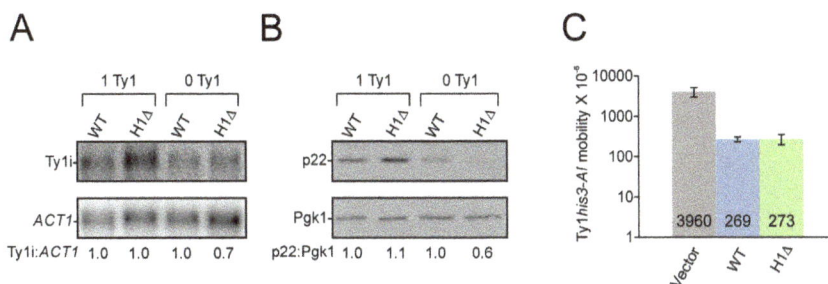

Figure 9. Effect of the hairpin H1 deletion on Ty1i RNA, p22 protein expression and Ty1*his3-AI* mobility. (**A**) Northern blotting of total RNA from the 1 Ty1 strain (DG2196) and 0 Ty1 strain (DG3582) containing either wild type (WT) pGPOLΔ or mutant pH1Δ plasmids. A [^{32}P]-labeled Ty1 riboprobe (nt 1266 to 1601) was used to detect Ty1i RNA. *ACT1* mRNA served as a loading control. Below are Ty1i:*ACT1* ratios as determined by phosphorimaging. (**B**) Whole cell extracts from strains used in (A) were immunoblotted with p18 antiserum to detect p22. Pgk1 served as a loading control. p22:Pgk1 ratios were determined by densitometry. (**C**) Quantitative Ty1*his3-AI* mobility assayed in the 1 Ty1 strain containing one genomic Ty1*his3-AI* element and empty vector, WT, or H1Δ plasmids. All strains were grown in glucose containing medium to repress *GAL1*-promoted Ty1 expression. Bars denote standard deviation.

We investigated the effect of H1Δ on Ty1i RNA level in a *S. paradoxus* strain with 1 chromosomal Ty1 element (DG2196; 1 Ty1) and the isogenic Ty1-less parent (DG3582; 0 Ty1) that contain WT pGPOLΔ or pH1Δ plasmids (Figure 9A). Northern blotting of total RNA from these strains showed no change in Ty1i RNA levels in the H1Δ mutant compared to the wild type (WT) plasmid in the 1 Ty1 strain. However, Ty1i H1Δ RNA levels decreased about 30% compared to WT Ty1i RNA in the 0 Ty1 strain (refer to Materials and Methods). These results suggest that the H1 hairpin may affect the stability of Ty1i RNA. In the 1 Ty1 strain, however, the defect in Ty1i H1Δ RNA stability was not evident. This may be due to additional Gag binding sites on Ty1i RNA that stabilize the transcript in the 1 Ty1 strain, as suggested by hydroxyl radical footprinting (Figure 8). Note that Gag binding sites C1081–C1090 remain intact in Ty1i H1Δ RNA and could function in vivo.

Total cell extracts from the same strains were subjected to Western analysis using an antiserum that detects p22 [14] (Figure 9B). The level of p22 remained about the same in the 1 Ty1 strain containing WT or H1Δ plasmids. In the 0 Ty1 strain, p22 decreased 43% (±12%) in the mutant pH1Δ when compared to WT pGPOLΔ. These results suggest that there is a correlation between p22 and Ty1i RNA levels (Figure 9A) in both strain backgrounds containing WT or H1Δ plasmids.

Finally, we asked if deleting the H1 hairpin from the Ty1i RNA affected Ty1 mobility (Figure 9C). A quantitative Ty1 mobility assay was performed in the 1 Ty1 yeast strain containing empty vector (Vector), WT or H1Δ plasmids. The single element in the 1 Ty1 strain is marked with the retrotransposition indicator gene *his3-AI* [33]. A Ty1*HIS3* genomic insertion that occurs following splicing of the *AI* (artificial intron) will complement the *HIS3* deletion mutation present in the strain. Therefore, the number of His$^+$ colonies generally reflect the level of Ty1 mobility. As expected for cells undergoing Ty1 CNC, the level of Ty1*his3-AI* mobility decreased about 15-fold from plasmid-based expression of p22 [13,14]. However, H1Δ and WT displayed similar levels of Ty1 mobility, suggesting that deleting the H1 hairpin does not affect Ty1 CNC despite the modest decrease in p22 observed in the 0 Ty1 strain (Figure 9B). Perhaps removing only one of the Gag binding sites in domain I of Ty1i RNA is not enough to affect CNC because Gag produced in the 1 Ty1 strain stabilizes Ty1 RNA through binding to other sites.

3.8. AUG1 is Exposed in a 3D Structural Model of Domain I RNA

Our Ty1i RNA structural and functional studies indicate that the 3D structure of domain I is important for efficient p22 translation. However, determining the 3D structure of RNA in solution is challenging. Therefore, we combined chemical probing experiments to map RNA secondary (Figure 3) and tertiary structures using RNAComposer [34]. To reveal the tertiary fold of domain I of AUG1AUG2 RNA and support RNAComposer predictions [36], we also used hydroxyl radicals to produce strand breaks. This approach allows one to map solvent exposed regions of the nucleic acid backbone. This analysis predicted >100 different 3D structures of domain I and clustered them based on their agreement with the hydroxyl radical cleavage data and the energy of the final RNA 3D structure. The structures that best-fit the hydroxyl radical cleavage data allowed us to explain the gain in SHAPE reactivity of H2 apical loop nucleotides upon S1 stem destabilization in the AUG1GCG2 and AUG1GUG2 RNA mutants. Our models suggest that the H2 hairpin stem bends due to the presence of an internal loop containing unpaired C1051 and A1063, which causes an apical loop of H2 to be positioned close to the 3-way junction. Thus, disruption of junction geometry due to S1 unwinding is likely to affect H2 apical loop reactivity. The best models shared the common feature of coaxial positioning of the S1 stem and H1 hairpin. Such an organization of the 3-way junction places AUG1 on the surface of the molecule between hairpins H1 and H2, and may contribute to AUG1's preferential use for initiating the translation of p22 (Figure 10).

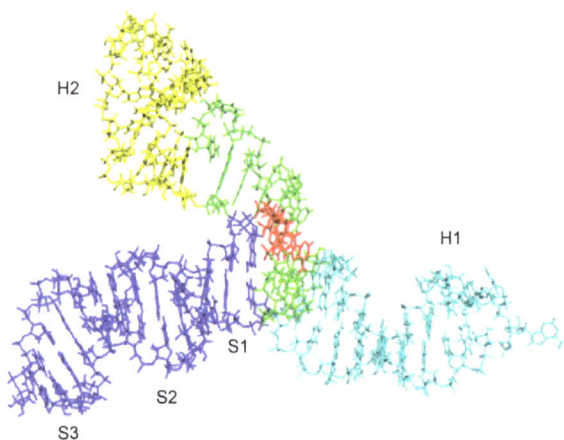

Figure 10. A 3D structure model of Ty1i RNA domain I. Structural elements are annotated: hairpin H1 (cyan), hairpin H2 (yellow), stem S1–3 (blue) and 3-way junction (green). AUG1 sequence is marked in red.

4. Conclusions

Translation initiation is the rate-limiting step of protein synthesis and is highly regulated by RNA binding factors and structural properties of the messenger RNA. This coordinated action allows cells to rapidly adapt to their environment without the need of de novo mRNA synthesis and transport from the nucleus to the cytoplasm [50]. In addition, a wide variety of viruses exploit variations in translation initiation to expand their coding capacity from a limited set of transcripts, including the use of alternative initiation codons and internal ribosome entry sites [17]. In the present work, we address how the Ty1 restriction factor p22 is translated from Ty1i RNA using a combination of structural and functional approaches. We show that two p22 initiation codons on Ty1i RNA are embedded in structural domain I, which is formed by an interaction between the 5′ UTR and the coding sequence. Our in vitro translation experiments show that both p22 initiation codons can be utilized but that AUG1 is used preferentially. We demonstrate that the structural integrity of Ty1i RNA is critical for the efficient expression of p22 from AUG1. Even small changes in the domain I sequence that disrupt its secondary and tertiary structure result in strong inhibition of p22 synthesis. Our studies have mapped two high affinity Ty1 Gag binding sites located in domain I of Ty1i RNA. Deletion of one of the binding sites leads to a decrease in the p22 level in vivo by destabilizing Ty1i RNA. Our work supports the hypothesis that structural motifs of domain I are not only important for the efficient translation of p22 protein but may also contribute to the stability of Ty1i RNA via interactions with Gag. Such interactions raise the possibility of an autogenous control loop where Gag positively controls the synthesis of p22, which in turn inhibits Gag function and mediates Ty1 CNC. However, more work will be required to understand how Gag binding to Ty1i RNA contributes to its stability.

Supplementary Materials: The following are available online at www.mdpi.com/1999-4915/9/4/74/s1, Figure S1: SHAPE reactivity AUG1GUG2 RNA mutant, Table S1: Primers used for construction of templates for in vitro transcription and reverse transcription, Table S2: Quantitation of the translation products from the gel in Figure 2, Data set S1: SHAPE data of AUG1AUG2 RNA.

Acknowledgments: We thank Agnieszka Kiliszek and Katarzyna Pachulska-Wieczorek for valuable discussions, and Jeremy Thorner for providing the Pgk1 antiserum. This work was supported by the Ministry of Science and Higher Education Poland [0492/IP1/2013/72], Foundation for Polish Science [HOMING PLUS/2012-6/12] (KJP), NIH grant GM095622 (DJG) and funds from UGARF (DJG). KJP also acknowledges support from the Ministry of Science and Higher Education Poland (MNiSW, fellowship for outstanding young scientists).

Author Contributions: L.B., M.B., A.S., D.J.G. and K.J.P. conceived and designed the experiments; L.B., M.B., A.S. performed the experiments; L.B., M.B., A.S., D.J.G. and K.J.P. analyzed the data; L.B., M.B., A.S., D.J.G. and K.J.P. wrote the paper.

Conflicts of Interest: The authors declare no conflict of interest.

References

1. Kim, J.M.; Vanguri, S.; Boeke, J.D.; Gabriel, A.; Voytas, D.F. Transposable elements and genome organization: A comprehensive survey of retrotransposons revealed by the complete *saccharomyces cerevisiae* genome sequence. *Genome Res.* **1998**, *8*, 464–478. [PubMed]

2. Curcio, M.J.; Lutz, S.; Lesage, P. The Ty1 ltr-retrotransposon of budding yeast. *Microbiol. Spectr.* **2015**, *3*, 1–35. [PubMed]

3. Feng, Y.X.; Moore, S.P.; Garfinkel, D.J.; Rein, A. The genomic RNA in Ty1 virus-like particles is dimeric. *J. Virol.* **2000**, *74*, 10819–10821. [CrossRef] [PubMed]

4. Belcourt, M.F.; Farabaugh, P.J. Ribosomal frameshifting in the yeast retrotransposon Ty: tRNAs induce slippage on a 7 nucleotide minimal site. *Cell* **1990**, *62*, 339–352. [CrossRef]

5. Roth, J.F.; Kingsman, S.M.; Kingsman, A.J.; Martin-Rendon, E. Possible regulatory function of the *saccharomyces cerevisiae* Ty1 retrotransposon core protein. *Yeast* **2000**, *16*, 921–932. [CrossRef]

6. Mellor, J.; Fulton, A.M.; Dobson, M.J.; Roberts, N.A.; Wilson, W.; Kingsman, A.J.; Kingsman, S.M. The Ty transposon of *saccharomyces cerevisiae* determines the synthesis of at least three proteins. *Nucleic Acids Res.* **1985**, *13*, 6249–6263. [CrossRef] [PubMed]

7. Pachulska-Wieczorek, K.; Le Grice, S.F.; Purzycka, K.J. Determinants of genomic RNA encapsidation in the *saccharomyces cerevisiae* long terminal repeat retrotransposons Ty1 and Ty3. *Viruses* **2016**, *8*. [CrossRef] [PubMed]

8. Bourc'his, D.; Bestor, T.H. Meiotic catastrophe and retrotransposon reactivation in male germ cells lacking Dnmt3L. *Nature* **2004**, *431*, 96–99. [CrossRef] [PubMed]

9. Yoder, J.A.; Walsh, C.P.; Bestor, T.H. Cytosine methylation and the ecology of intragenomic parasites. *Trends Genet.* **1997**, *13*, 335–340. [CrossRef]

10. Harris, R.S.; Dudley, J.P. Apobecs and virus restriction. *Virology* **2015**, *479–480*, 131–145. [CrossRef] [PubMed]

11. Drinnenberg, I.A.; Fink, G.R.; Bartel, D.P. Compatibility with killer explains the rise of RNAi-deficient fungi. *Science* **2011**, *333*, 1592. [CrossRef] [PubMed]

12. Drinnenberg, I.A.; Weinberg, D.E.; Xie, K.T.; Mower, J.P.; Wolfe, K.H.; Fink, G.R.; Bartel, D.P. RNAi in budding yeast. *Science* **2009**, *326*, 544–550. [CrossRef] [PubMed]

13. Garfinkel, D.J.; Nyswaner, K.; Wang, J.; Cho, J.Y. Post-transcriptional cosuppression of Ty1 retrotransposition. *Genetics* **2003**, *165*, 83–99. [PubMed]

14. Nishida, Y.; Pachulska-Wieczorek, K.; Blaszczyk, L.; Saha, A.; Gumna, J.; Garfinkel, D.J.; Purzycka, K.J. Ty1 retrovirus-like element gag contains overlapping restriction factor and nucleic acid chaperone functions. *Nucleic Acids Res.* **2015**, *43*, 7414–7431. [CrossRef] [PubMed]

15. Saha, A.; Mitchell, J.A.; Nishida, Y.; Hildreth, J.E.; Ariberre, J.A.; Gilbert, W.A.; Garfinkel, D.J. A trans-dominant form of gag restricts Ty1 retrotransposition and mediates copy number control. *J. Virol.* **2015**, *89*, 3922–3938. [CrossRef] [PubMed]

16. Tucker, J.M.; Larango, M.E.; Wachsmuth, L.P.; Kannan, N.; Garfinkel, D.J. The Ty1 retrotransposon restriction factor p22 targets gag. *PLoS Genet.* **2015**, *11*, e1005571. [CrossRef] [PubMed]

17. Firth, A.E.; Brierley, I. Non-canonical translation in RNA viruses. *J Gen. Virol.* **2012**, *93*, 1385–1409. [CrossRef] [PubMed]

18. Bolinger, C.; Boris-Lawrie, K. Mechanisms employed by retroviruses to exploit host factors for translational control of a complicated proteome. *Retrovirology* **2009**, *6*, 8. [CrossRef] [PubMed]

19. Pfingsten, J.S.; Kieft, J.S. RNA structure-based ribosome recruitment: Lessons from the dicistroviridae intergenic region ireses. *RNA* **2008**, *14*, 1255–1263. [CrossRef] [PubMed]

20. Kozak, M. Circumstances and mechanisms of inhibition of translation by secondary structure in eucaryotic mRNArs. *Mol. Cell. Biol.* **1989**, *9*, 5134–5142. [CrossRef] [PubMed]

21. Sagliocco, F.A.; Vega Laso, M.R.; Zhu, D.; Tuite, M.F.; McCarthy, J.E.; Brown, A.J. The influence of 5′-secondary structures upon ribosome binding to mRNA during translation in yeast. *J Biol. Chem.* **1993**, *268*, 26522–26530. [PubMed]

22. Vega Laso, M.R.; Zhu, D.; Sagliocco, F.; Brown, A.J.; Tuite, M.F.; McCarthy, J.E. Inhibition of translational initiation in the yeast *saccharomyces cerevisiae* as a function of the stability and position of hairpin structures in the mRNA leader. *J. Biol. Chem.* **1993**, *268*, 6453–6462. [PubMed]

23. Babendure, J.R.; Babendure, J.L.; Ding, J.H.; Tsien, R.Y. Control of mammalian translation by mRNA structure near caps. *RNA* **2006**, *12*, 851–861. [CrossRef] [PubMed]

24. Kozak, M. Context effects and inefficient initiation at non-AUG codons in eucaryotic cell-free translation systems. *Mol. Cell. Biol.* **1989**, *9*, 5073–5080. [CrossRef] [PubMed]

25. Kozak, M. Downstream secondary structure facilitates recognition of initiator codons by eukaryotic ribosomes. *Proc. Natl. Acad. Sci. USA* **1990**, *87*, 8301–8305. [CrossRef] [PubMed]

26. Kochetov, A.V.; Palyanov, A.; Titov, I.I.; Grigorovich, D.; Sarai, A.; Kolchanov, N.A. AUG_hairpin: Prediction of a downstream secondary structure influencing the recognition of a translation start site. *BMC Bioinform.* **2007**, *8*, 318. [CrossRef] [PubMed]

27. Kozak, M. Influences of mRNA secondary structure on initiation by eukaryotic ribosomes. *Proc. Natl. Acad. Sci. USA* **1986**, *83*, 2850–2854. [CrossRef] [PubMed]

28. Blaszczyk, L.; Ciesiolka, J. Secondary structure and the role in translation initiation of the 5′-terminal region of p53 mRNA. *Biochemistry* **2011**, *50*, 7080–7092. [CrossRef] [PubMed]

29. Gorska, A.; Blaszczyk, L.; Dutkiewicz, M.; Ciesiolka, J. Length variants of the 5′ untranslated region of p53 mRNA and their impact on the efficiency of translation initiation of p53 and its n-truncated isoform deltanp53. *RNA Biol.* **2013**, *10*, 1726–1740. [CrossRef] [PubMed]

30. Purzycka, K.J.; Legiewicz, M.; Matsuda, E.; Eizentstat, L.D.; Lusvarghi, S.; Saha, A.; Le Grice, S.F.; Garfinkel, D.J. Exploring Ty1 retrotransposon RNA structure within virus-like particles. *Nucleic Acids Res.* **2013**, *41*, 463–473. [CrossRef] [PubMed]
31. Pachulska-Wieczorek, K.; Blaszczyk, L.; Biesiada, M.; Adamiak, R.W.; Purzycka, K.J. The matrix domain contributes to the nucleic acid chaperone activity of HIV-2 Gag. *Retrovirology* **2016**, *13*, 18. [CrossRef] [PubMed]
32. Purzycka, K.J.; Pachulska-Wieczorek, K.; Adamiak, R.W. The in vitro loose dimer structure and rearrangements of the HIV-2 leader RNA. *Nucleic Acids Res.* **2011**, *39*, 7234–7248. [CrossRef] [PubMed]
33. Curcio, M.J.; Garfinkel, D.J. Single-step selection for Ty1 element retrotransposition. *Proc. Natl. Acad. Sci. USA* **1991**, *88*, 936–940. [CrossRef] [PubMed]
34. Popenda, M.; Szachniuk, M.; Antczak, M.; Purzycka, K.J.; Lukasiak, P.; Bartol, N.; Blazewicz, J.; Adamiak, R.W. Automated 3D structure composition for large RNAs. *Nucleic Acids Res.* **2012**, *40*, e112. [CrossRef] [PubMed]
35. RNAComposer. Automated RNA Structure 3D Modeling Server. Available online: http://rnacomposer.ibch. poznan.pl/ (accessed on 27 October 2016).
36. Biesiada, M.; Purzycka, K.J.; Szachniuk, M.; Blazewicz, J.; Adamiak, R.W. Automated RNA 3D structure prediction with RNAcomposer. *Methods Mol. Biol.* **2016**, *1490*, 199–215. [PubMed]
37. Wilkinson, K.A.; Merino, E.J.; Weeks, K.M. Selective 2'-hydroxyl acylation analyzed by primer extension (shape): Quantitative RNA structure analysis at single nucleotide resolution. *Nat. Protoc.* **2006**, *1*, 1610–1616. [CrossRef] [PubMed]
38. Deigan, K.E.; Li, T.W.; Mathews, D.H.; Weeks, K.M. Accurate shape-directed RNA structure determination. *Proc. Natl. Acad. Sci. USA* **2009**, *106*, 97–102. [CrossRef] [PubMed]
39. Reuter, J.S.; Mathews, D.H. RNAstructure: Software for RNA secondary structure prediction and analysis. *BMC Bioinform.* **2010**, *11*, 129. [CrossRef] [PubMed]
40. Tijerina, P.; Mohr, S.; Russell, R. DMS footprinting of structured RNAs and RNA-protein complexes. *Nat. Protoc.* **2007**, *2*, 2608–2623. [CrossRef] [PubMed]
41. Pachulska-Wieczorek, K.; Purzycka, K.J.; Adamiak, R.W. New, extended hairpin form of the TAR-2 RNA domain points to the structural polymorphism at the 5' end of the HIV-2 leader RNA. *Nucleic Acids Res.* **2006**, *34*, 2984–2997. [CrossRef] [PubMed]
42. Woodson, S.A.; Koculi, E. Analysis of RNA folding by native polyacrylamide gel electrophoresis. *Methods Enzymol.* **2009**, *469*, 189–208. [PubMed]
43. Araujo, P.R.; Yoon, K.; Ko, D.; Smith, A.D.; Qiao, M.; Suresh, U.; Burns, S.C.; Penalva, L.O. Before it gets started: Regulating translation at the 5' UTR. *Comp. Funct. Genom.* **2012**, *2012*, 475731. [CrossRef] [PubMed]
44. Mathews, D.H. RNA secondary structure analysis using RNAstructure. *Curr. Protoc. Bioinform.* **2014**, *46*. [CrossRef]
45. Huang, Q.; Purzycka, K.J.; Lusvarghi, S.; Li, D.; Legrice, S.F.; Boeke, J.D. Retrotransposon Ty1 RNA contains a 5'-terminal long-range pseudoknot required for efficient reverse transcription. *RNA* **2013**, *19*, 320–332. [CrossRef] [PubMed]
46. Dikstein, R. Transcription and translation in a package deal: The tisu paradigm. *Gene* **2012**, *491*, 1–4. [CrossRef] [PubMed]
47. Kozak, M. A short leader sequence impairs the fidelity of initiation by eukaryotic ribosomes. *Gene Expr.* **1991**, *1*, 111–115. [PubMed]
48. Kozak, M. Pushing the limits of the scanning mechanism for initiation of translation. *Gene* **2002**, *299*, 1–34. [CrossRef]
49. Garfinkel, D.J.; Tucker, J.M.; Saha, A.; Nishida, Y.; Pachulska-Wieczorek, K.; Błaszczyk, L.; Purzycka, K.J. A self-encoded capsid derivative restricts Ty1 retrotransposition in Saccharomyces. *Curr. Genet.* **2015**, 1–9. [CrossRef] [PubMed]
50. Sonenberg, N.; Hinnebusch, A.G. Regulation of translation initiation in eukaryotes: Mechanisms and biological targets. *Cell* **2009**, *136*, 731–745. [CrossRef] [PubMed]

viruses

MDPI

Review

Reverse Transcription in the *Saccharomyces cerevisiae* Long-Terminal Repeat Retrotransposon Ty3

Jason W. Rausch, Jennifer T. Miller and Stuart F. J. Le Grice *

Reverse Transcriptase Biochemistry Section, Basic Research Laboratory, Frederick National Laboratory for
Cancer Research, Frederick, MD 21702, USA; rauschj@mail.nih.gov (J.W.R.); millerj@mail.nih.gov (J.T.M.)
* Correspondence: legrices@mail.nih.gov; Tel.: +1-301-846-5256

Academic Editors: David J. Garfinkel and Katarzyna J. Purzycka
Received: 2 February 2017; Accepted: 7 March 2017; Published: 15 March 2017

Abstract: Converting the single-stranded retroviral RNA into integration-competent double-stranded DNA is achieved through a multi-step process mediated by the virus-coded reverse transcriptase (RT). With the exception that it is restricted to an intracellular life cycle, replication of the *Saccharomyces cerevisiae* long terminal repeat (LTR)-retrotransposon Ty3 genome is guided by equivalent events that, while generally similar, show many unique and subtle differences relative to the retroviral counterparts. Until only recently, our knowledge of RT structure and function was guided by a vast body of literature on the human immunodeficiency virus (HIV) enzyme. Although the recently-solved structure of Ty3 RT in the presence of an RNA/DNA hybrid adds little in terms of novelty to the mechanistic basis underlying DNA polymerase and ribonuclease H activity, it highlights quite remarkable topological differences between retroviral and LTR-retrotransposon RTs. The theme of overall similarity but distinct differences extends to the priming mechanisms used by Ty3 RT to initiate (−) and (+) strand DNA synthesis. The unique structural organization of the retrotransposon enzyme and interaction with its nucleic acid substrates, with emphasis on polypurine tract (PPT)-primed initiation of (+) strand synthesis, is the subject of this review.

Keywords: retrotransposon; Ty3; reverse transcriptase; reverse transcription; ribonuclease H (RNase H); DNA polymerase; retroelement

1. Introduction

Central to the propagation of retroviruses and long terminal repeat (LTR)-retrotransposons is the conversion of their single-stranded RNA genome into integration-competent double-stranded DNA, a multi-step process mediated by the element-encoded reverse transcriptase (RT) [1]. Crucial steps in this process involve the use of RNA primers to initiate synthesis of the (−) and (+) strand DNAs (a host-coded transfer RNA (tRNA) and the element-encoded polypurine tract (PPT), respectively). Our understanding of these events has come almost exclusively from retroviruses where, over some 50 years, the field has witnessed a progression from the discovery of an enzyme capable of synthesizing DNA on an RNA template [2,3] to high resolution X-ray structures for human immunodeficiency virus type 1 (HIV-1) RT that have proven instrumental to the success of combination antiviral therapy to stem HIV infection and the progression of acquired immunodeficiency syndrome (AIDS) [1].

Based on literature that has been amassed on RT from human, avian and murine retroviruses, it might be considered reasonable to assume that counterpart enzymes of transposable elements (e.g., *Drosophila* (*copia*) and *Saccharomyces cerevisiae* (Ty1 and Ty3)), as well as their cognate nucleic acid substrates, are merely minor variations of a common theme. However, the observation that (a) Ty1 and Ty3 RTs use a bipartite primer binding site (PBS); (b) the *Schizosaccharomyces pombe* element Tf1 uses a tRNA-independent mechanism; and (c) a "half-tRNA" is employed by *Drosophila melanogaster copia* to initiate (−) strand DNA synthesis [4] suggests their respective polymerases might also not share the topological features of HIV-1 RT. This issue is highlighted by structural data for several monomeric retroviral and retrotransposon RTs such as the gammaretroviruses xenotropic murine leukemia virus-related virus (XMRV) and Moloney murine leukemia viruses, mouse mammary tumor virus, simian foamy virus, bovine leukemia virus, and the Tf1 element [5–10]. As the third RNA-dependent DNA polymerase to be crystallized in the presence of an RNA/DNA hybrid, the goal of data presented in this review is to illustrate the unique topological complexity of Ty3 RT and point out to the reader that our understanding of reverse transcription should be the consequence of comparative studies and not simply those from a single enzyme.

2. Reverse Transcription Overview

Ty3 RT performs a series of orchestrated events to convert the diploid plus (+) stranded retrotransposon RNA into double-stranded DNA (dsDNA) that is subsequently integrated into the host cell genome (Figure 1). Minus (−) strand DNA synthesis initiates from the 3′-end of a host-derived tRNA hybridized to a bipartite primer binding site (PBS) and continues until the 5′-end of the genome is reached (Figure 1B–D). RT-associated RNase H activity then hydrolyzes the 5′-terminal repeat (R) and U5 segments of the RNA template, allowing transfer of the nascent (−) strong stop DNA (ssDNA) to the 3′-terminal R segment (Figure 1D,E). After the template switch, minus (−) strand DNA synthesis proceeds with concomitant RNase H-mediated degradation of viral RNA, leaving a small RNase H-resistant purine-rich RNA fragment (polypurine tract, or PPT) hybridized to the nascent DNA (Figure 1F,G). In contrast to retroviruses and Ty1, no central PPT has been identified for Ty3. The Ty3 PPT fragment primes (+) strand DNA synthesis in a manner that diverges somewhat from the equivalent event in retroviruses. Through a mechanism that will be discussed in more detail below, the (+) strong stop DNA generated from a second PPT priming event is transferred to the 3′-end of the nascent (−) DNA by virtue of the terminal repeat (R) sequences. Once both the (+) and (−) strands are filled out, the final dsDNA contains a repeated U3-R-U5 sequence flanking the coding regions of the retrotransposon genome (Figure 1G–M).

2.1. Minus (−) Strand Initiation and tRNA-Retrotransposon RNA Interactions

Minus (−) strand DNA synthesis in Ty3 is primed by host tRNA$_i^{Met}$, the same species utilized by the distantly-related Ty1 and Ty5 retrotransposons [11]. Interestingly, while Ty3 and Ty1 prime from the native 3′-end of the tRNA, Ty5 RT initiates from a 3′-end produced by host cell RNase P-mediated internal cleavage within the anticodon loop [12–14]. Also, both Ty3 and Ty1 utilize a bipartite PBS, although the details of how the PBS is divided and where the segments reside in their respective RNA genomes differs between the two elements. Like those of retroviruses, the PBS of Ty1 is contained entirely within U5, with the two segments separated by a relatively small internal loop. In contrast, the 5′ and 3′ segments of the bipartite Ty3 PBS are separated by ~4800 nt and reside in the 5′ (PBS) and 3′ (U3) untranslated regions (UTRs), respectively. To form a DNA synthesis-competent initiation complex, the acceptor stem and TΨC arm of tRNA$_i^{Met}$ hybridize to the 5′ and 3′ components of the PBS while the D arm interacts with viral RNA in U3 [15,16]. Such intricate interactions to establish (−) strand initiation complexes are a common requirement for many retroelements, including retroviruses. For instance, mutational analyses of HIV-1, feline immunodeficiency virus (FIV) and Rous sarcoma virus (RSV) complexes indicate that base pairing between tRNA and viral RNA sequences outside of the PBS support an efficient transition from the initiation to elongation phase of DNA synthesis [17–22].

Ty3 nucleocapsid protein (NC) is produced by proteolytic cleavage of the *GAG3* (CA-SP-NC) precursor [23]. Ty3 NC has a single zinc finger, the highly-basic N-terminal domain of which contributes to nucleic acid binding efficiency [13,24], facilitating annealing of tRNA$_i^{Met}$ to the PBS, formation of ribonucleoprotein complexes, and genomic RNA dimerization. Deletion analysis has determined that these NC functions are more dependent on the basic region than the zinc finger [15]. Together with tRNA$_i^{Met}$-PBS hybridization, Ty3 NC enables initiation complex dimerization by promoting interstrand base pairing between 12 nt G:C rich palindromic sequences at the tRNA 5′-ends [13].

One study also suggests that a global complex in which viral RNA 5′ and 3′ termini are brought into proximity may be stabilized by a transient covalent linkage between the two ends, as knockdown mutations in the lariat debranching enzyme Dbr1 have significantly decreased levels of Ty3 cDNA accumulation [25].

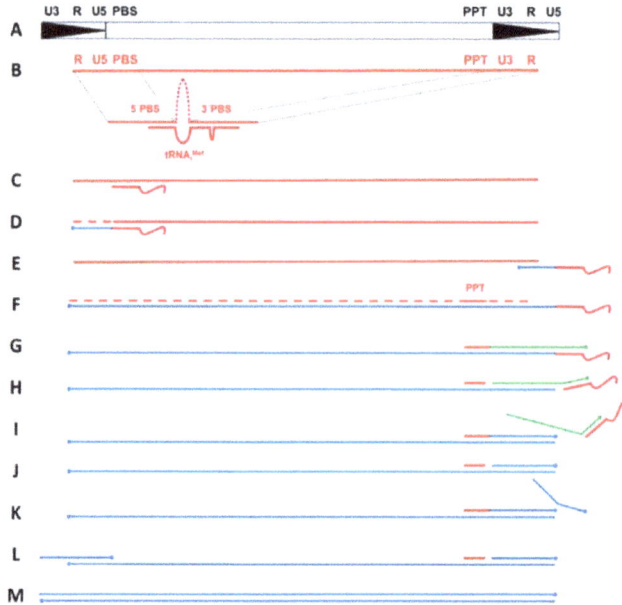

Figure 1. Ty3 Reverse Transcription Cycle. (**A**) Structure of the double stranded preintegrative Ty3 DNA (black). U3, unique 3′ sequence; R, repeat sequence; U5, unique 5′ sequence; PBS, primer binding site; PPT, polypurine tract; (**B**) Genomic RNA is depicted in red. The bipartite nature of the PBS comprises sequences from both the 5′ PBS and the 3′ U3 regions; (**C**) Simplified initiation complex excluding the transfer RNA (tRNA) 5′ terminal nucleotides; (**D**) (−) strand strong stop synthesis, with concomitant degradation of genomic RNA by RNase H. Newly synthesized (−) strand DNA is shown in blue; (**E**) (−) strand transfer; (**F**) (−) strand synthesis and concomitant degradation of genomic RNA by RNase H; (**G**) (+) strand synthesis initiates from the PPT and extends into tRNA. Nascent (+) strand DNA is shown in green; (**H**) PPT is re-cleaved from (+) strand DNA and tRNA is cleaved from (−) strand DNA by RNase H; (**I**) Second (+) strand DNA, indicated in blue, displaces first; (**J**) PPT is again cleaved; (**K**) Third (+) strand synthesis initiates, and displaces second (+) strand; (**L**) Second (+) strand transfers to 3′-end of (−) DNA and PPT is cleaved; (**M**) Synthesis of both (+) and (−) strands is completed.

2.2. Plus (+) Strand Initiation, (+) sssDNA Synthesis, and (+) Strand Transfer

Plus-strand synthesis in Ty3 initiates from a PPT RNA fragment located just upstream of U3. However, in Ty3 and Ty1, this PPT appears to prime DNA synthesis more than once. This revelation came from experiments in which a mutant tRNA was used to prime minus-strand initiation, yet this change was not reflected in the PBS region of Ty1 or Ty3 DNA following retrotransposition [26].

In these experiments, researchers utilized a mutant yeast strain devoid of any endogenous tRNA$_i^{Met}$ genes but expressing a similar mutant tRNA containing a nucleotide substitution in the anti-PBS sequence [27]. After performing a Ty3-specific integration assay, progeny retrotransposon DNA was sequenced and did not contain the mutation, indicating the genomic PBS sequence could not be derived from reverse transcription of the (−) DNA-priming tRNA, as is the case in retroviruses.

The authors proposed the alternative PPT recycling mechanism shown schematically in Figure 1. In this process, (+)-DNA synthesis initiates from the PPT and terminates after reverse transcribing 12 nt of the (−)-strand priming tRNA (Figure 1G). RT then separates the tRNA from the (−) DNA template by cleaving at or near the tRNA-DNA junction. RT also cleaves at the junction between the PPT and nascent (+) DNA, after which synthesis of a second (+) strand initiates from the regenerated 3′PPT primer, displaces the first (+) strand strong stop DNA (sssDNA), and terminates at the end of U5 (since the tRNA has been removed from the (−) DNA template) (Figure 1H,I). Finally, a third cleavage of the PPT allows re-initiation of a third (+) DNA synthesis product, resulting in displacement of the second (+) sssDNA, and making it available for hybridization to the complementary R and U5 sequences at the (−) DNA 3′ terminus (Figure 1J–L). Re-initiation of (−) DNA synthesis from the transferred strand completes the (+) strand transfer process (Figure 1L,M).

As the alternative (+) DNA synthesis mechanism would suggest, dead end (+) sssDNA products have been found to accumulate to high levels in Ty3 virus like particles [28]. This observation, together with finding that the PBS sequence is not preserved by reverse transcription of the tRNA 3′ terminus, lends support to this distinctive and intriguing model of (+) strand synthesis and strand transfer.

2.3. Involvement of Ty3 Integrase

Ty3 integrase (IN) is produced by proteolytic cleavage of the polyprotein precursor GAG3-POL3 (PR-J-RT-IN) [23,29]. To determine whether this enzyme might affect stages of retrotransposition outside of integration, researchers substituted alanine for charged non-catalytic residues in both the N- and C-terminal domains of Ty3 IN and studied the effects in vivo. One class of such mutations that reduced steady state levels of IN in cells also produced a correlative decrease in accumulated cDNA. Similarly, mutant virus-like particles (VLPs) contained less primer tRNA and produced less (−) sssDNA in exogenous RT assays, suggesting IN may contribute a stimulatory role at early stages of reverse transcription. Trans-complementation with a capsid (CA)-RT-IN, but not a CA-IN construct, rescued cDNA production, indicating that the stimulatory effects of IN on cDNA synthesis may be mediated by close association of this enzyme with RT [30]. Ty1 experiments in which native IN was provided in trans yielded similar results wherein trans-complementation of IN alone failed to rescue reverse transcription defects in an IN-deficient Ty1 model system [31]. Taken together, these studies suggest that the mechanism of activating initiation of (−) DNA synthesis by association of IN with RT may be common among retrotransposons.

3. Ty3 RT Structural Organization and Biochemical Characterization

The reverse transcription process has been thoroughly characterized for several retroviruses and LTR-containing retrotransposons. In contrast, high resolution structural details on their associated RTs have been limited largely to the HIV-1 enzyme as a consequence of its central role as an antiviral target [32]. In the absence or presence of its nucleic acid substrate, HIV-1 RT is organized as an asymmetric heterodimer of 66 and 51 kDa subunits (p66 and p51, respectively) derived from the same gene, but differing in that p51 lacks the ~15 kDa, C-terminal RNase H domain as a consequence of processing by the virus-coded protease [33]. Similar to other nucleic acid polymerases, p66 subdomains were designated "fingers", "palm", and "thumb", which were tethered to the C-terminal RNase H domain via a "connection" subdomain. Alternative folding of the p51 subunit positioned the connection between its fingers and palm, thereby occluding its DNA polymerase active site [34]. The lack of a p51-associated RNase H domain thus indicated that both the polymerizing and hydrolytic activities of HIV-1 RT were a property of the p66 subunit.

Later studies with RT from the gammaretrovirus xenotropic murine leukemia virus-related virus (XMRV) [5] demonstrated a monomeric organization in the absence and presence of nucleic acid substrate, providing a second example of a retroviral polymerase whose dual enzymatic functions reside on the same subunit. The availability of high resolution structures for two retroviral enzymes in the presence of an RNA/DNA hybrid thus predicted that their LTR-retrotransposon counterpart would assume one of these two configurations. Initial clues that this might not be so simple came from phylogenetic studies indicating that LTR-retrotranspon RT lacks a "connection" subdomain (i.e., its RNase H and DNA polymerase domains domain were juxtaposed) [35]. Initial biochemical characterization of recombinant Ty3 RT indicated that, following gel permeation chromatography, DNA polymerase activity was associated with a polypeptide that migrated consistent with 55 kDa monomer [36]. However, when the same analysis was conducted in the presence of nucleic acid, the migration properties of the nucleoprotein complex, 125 kDa, suggested the intriguing notion of substrate-dependent dimerization [37], in this case a homodimer. However, in contrast to HIV-1 RT, the Ty3 homodimer would retain two copies of the C-terminal RNase domain, raising speculation that both might exhibit activity. Our high resolution structure of Ty3 RT containing an RNA/DNA hybrid derived from its PPT answered this question, while at the same time it also demonstrated a uniquely versatile enzyme with respect to subunit topology.

As depicted in Figure 2, Ty3 RT is an asymmetric homodimer comprised of subunits we designated A and B. In contrast to the previous studies of Sarafianos et al. [38], but in keeping with our own data for HIV-1 [39,40] and XMRV RT [5], the RNA/DNA hybrid assumes a more A-like configuration, displaying no steric clashes between O2′ and O4′ oxygens of adjacent riboses of the RNA strand. Although lacking a connection subdomain, the fingers, palm, thumb, and RNase H domain of Ty3 RT subunit A are topologically similar to those of HIV-1 RT p66. In addition to crystallographic data in the presence of an RNA/DNA hybrid, ascribing DNA polymerase function exclusively to subunit A was based on the observation that alternative folding positioned the subunit B RNase H domain between its fingers and palm. Thus, despite major structural differences between HIV-1 and Ty3 RT, they share the common property that alternative folding of the two subunits occludes one of the DNA polymerase active sites. A summary of amino acid contacts supported by subunits A and B is illustrated in Figure 3.

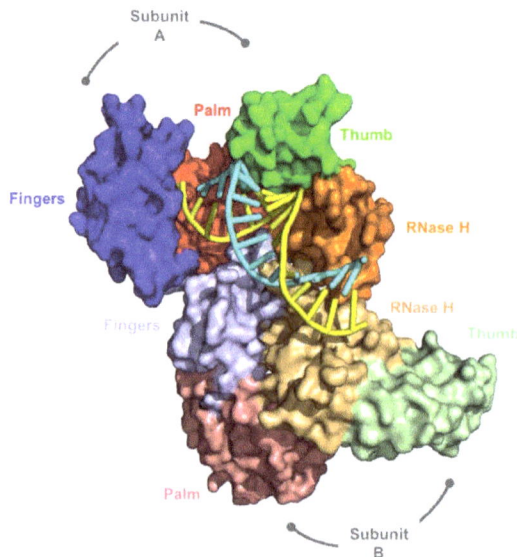

Figure 2. Structure of the asymmetric Ty3 RT homodimer in complex with its PPT-containing RNA/DNA hybrid. DNA and RNA strands of the cartoon representation are denoted in cyan and yellow, respectively. Subunit domains are color coded blue, red, green, and orange for fingers, palm, thumb, and RNase H, respectively, and the darker shading represents subunit A. Note the absence of a connection subdomain, a significant contrast between retroviral and LTR-retrotransposon RTs. Adapted from [37].

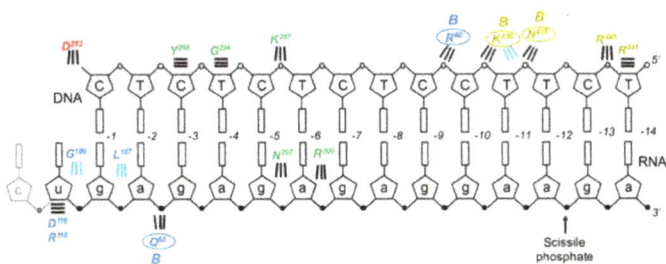

Figure 3. Contacts between Ty3 RT subunits A and B and the PPT-containing RNA/DNA hybrid. Color coding is consistent with subdomain designation of Figure 2, and DNA and RNA nucleotides are denoted in capital and small letters, respectively. The scissile PPT/U3 junction has been indicated, and base numbering is relative to substrate bound at the DNA polymerase active site Subunit B contacts are denoted "B" and circled. Parallel horizontal lines indicate van der Waals interactions. Diagonal and vertical lines indicate interactions mediated by the protein backbone (cyan) or side chains (black).

3.1. DNA Polymerase Active Site Residues

As originally identified by homology with HIV-1 RT, D151, D213, and D214 are housed in the palm subdomain and comprise the catalytic triad of the -D-(aa)$_n$-Y-L-D-D- DNA polymerase active site of Ty3 RT [41] (Figure 4). These residues were mutated to either asparagine or glutamate and the effects on enzyme function were determined in vitro in the context of purified enzyme as well as transposition activity in *S. cerevisiae*. D151N and D213N substitutions eliminated both RNA-dependent and DNA-dependent DNA polymerase activities, whereas activity was retained in D214N and D214E mutants (although enzyme processivity was substantially reduced). D151E mutants were likewise devoid of polymerase activity, although D213E was partially tolerated. Reduced pyrophosphorolysis activity was found to parallel DNA polymerase activity deficits, and none of these mutants were substantially rescued by substituting MnCl$_2$ for MgCl$_2$ in enzyme assays. Quantitative kinetic analysis indicated that the principle effects of these mutations were on turnover and processivity rather than substrate binding.

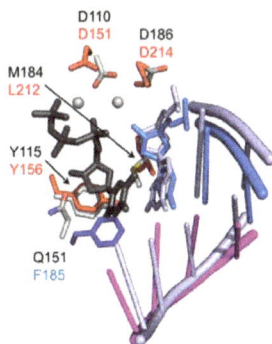

Figure 4. Alignment of the DNA polymerase active sites of Ty3 (PDB ID 4OL8, REF) and HIV-1 RT (PDB ID:1RTD). Carbon atoms of select Ty3 RT residues are shown in red (palm) and blue (fingers), and those of HIV-1 residues are in grey. The two catalytic metal ions and incoming dTTP are shown in grey and dark grey, respectively. Both HIV-1 DNA strands are shown as a light blue ladder, and the RNA template and DNA primer bound by Ty3 RT are shown in magenta and marine, respectively. The 3'-terminal nucleotides in both DNA primer strands are shown in stick form, and the stick radius of the incoming dTTP has been slightly expanded for contrast. Adapted from [37].

In vitro, D151E RT was only 2% active relative to the wild type enzyme. All other mutants were at least 25% active, indicating that they were not structurally compromised and still capable of substrate binding. Both wild type and mutant enzymes retained the precision of RNase H activity, indicating that active site residues do not affect positioning of the enzyme on the substrate. In vivo, all mutations proved lethal for transposition. Taken together, these results suggested that D151 and D213 were required for coordination of the catalytically essential divalent Mg^{++}, while D214 may stabilize the polymerase activation complex or otherwise facilitate catalytic chemistry. The Ty3 RT-RNA/DNA co-crystal structure also shows that, in addition to its role in metal ion chelation, the D213 side chain also contacts the 3'-terminal nucleotide of the DNA primer [37].

3.2. Thumb Subdomain Residues Contacting Nucleic Acid

In retroviral RTs and other DNA polymerases, the thumb subdomain is flexible and, in the context of an active polymerase domain, functions both in substrate binding and translocation during DNA synthesis [42]. Numerous residues in the Ty3 subunit A thumb contact either the primer or template strand in the RT-RNA/DNA co-crystal [37]. Specifically, DNA primer nucleotides at positions −3 to −5 form backbone contacts with thumb residues Y298, G294, and K287, respectively, while N297 and R300 contact the 2'OH moiety of the RNA strand at positions −5 and −6. Equivalent residues in the B subunit do not contact nucleic acid, as the thumb subdomain is displaced from the palm and rotated relative to the RNase H domain. Before the high resolution crystal structure became available, thumb residues proposed to interact with the nucleic acid substrate were identified by homology to the equivalent domain in HIV-1 RT [43]. On this basis, residues Q290, F292, G294, N297, and Y298 were subjected to mutational and biochemical analysis to characterize their roles in enzyme function.

A novel assay developed for this study utilized duplex DNA substrates containing serial locked nucleic acid (LNA) substitutions in either the primer or template strand [43,44]. Because LNA can only assume an RNA-like C3'-endo sugar pucker and contains a methylene bridge between ribose 2'-O and 4'-C atoms, its introduction into DNA creates a localized steric barrier to polymerase binding and/or translocation. Moreover, because only the ribose groups of LNAs are chemically modified, measuring the efficiency of single nucleotide incorporation in these substituted substrates can be exploited to determine contact sites between the enzyme and sugar-phosphate backbone irrespective of nucleoside base identity.

In this assay, LNA substitutions at either position −3 or −4 in the DNA primer strand or position −6 or −7 in the DNA template strand impaired single nucleotide incorporation, indicating the importance of enzyme-nucleic acid contacts at these sites for proper substrate binding. This finding was corroborated by parallel assays in which a basic nucleoside analogs were serially substituted into nucleic acid substrates, and is in remarkable agreement with the high resolution Ty3 RT-RNA/DNA co-crystal structure published nine years later [37]. Analysis of Ty3 RT thumb mutants using this assay indicated that subunit A residues G294, N297, and Y298 contact the DNA substrate at or near the sites indicated in the co-crystal structure. Perhaps the most remarkable finding was the compensatory interaction between the Y298A mutant and the DNA substrate with an LNA substitution at primer nucleotide −3. Primer extension activity of this mutant was substantially greater than that of wild type Ty3 RT, indicating a reciprocally favorable binding interaction between the smaller Ala side chain and the bulky modified nucleoside.

The important contribution of thumb contacts to Ty3 RT function was further established by more conventional biochemical assays [43]. Higher rates of dissociation from duplex DNA substrates were measured in steady-state kinetic assays, while mutants containing G294, N297, or Y298 substitutions exhibited reduced RNase H activity.

3.3. A Single Subunit of the Ty3 RT Asymmetric Homodimer Contributes to RNase H Activity

Although contacts with the DNA strand of the RNA/DNA hybrid could be identified for both RNase H domains in the crystal structure, neither RNase H active site was in the vicinity of the

RNA scissile bond. Since simple site-directed mutagenesis would duplicate any modification in both subunits, the origin of RNase H activity was determined using a novel phenotypic mixing strategy in which the nucleoprotein complex was reconstituted with selectively-deficient Ty3 RT monomers.

Residue D426 constitutes one of the catalytically critical residues of the RNase H domain, and its replacement with asparagine (N426) was shown to eliminate RNase H activity [45]. The capacity of this variant to dimerize, however, appears to be unaffected, as the D426N enzyme was fully functional as a DNA polymerase. In contrast, R140 and R203 of Ty3 RT subunit A localize to the dimerization interface, suggesting that mutating these residues might impair dimerization, and hence enzyme function. Indeed, an R140A/R203A double mutant was defective in both DNA polymerase and RNase H activities, presumably reflecting a failure to dimerize. It is important to note that these mutations only prevent dimerization when present in the context of the A subunit; in the B subunit, residues R140 and R203 do not appear to be directly involved in dimerization or any other aspect of RT function.

The possible complementation outcomes of the mixing of D426N and R140A/R203A Ty3 RT monomers are depicted in Figure 5. In brief, the only way for these variants to combine to form an active dimer with RNase H activity would be if (i) mutants D426N and R140A/R203A occupied the subunit A and B positions, respectively; and (ii) the RNase H domain of subunit B confers RNase H activity to Ty3 RT. This was indeed what we observed experimentally [37], demonstrating that DNA polymerase and RNase H activity are exclusive to the A and B subunits of Ty3 RT, respectively. An unresolved question, however, was the conformational change necessary to position the subunit B active site in the vicinity of the scissile bond of the RNA backbone. Although located closer to the scissile phosphate, the subunit B RNase H domain (and thumb subdomain) would be required to move ~40 Å, a translation molecular modeling suggests could be accommodated for without invoking steric clashes. In summary, although the active site residues of DNA polymerase domains of lentiviral, gammaretroviral, and LTR-retrotransposon RTs are well conserved, the major differences they exhibit in the topology of their RNase H domains possibly reflect an intricate evolutionary mechanism whereby cellular RNases H were sequestered by the retroviral polymerase into bifunctional enzymes.

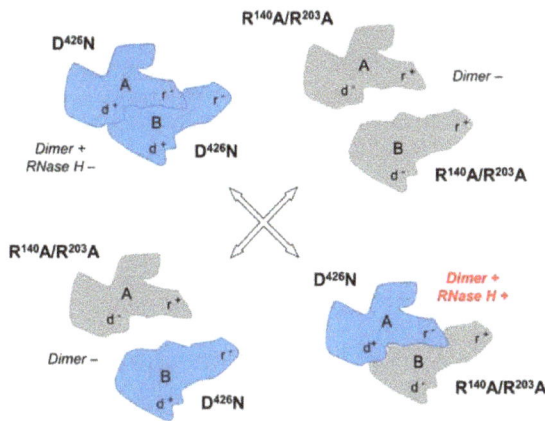

Figure 5. Phenotypic mixing strategy to determine the RNase H-competent Ty3 RT subunit. RNase H defective (D426N) and dimerization defective (R140A/R203A) mutant monomers are indicated in blue and grey, respectively. Notations d^+ and d^- indicate a dimerization-competent and dimerization-incompetent subunit interface, while r^+ and r^- denote RNase H-competent and RNase H-incompetent, respectively. Note that the d^- mutant only prevents dimerization when in the A subunit position. When purified mutants are mixed, RNase H activity is only recovered in a reconstituted dimer whose subunit B contributes to RNase H activity.

3.4. RNase H Domain Structure

Retroviral, bacterial, human H1, and Ty3 RNase H enzymes/domains adopt a common "RNase H fold" characterized by a 5-stranded β-sheet flanked by 2–3 α-helices on one side and one on the other [46]. Aside from their positioning relative to nucleic acid substrate and the Ty3 fingers, palm, and thumb subdomains, the Ty3 RNase H domain differs from the retroviral and RNase H1 counterparts in the length of the first β-strand (~10 residues shorter for Ty3 RT) and arrangement of α-helices between β-strands 4 and 5. Secondary, tertiary, and quaternary structures of Ty3 RNase H domains also resemble the connection subdomains of closely related retroviral enzymes, although the latter elements lack the functional catalytic residues [35].

Critical active site residues of the Ty3 RNase H domain are D358, E401, D426, and D469 [37,45]. These residues are superimposable with their counterparts in cellular and retroviral enzymes (Figure 6), suggesting they support a common catalytic mechanism. In biochemical assays, D358N, E401Q, and D426N substitutions eliminated RNase H activity while a D469N mutation led to its reduction [45]. The diminished effects of the D469N mutation were consistent with a prior study of the homologous residue in HIV-1 RT as well as the distinct role this acidic residue is purported to play in the 2-metal ion catalyzed model of RNase H-mediated RNA cleavage [47,48]. One distinct feature of the Ty3 RT domains is the reduced size of a loop located proximal to the active site in cellular and retroviral enzymes. As this loop harbors a conserved histidine residue (H264 in human RNase, H1 and H539 in HIV-1 RNase H) that is proposed to facilitate product dissociation following hydrolysis [47,49], its absence in the Ty3 enzyme may reduce catalytic turnover relative to the human and retroviral counterparts.

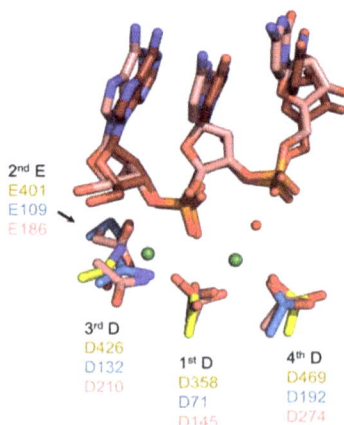

Figure 6. Alignment of RNase H active sites from Ty3 RT (PDB ID 4OL8, REF), *Bacillus halodurans* RNase H1 (PDB ID: 1ZB1, REF), and human RNase H1 (PDB ID: 2QK9, REF). Residue carbon atoms are shown in yellow, blue, and salmon, respectively. RNA strands from human and bacterial RNases H1 are shown in salmon and red, and two catalytic Mg^{++} ions from the Bh-RNase H1 structure are depicted as green spheres. The attacking nucleophilic water is shown as a red sphere.

In the co-crystal structure containing an RNA/DNA hybrid, subunit A RNase H residues R441 and R445 make backbone contacts with the DNA at positions −13/−14, while subunit B residues N435 and K436 make contacts between positions −10/−11 [37]. The functional role these residues play in substrate binding and/or RNase H activity of Ty3 RT is unclear, since neither subunit is positioned for cleavage in the crystallized complex. Conversely, because homologs of Ty3 residues R473 and Y459 in HIV-1 have been shown to interact with the backbone of the RNA strand in an HIV RT-RNA/DNA co-crystal, these residues might be expected to play a similar role in a "cleavage-ready" Ty3 RT complex.

R473 is well conserved among Gypsy retroelements, while mutating Y459 greatly reduces RNase H activity [45].

Homology modeling of a productive Ty3 RNase H-RNA/DNA complex indicates that a number of contacts observed to occur between cellular and retroviral RNases H and their RNA/DNA hybrid substrates would likely be missing. For example, C-terminal residues of β1' in bacterial and human RNase H1 mediate contacts with 2'-OH groups on the 3' side of active site that have been postulated as important determinants of substrate specificity [47]. Since this β-sheet is ~10 residues shorter in Ty3 RT, no such 2'-OH interactions could be established in a homologous complex. Similarly, there appears to be no Ty3 homologs of cellular and retroviral RNase H residues shown to contact the minor groove side of substrate bases (e.g., E449, N474, and Q475 of HIV-1 RNase H) [39,49]. Finally, conserved residues of the phosphate binding pocket—a motif critical for substrate recognition and DNA deformation in hybrid duplexes—have no clear homologs in the Ty3 RNase H domain [47]. Taken together, these observations suggest that, although the active site of Ty3 RNase H likely functions through a very similar mechanism to cellular enzymes, the mode of RNA-DNA binding involves fewer contacts with nucleic acid, and in particular with the DNA strand.

4. Structural Determinants of PPT Cleavage by Ty3 RT-Associated RNase H

(+) strand DNA synthesis in LTR-retrotransposons from an RNase H-resistant PPT-containing RNA/DNA hybrid parallels mechanisms established for retroviruses. In brief, this involves (i), exposure of the PPT 3'-OH in the RNA/DNA replication intermediate; (ii), initiation of (+) strand DNA-dependent DNA synthesis; and (iii), precise removal of the RNA primer from the RNA-DNA chimera. Curiously, however, the Ty3 PPT sequence, 5'-G-A-G-A-G-G-A-G-A-G-G-A-A-3' differs from its retroviral counterparts, which in general have a more homopolymeric organization (e.g., 5'-A-A-A-A-G-A-A-A-A-G-G-G-G-G-G-3' for HIV-1). In addition, the Ty3 and HIV PPTs differ in length (12 nt and 15 nt, respectively). Despite this, model systems mimicking Ty3 PPT primer selection and its release from nascent (+) strand DNA demonstrate a high degree of precision (Figure 7), while in a heterologous system, Ty3 RT fails to recognize the HIV PPT/U3 junction [50]. Together, these observations suggest a mechanistically appropriate "fit" between the retroviral or retrotransposon polymerase and its cognate PPT drives cleavage specificity. Nucleic acid interference experiments, in combination with nuclear magnetic resonance (NMR) spectroscopy, have provided important insights into the structural basis for Ty3 PPT cleavage specificity.

The nonpolar pyrimidine mimic, 2,4-difluoro-5-methylbenzene deoxynucleoside (F, Figure 8) is isosteric with thymine, but has severely reduced hydrogen bonding capacity [51]. Its strategic insertion into the DNA strand of a Ty3 PPT RNA/DNA hybrid provided a unique means of assessing the role of hydrogen bonding without invoking major steric clashes. Most prominent among the outcomes of this strategy was the observation that a tandem $-1/-2$ T $\rightarrow$ F substitution quantitatively relocated cleavage specificity ~11 bp downstream (i.e., to positions +10 and +11, Figure 8). Although some specificity for the PPT/U3 junction was retained, additional dual substitutions likewise re-directed the RNase H catalytic site some 10–12 bp downstream [50]. Since the position of cleavage defined the disposition of the Ty3 RNase H domain on the hybrid, mutagenesis data indicated that local T $\rightarrow$ F-induced flexibility was "sensed" and sequestered by a structural component of Ty3 RT, leading to re-positioning of the RNase H active site. Crystallographic evidence with HIV-1 RT had suggested that several residues of its p66 thumb that were in close contact with the nucleic acid substrate could assume the role of a sensor of nucleic acid configuration [43]. Preliminary studies on Ty3 RNase H activity indicated its DNA polymerase and RNase H active sites were separated by ~13 bp of RNA/DNA hybrid [45], predicting a shorter separation distance between its thumb and RNase H domain. As indicated in Figure 3, this distance is ~10 bp, supporting such a sensor role for the subunit A thumb.

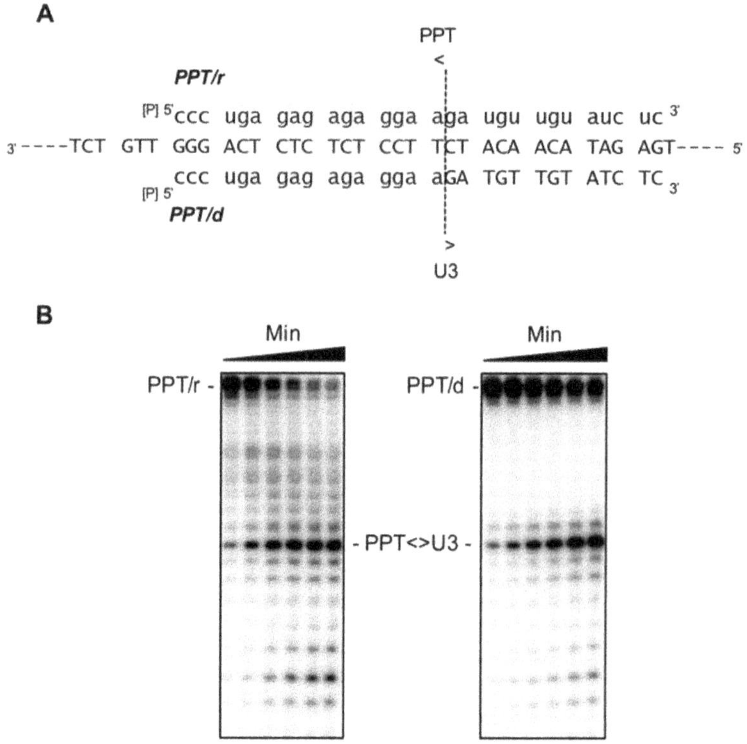

A

PPT
<
PPT/r
[P] 5'ccc uga gag aga gga aga ugu ugu auc uc 3'
3'----TCT GTT GGG ACT CTC TCT CCT TCT ACA ACA TAG AGT---- 5'
ccc uga gag aga gga aGA TGT TGT ATC TC 3'
[P] 5'
PPT/d
>
U3

B

Min   Min

PPT/r -   PPT/d -

- PPT<>U3 -

Figure 7. (**A**) Model RNA/DNA hybrids to illustrate the specificity of cleavage at the Ty3 PPT/U3 junction. A hybrid containing the "all-RNA" strand, PPT/r, mimics selection of the PPT 3'-OH from the RNA/DNA replication mediate during (−) strand DNA synthesis, while a hybrid containing the RNA-DNA chimera, PPT/d, mimics release of the PPT 3'-OH from nascent DNA, an obligate step following initiation of (+) strand DNA synthesis; (**B**) experimental data. For both model substrates, the position of the PPT/U3 junction has been indicated. Adapted from [50].

In an effort to correlate these findings with the selection of the PPT primer 3'-OH in vivo, pyrimidine isostere experiments raised the possibility that local anomalies in nucleic acid geometry, either at or upstream of the scissile junction, might also serve as recognition signals for RT positioning. A clue to this possibility was provided by NMR studies, which indicated an A- to B-transition in the +1rG sugar pucker at the Ty3 PPT/U3 junction [52]. Structurally, this local alteration in sugar pucker would alter the backbone conformation of the RNA/DNA hybrid, creating both a local distortion and, potentially, more long range kinking of the helix. An NMR structure of the junction formed at the HIV-1 (−) strand initiation site has also revealed a deoxyribose sugar switch one base step away from the junction between the tRNA primer and nascent (−) strand DNA [53]. Thus, sugar pucker switches may provide a common mechanism that contributes towards aligning RNA/DNA hybrids for correct cleavage at the RNase H active site.

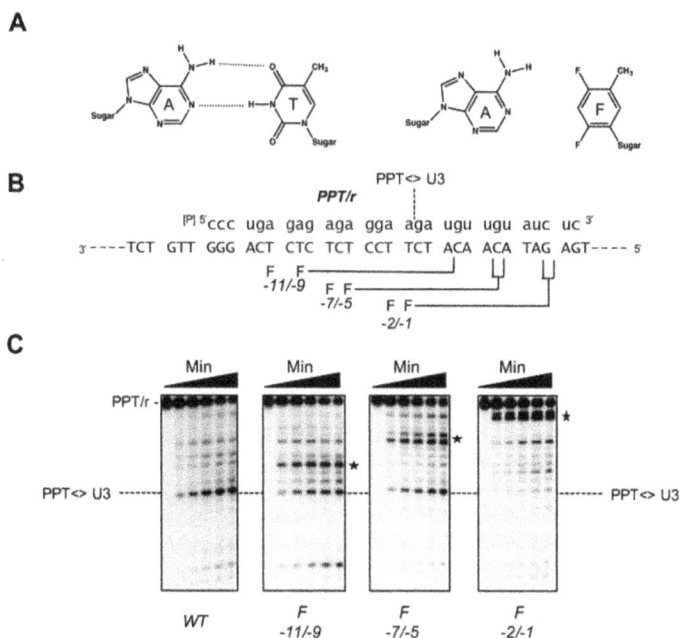

Figure 8. Modulation of Ty3 PPT cleavage by targeted insertion of non-polar pyrimidine isosteres. (A) Representation of an A:T base pair and its A:F counterpart; (B) Model Ty3 RNA/DNA hybrid and a summary of pyrimidine isostere mutagenesis. DNA and RNA strands are depicted in capital and small letters, respectively, and the scissile PPT/U3 junction is indicated. Base-pair numbering is relative to the PPT/U3 junction (i.e., the last base of the PPT is denoted −1). Sites of cleavage relative to the position of T-F modification in the DNA strand are indicated; (C) experimental data. WT, unmodified hybrid, indicating cleavage at the PPT/U3 junction. For additional panels, the position of T-F modification in the DNA strand are indicated, and the asterisk illustrates the relocated RNase H cleavage in response to these modifications. Adapted from [50,51].

Finally, as another example of subtle mechanistic differences in RTs that catalyze common steps in reverse transcription, pyrimidine isostere insertions into the DNA strand of the HIV-1 PPT have been demonstrated to similarly re-align the RNase H active site, but in this case 3–4 bp from their sites of insertion [54]. An HIV RT motif that might respond to structural anomalies is the "RNase H primer grip" (alternatively designated the phosphate binding pocket) which interacts with nucleic acid ~5 bp from the RNase H active site [38].

5. Conclusions and Perspectives

While the Ty3 lifecycle and RT structure share many of the features common among retroelements, numerous unique aspects of Ty3 have been highlighted in this review. The cognate minus strand primer tRNA hybridizes to distinct segments of Ty3 PBS separated by ~4800 nt in the genomic sequence, plus strand synthesis initiates multiple times from the PPT in a single reverse transcription cycle, and the PBS sequence is not perpetuated by reverse transcription of tRNA. Moreover, the RNase H domains of Ty3 RT are homologous to retroviral connection subdomains in both sequence and structural organization, and the DNA polymerase and RNase H activities of the enzyme are catalyzed by different subunits of an asymmetric homodimer. Such findings highlight not only the evolutionary commonalities and divergences among retroelements, but also the value of comparative studies in biological and biochemical research.

Acknowledgments: J.W.R., J.T.M., and S.F.J.L.G. are supported by the Intramural Research Program of the National Cancer Institute, National Institutes of Health, and Department of Health and Human Services.

Author Contributions: J.W.R., J.T.M, and S.F.J.L.G. wrote, organized and edited this manuscript.

Conflicts of Interest: The authors declare no conflict of interest. The funding sponsors had no role in the design of the study; in the collection, analyses, or interpretation of data; in the writing of the manuscript, and in the decision to publish the results.

References

1. Telesnitsky, A.; Goff, S.P. Reverse transcriptase and the generation of retroviral DNA. In *Retroviruses*; Coffin, J.M., Hughes, S.H., Varmus, H.E., Eds.; Cold Spring Harbor Laboratory Press: Cold Spring Harbor, NY, USA, 1997; pp. 121–160.
2. Baltimore, D. RNA-dependent DNA polymerase in virions of RNA tumour viruses. *Nature* **1970**, *226*, 1209–1211. [CrossRef] [PubMed]
3. Temin, H.M.; Mizutani, S. RNA-dependent DNA polymerase in virions of rous sarcoma virus. *Nature* **1970**, *226*, 1211–1213. [CrossRef] [PubMed]
4. Le Grice, S.F. "In the beginning": Initiation of minus strand DNA synthesis in retroviruses and LTR-containing retrotransposons. *Biochemistry* **2003**, *42*, 14349–14355. [CrossRef] [PubMed]
5. Nowak, E.; Potrzebowski, W.; Konarev, P.V.; Rausch, J.W.; Bona, M.K.; Svergun, D.I.; Bujnicki, J.M.; Le Grice, S.F.; Nowotny, M. Structural analysis of monomeric retroviral reverse transcriptase in complex with an RNA/DNA hybrid. *Nucleic Acids Res.* **2013**, *41*, 3874–3887. [CrossRef] [PubMed]
6. Kirshenboim, N.; Hayouka, Z.; Friedler, A.; Hizi, A. Expression and characterization of a novel reverse transcriptase of the LTR retrotransposon Tf1. *Virology* **2007**, *366*, 263–276. [CrossRef] [PubMed]
7. Benzair, A.B.; Rhodes-Feuillette, A.; Emanoil-Ravicovitch, R.; Peries, J. Reverse transcriptase from simian foamy virus serotype 1: Purification and characterization. *J. Virol.* **1982**, *44*, 720–724. [PubMed]
8. Das, D.; Georgiadis, M.M. The crystal structure of the monomeric reverse transcriptase from moloney murine leukemia virus. *Structure* **2004**, *12*, 819–829. [CrossRef] [PubMed]
9. Perach, M.; Hizi, A. Catalytic features of the recombinant reverse transcriptase of bovine leukemia virus expressed in bacteria. *Virology* **1999**, *259*, 176–189. [CrossRef] [PubMed]
10. Taube, R.; Loya, S.; Avidan, O.; Perach, M.; Hizi, A. Reverse transcriptase of mouse mammary tumour virus: Expression in bacteria, purification and biochemical characterization. *Biochem. J.* **1998**, *329 Pt 3*, 579–587. [CrossRef] [PubMed]
11. Sandmeyer, S.; Patterson, K.; Bilanchone, V. Ty3, a position-specific retrotransposon in budding yeast. *Microbiol. Spectr.* **2015**, *3*. MDNA3-0057-2014. [CrossRef] [PubMed]
12. Friant, S.; Heyman, T.; Wilhelm, M.L.; Wilhelm, F.X. Extended interactions between the primer tRNAi(met) and genomic RNA of the yeast Ty1 retrotransposon. *Nucleic Acids Res.* **1996**, *24*, 441–449. [CrossRef] [PubMed]
13. Gabus, C.; Ficheux, D.; Rau, M.; Keith, G.; Sandmeyer, S.; Darlix, J.L. The yeast Ty3 retrotransposon contains a 5′-3′ bipartite primer-binding site and encodes nucleocapsid protein NCp9 functionally homologous to HIV-1 NCp7. *EMBO J.* **1998**, *17*, 4873–4880. [CrossRef] [PubMed]
14. Ke, N.; Gao, X.; Keeney, J.B.; Boeke, J.D.; Voytas, D.F. The yeast retrotransposon Ty5 uses the anticodon stem-loop of the initiator methionine tRNA as a primer for reverse transcription. *RNA* **1999**, *5*, 929–938. [CrossRef] [PubMed]
15. Cristofari, G.; Gabus, C.; Ficheux, D.; Bona, M.; Le Grice, S.F.; Darlix, J.L. Characterization of active reverse transcriptase and nucleoprotein complexes of the yeast retrotransposon Ty3 in vitro. *J. Biol. Chem.* **1999**, *274*, 36643–36648. [CrossRef] [PubMed]
16. Friant, S.; Heyman, T.; Bystrom, A.S.; Wilhelm, M.; Wilhelm, F.X. Interactions between Ty1 retrotransposon RNA and the T and D regions of the tRNA(imet) primer are required for initiation of reverse transcription in vivo. *Mol. Cell. Biol.* **1998**, *18*, 799–806. [CrossRef] [PubMed]
17. Lanchy, J.M.; Keith, G.; Le Grice, S.F.; Ehresmann, B.; Ehresmann, C.; Marquet, R. Contacts between reverse transcriptase and the primer strand govern the transition from initiation to elongation of HIV-1 reverse transcription. *J. Biol. Chem.* **1998**, *273*, 24425–24432. [CrossRef] [PubMed]

18. Liu, S.; Harada, B.T.; Miller, J.T.; Le Grice, S.F.; Zhuang, X. Initiation complex dynamics direct the transitions between distinct phases of early HIV reverse transcription. *Nat. Struct. Mol. Biol.* **2010**, *17*, 1453–1460. [CrossRef] [PubMed]

19. Beerens, N.; Berkhout, B. The tRNA primer activation signal in the human immunodeficiency virus type 1 genome is important for initiation and processive elongation of reverse transcription. *J. Virol.* **2002**, *76*, 2329–2339. [CrossRef] [PubMed]

20. Isel, C.; Westhof, E.; Massire, C.; Le Grice, S.F.; Ehresmann, B.; Ehresmann, C.; Marquet, R. Structural basis for the specificity of the initiation of HIV-1 reverse transcription. *EMBO J.* **1999**, *18*, 1038–1048. [CrossRef] [PubMed]

21. Miller, J.T.; Ehresmann, B.; Hubscher, U.; Le Grice, S.F. A novel interaction of tRNA(Lys,3) with the feline immunodeficiency virus RNA genome governs initiation of minus strand DNA synthesis. *J. Biol. Chem.* **2001**, *276*, 27721–27730. [CrossRef] [PubMed]

22. Aiyar, A.; Ge, Z.; Leis, J. A specific orientation of RNA secondary structures is required for initiation of reverse transcription. *J. Virol.* **1994**, *68*, 611–618. [PubMed]

23. Kirchner, J.; Sandmeyer, S. Proteolytic processing of Ty3 proteins is required for transposition. *J. Virol.* **1993**, *67*, 19–28. [PubMed]

24. Cristofari, G.; Ficheux, D.; Darlix, J.L. The gag-like protein of the yeast Ty1 retrotransposon contains a nucleic acid chaperone domain analogous to retroviral nucleocapsid proteins. *J. Biol. Chem.* **2000**, *275*, 19210–19217. [CrossRef] [PubMed]

25. Karst, S.M.; Rutz, M.L.; Menees, T.M. The yeast retrotransposons Ty1 and Ty3 require the RNA lariat debranching enzyme, Dbr1p, for efficient accumulation of reverse transcripts. *Biochem. Biophys. Res. Commun.* **2000**, *268*, 112–117. [CrossRef] [PubMed]

26. Lauermann, V.; Boeke, J.D. The primer tRNA sequence is not inherited during Ty1 retrotransposition. *Proc. Natl. Acad. Sci. USA* **1994**, *91*, 9847–9851. [CrossRef] [PubMed]

27. Lauermann, V.; Boeke, J.D. Plus-strand strong-stop DNA transfer in yeast Ty retrotransposons. *EMBO J.* **1997**, *16*, 6603–6612. [CrossRef] [PubMed]

28. Pochart, P.; Agoutin, B.; Rousset, S.; Chanet, R.; Doroszkiewicz, V.; Heyman, T. Biochemical and electron microscope analyses of the DNA reverse transcripts present in the virus-like particles of the yeast transposon Ty1. Identification of a second origin of Ty1DNA plus strand synthesis. *Nucleic Acids Res.* **1993**, *21*, 3513–3520. [CrossRef] [PubMed]

29. Nymark-McMahon, M.H.; Sandmeyer, S.B. Mutations in nonconserved domains of Ty3 integrase affect multiple stages of the ty3 life cycle. *J. Virol.* **1999**, *73*, 453–465. [PubMed]

30. Nymark-McMahon, M.H.; Beliakova-Bethell, N.S.; Darlix, J.L.; Le Grice, S.F.; Sandmeyer, S.B. Ty3 integrase is required for initiation of reverse transcription. *J. Virol.* **2002**, *76*, 2804–2816. [CrossRef] [PubMed]

31. Wilhelm, M.; Wilhelm, F.X. Cooperation between reverse transcriptase and integrase during reverse transcription and formation of the preintegrative complex of Ty1. *Eukaryot. Cell* **2006**, *5*, 1760–1769. [CrossRef] [PubMed]

32. Le Grice, S.F.J. Human immunodeficiency virus reverse transcriptase: 25 years of research, drug discovery, and promise. *J. Biol. Chem.* **2012**, *287*, 40850–40857. [CrossRef] [PubMed]

33. Mous, J.; Heimer, E.P.; Le Grice, S.F. Processing protease and reverse transcriptase from human immunodeficiency virus type I polyprotein in *Escherichia coli*. *J. Virol.* **1988**, *62*, 1433–1436. [PubMed]

34. Kohlstaedt, L.A.; Wang, J.; Friedman, J.M.; Rice, P.A.; Steitz, T.A. Crystal Structure at 3.5 A resolution of HIV-1 reverse transcriptase complexed with an inhibitor. *Science* **1992**, *256*, 1783–1790. [CrossRef] [PubMed]

35. Malik, H.S.; Eickbush, T.H. Phylogenetic analysis of ribonuclease H domains suggests a late, chimeric origin of LTR retrotransposable elements and retroviruses. *Genome Res.* **2001**, *11*, 1187–1197. [CrossRef] [PubMed]

36. Rausch, J.W.; Grice, M.K.; Henrietta, M.; Nymark, M.; Miller, J.T.; Le Grice, S.F. Interaction of p55 reverse transcriptase from the *Saccharomyces cerevisiae* retrotransposon Ty3 with conformationally distinct nucleic acid duplexes. *J. Biol. Chem.* **2000**, *275*, 13879–13887. [CrossRef] [PubMed]

37. Nowak, E.; Miller, J.T.; Bona, M.K.; Studnicka, J.; Szczepanowski, R.H.; Jurkowski, J.; Le Grice, S.F.; Nowotny, M. Ty3 reverse transcriptase complexed with an RNA-DNA hybrid shows structural and functional asymmetry. *Nat. Struct. Mol. Biol.* **2014**, *21*, 389–396. [CrossRef] [PubMed]

38. Sarafianos, S.G.; Das, K.; Tantillo, C.; Clark, A.D., Jr.; Ding, J.; Whitcomb, J.M.; Boyer, P.L.; Hughes, S.H.; Arnold, E. Crystal structure of HIV-1 reverse transcriptase in complex with a polypurine tract RNA:DNA. *EMBO J.* **2001**, *20*, 1449–1461. [CrossRef] [PubMed]

39. Lapkouski, M.; Tian, L.; Miller, J.T.; Le Grice, S.F.; Yang, W. Complexes of HIV-1 RT, NNRTI and RNA/DNA hybrid reveal a structure compatible with RNA degradation. *Nat. Struct. Mol. Biol.* **2013**, *20*, 230–236. [CrossRef] [PubMed]

40. Lapkouski, M.; Tian, L.; Miller, J.T.; Le Grice, S.F.; Yang, W. Reply to "Structural requirements for RNA degradation by HIV-1 reverse transcriptase". *Nat. Struct. Mol. Biol.* **2013**, *20*, 1342–1343. [CrossRef] [PubMed]

41. Bibillo, A.; Lener, D.; Klarmann, G.J.; Le Grice, S.F. Functional roles of carboxylate residues comprising the DNA polymerase active site triad of Ty3 reverse transcriptase. *Nucleic Acids Res.* **2005**, *33*, 171–181. [CrossRef] [PubMed]

42. Sawaya, M.R.; Pelletier, H.; Kumar, A.; Wilson, S.H.; Kraut, J. Crystal structure of rat DNA polymerase beta: Evidence for a common polymerase mechanism. *Science* **1994**, *264*, 1930–1935. [CrossRef] [PubMed]

43. Bibillo, A.; Lener, D.; Tewari, A.; Le Grice, S.F. Interaction of the Ty3 reverse transcriptase thumb subdomain with template-primer. *J. Biol. Chem.* **2005**, *280*, 30282–30290. [CrossRef] [PubMed]

44. Koshkin, A.A.; Singh, S.K.; Nielsen, P.; Rajwanshi, V.K.; Kumar, R.; Meldgaard, M.; Olsen, C.E.; Wengel, J. LNA (locked nucleic acids): Synthesis of the adenine, cytosine, guanine, 5-methylcytosine, thymine and uracil bicyclonucleoside monomers, oligomerisation, and unprecedented nucleic acid recognition. *Tetrahedron* **1998**, *54*, 3607–3630. [CrossRef]

45. Lener, D.; Budihas, S.R.; Le Grice, S.F. Mutating conserved residues in the ribonuclease H domain of Ty3 reverse transcriptase affects specialized cleavage events. *J. Biol. Chem.* **2002**, *277*, 26486–26495. [CrossRef] [PubMed]

46. Nowotny, M. Retroviral integrase superfamily: The structural perspective. *EMBO Rep.* **2009**, *10*, 144–151. [CrossRef] [PubMed]

47. Nowotny, M.; Gaidamakov, S.A.; Crouch, R.J.; Yang, W. Crystal structures of RNase H bound to an RNA/DNA hybrid: Substrate specificity and metal-dependent catalysis. *Cell* **2005**, *121*, 1005–1016. [CrossRef] [PubMed]

48. Rausch, J.W.; Le Grice, S.F. Substituting a conserved residue of the ribonuclease H domain alters substrate hydrolysis by retroviral reverse transcriptase. *J. Biol. Chem.* **1997**, *272*, 8602–8610. [CrossRef] [PubMed]

49. Nowotny, M.; Gaidamakov, S.A.; Ghirlando, R.; Cerritelli, S.M.; Crouch, R.J.; Yang, W. Structure of human RNase H1 complexed with an RNA/DNA hybrid: Insight into HIV reverse transcription. *Mol. Cell* **2007**, *28*, 264–276. [CrossRef] [PubMed]

50. Lener, D.; Kvaratskhelia, M.; Le Grice, S.F. Nonpolar thymine isosteres in the Ty3 polypurine tract DNA template modulate processing and provide a model for its recognition by Ty3 reverse transcriptase. *J. Biol. Chem.* **2003**, *278*, 26526–26532. [CrossRef] [PubMed]

51. Guckian, K.M.; Krugh, T.R.; Kool, E.T. Solution structure of a nonpolar, non-hydrogen-bonded base pair surrogate in DNA. *J. Am. Chem. Soc.* **2000**, *122*, 6841–6847. [CrossRef] [PubMed]

52. Yi-Brunozzi, H.Y.; Brabazon, D.M.; Lener, D.; Le Grice, S.F.; Marino, J.P. A ribose sugar conformational switch in the LTR-retrotransposon Ty3 polypurine tract-containing RNA/DNA hybrid. *J. Am. Chem. Soc.* **2005**, *127*, 16344–16345. [CrossRef] [PubMed]

53. Szyperski, T.; Gotte, M.; Billeter, M.; Perola, E.; Cellai, L.; Heumann, H.; Wuthrich, K. NMR structure of the chimeric hybrid duplex r(gcaguggc).R(gcca)d(CTGC) comprising the tRNA-DNA junction formed during initiation of HIV-1 reverse transcription. *J. Biomol. NMR* **1999**, *13*, 343–355. [CrossRef] [PubMed]

54. Rausch, J.W.; Qu, J.; Yi-Brunozzi, H.Y.; Kool, E.T.; Le Grice, S.F. Hydrolysis of RNA/DNA hybrids containing nonpolar pyrimidine isosteres defines regions essential for HIV type 1 polypurine tract selection. *Proc. Natl. Acad. Sci. USA* **2003**, *100*, 11279–11284. [CrossRef] [PubMed]

viruses

MDPI

Review

Cross-Regulation between Transposable Elements and Host DNA Replication

Mikel Zaratiegui

Department of Molecular Biology and Biochemistry, Rutgers, the State University of New Jersey, 604 Allison Rd, Nelson Biolabs A133, Piscataway, NJ 08854, USA; zaratiegui@dls.rutgers.edu; Tel.: +1-848-445-1497

Academic Editors: David J. Garfinkel and Katarzyna J. Purzycka
Received: 1 February 2017; Accepted: 15 March 2017; Published: 21 March 2017

Abstract: Transposable elements subvert host cellular functions to ensure their survival. Their interaction with the host DNA replication machinery indicates that selective pressures lead them to develop ancestral and convergent evolutionary adaptations aimed at conserved features of this fundamental process. These interactions can shape the co-evolution of the transposons and their hosts.

Keywords: transposable elements; DNA replication; replication fork; transcription; genome integrity

---

1. Introduction

Transposable elements (TE) are ubiquitous in the tree of life. They have colonized almost all genomes sequenced to date, throughout eukaryotic, prokaryotic and archaeal domains. TE maintain their presence in the host genome by increasing their copy number via transposition, and colonize new genomes through horizontal transfer. Through these activities, TE exert a major influence in the evolution of the species.

Like viruses, TE are molecular parasitic elements that contain few genes, and they must condense multiple activities to subvert cellular functions to enable their continued presence in the host genome. This paucity of genetic payload leads molecular parasites to focus their intervention towards very fundamental cellular processes. As a consequence, research into viruses has led to some of the most seminal discoveries in molecular biology, such as the mechanisms of eukaryotic DNA replication, mRNA processing and many others. Similarly, the investigation of the transcriptional and post-transcriptional regulation of eukaryotic TE has been very fruitful, advancing our understanding of transcription and chromatin dynamics [1,2].

The equally fundamental process of DNA replication is another point of interaction between parasites and their hosts. The potential of TE to influence host genome stability and evolution make this problem a subject of particular interest, because it could have direct implications in the etiology of diseases like cancer and aging. Since the influence of host DNA replication extends across both type I retrotransposons and type II DNA transposons, it is worthwhile to discuss them together. The purpose of this review is to summarize the current evidence of TE influence on host DNA replication and vice versa, and to speculate on the potential selection pressures that shape its evolution.

2. DNA Transposon Duplication

Type II elements, also known as DNA transposons, do not generate an RNA transposition intermediate, and they must rely on the host DNA replication machinery to increase their copy number. One way to do this is through a partial transposition in which a single strand of the donor copy is inserted in a target site, leaving DNA replication to generate the complementary strand of both the donor copy and the new insertion. The Mu phage and the bacterial Tn3 family of transposons can undergo such a replicative transposition through single-stranded donor cleavage and strand transfer into the target site, yielding a θ structure known as the Shapiro intermediate [3] (Figure 1). Subsequent DNA replication duplicates the joint insertion into a co-integrate, doubling the copy number. Similarly, the concerted model of Helitron transposition starts with single-stranded cleavage and 5′ strand transfer, followed by strand displacement of the transposed strand by replication from the free 3′ OH of the donor [4]. The displaced strand is cleaved and joined with the 5′ end of the target nick, leaving it as a heteroduplex that resolves by passive DNA replication, generating a new copy of the Helitron in one of the daughter strands.

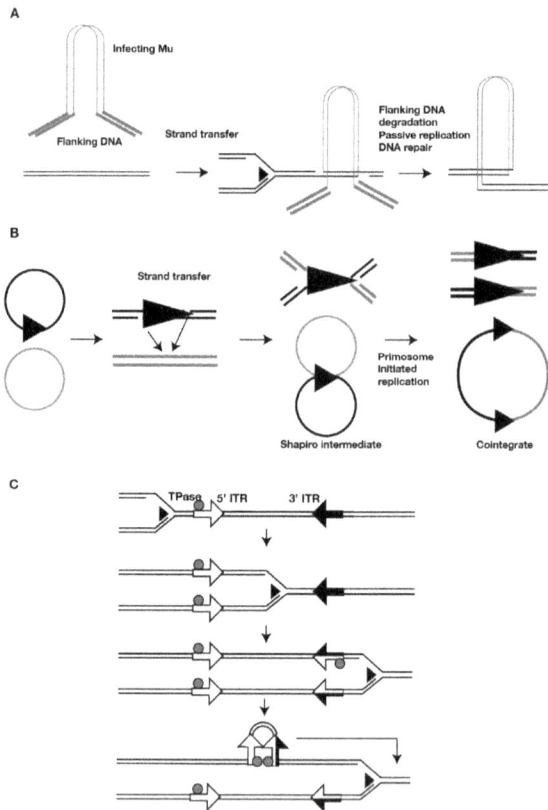

Figure 1. Replication of type II DNA transposons. (**A**) Non-replicative transposition of Mu after infection. The Mu phage and flanking DNA are injected into the host. Cleavage and strand transfer join the Mu phage DNA to the target site, leaving single-stranded gaps. Upon arrival of a replication fork the flanking DNA is degraded, and the gaps create a double stranded end create a double stranded end. Both gaps are simultaneously filled by passive DNA replication, yielding a mature prophage; (**B**) Replicative transposition of Mu in the lytic phase. Strand transfer of the prophage into the target site create a Θ-shaped Shapiro intermediate, with the Mu element flanked by fork-like structures. Primosome-started replication at these structures duplicate the Mu element in a joined cointegrate; (**C**) Control of activator/dissociator (Ac/Ds) transposition by replication fork passage. Methylation at the inverted terminal repeats (ITRs) is depicted as filled arrows. Hemimethylated ITR depicted as half-filled arrows, with the filled portion indicating the methylated strand. Replication of the methylated 3′ ITR yields two hemimethylated daughter ITR, only one of which binds the transposase (TPase), determining which of the two daughter elements can assemble the transpososome.

Both of these strategies rely on DNA replication for completion of the transposition. The initial integration of infecting Mu (lysogenic stage) is a non-replicative transposition but it nevertheless depends on passive DNA replication for completion. In this case a native host-initiated DNA replication fork that encounters the transpososome directs the degradation of flanking DNA that usually accompanies the injected Mu phage DNA, and repairs the gaps resulting from the staggered nicks in the insertion site, resulting in a mature prophage [5] (Figure 1A). In contrast, the Mu phage in its lytic stage completes each new replicative transposition by the assembly of a primosome dependent "restart" replisome at the fork structures created by strand transfer [6] (Figure 1B). The switch from transposition to replication is coordinated by the transpososome in collaboration with host factors, and in this case, it can be said that the transposon machinery initiates replication.

The more prevalent canonical "cut-and-paste" transposition mechanism can also take advantage of passive DNA replication to increase copy number, by directing mobilization of a copy from one of the daughter chromatids generated after passage of the replication fork into an unreplicated region of the host genome. The transposition machinery can sense the replicated/unreplicated status of an insertion by measuring the level of DNA methylation of the two DNA strands. Nascent DNA is unmethylated, and DNA replication leaves transiently hemimethylated sites that are subsequently restored to full methylation by the maintenance DNA methyltransferases. A methylation-sensing mechanism has been demonstrated in multiple bacterial transposons and in the maize transposons activator/dissociator (Ac/Ds) (Figure 1C).

Ac/Ds transposes during DNA replication. Only one of the two daughter elements becomes active, and can transpose ahead of the replication fork to create a new insertion [7–9]. The cause of both the S-phase activity and the "chromatid selectivity" of mobilization was traced to the methylation status of the transposase (TPase) binding sites in the inverted terminal repeats (ITRs) [10]. The Ac TPase binds with strongly differential affinity depending on which strand is hemimethylated [11]. Since only the 3' ITR presents high levels of methylation [12], passage of the fork leaves one daughter element that allows TPase binding at both ends because its 3' ITR shows permissive hemimethylation. This element can actively mobilize, but the other daughter element, with non-permissive hemimethylation in the 3' ITR, remains inactive [10].

Replication fork passage controls Tn10/IS10 transposition and chromatid selectivity by a very similar mechanism, utilizing the Dam methylation motif to control binding of the TPase. In addition, hemimethylation of the Dam motifs allow binding of the RNA polymerase and transcription of the TPase gene, further coupling transposition to DNA replication [13].

Type II transposons are much less prevalent in mammals, and a potential role of replication fork passage sensing is yet to be demonstrated. The human Tc1/mariner family element HsMar1 represents a potential example. HsMar1 transpososome formation is sensitive to DNA topology, and is enhanced by negatively supercoiled DNA that could occur in the wake of the replication fork [14,15]. Besides DNA methylation, other epigenetic marks that exhibit slow re-establishment in the wake of the fork, such as Histone 4 Lysine 20 methylation [16], may also regulate transposon activity in eukaryotes [17].

3. Role of the Replication Fork in Transposon Target-Site Selection

The involvement of replication forks extends into the insertion stage of mobilization. The functional implications are difficult to gauge because they often involve essential cellular functions. Nevertheless, evidence from TE representing multiple classes of elements, both type I and type II, point to a direct role of replication fork dynamics in target site selection and the nucleic acid transactions that underlie insertion.

One point of cross-talk between transposons and the replication fork that extends across type I and type II elements in eukaryotes and prokaryotes is the interaction between the transposition machinery and the sliding clamps that coordinate replisome function. This activity is a universal requirement for processive DNA replication, and is carried out by proteins showing the DNA clamp fold, which forms multimers that encircle double stranded DNA [18]. The bacterial sliding clamp is a homodimer of the beta subunit of DNA polymerase III (Pol III; β-clamp), while in Archaea and Eukaryotes it is formed by a homotrimer of the proliferating cell nuclear antigen (PCNA).

Sliding clamps recruit a myriad of proteins involved in DNA replication, DNA repair and, in the case of PCNA, chromatin assembly. The majority of these interactors bind via hydrophobic pockets on the advancing face of the sliding clamp, gaining access to the primer terminus of the nascent DNA. These conserved hydrophobic domains recruit proteins sporting consensus binding motifs: β-clamp interactors show QxxL(x)F or QL(S/D)LF, and PCNA interactors show a remarkably similar sequence known as PCNA Interacting Protein motif (PIP-box: Qxx[I/L/M]xxF[F/Y]). The first transposon protein observed to interact with a sliding clamp was the *Drosophila melanogaster* type II POGO TPase, which was identified as a PCNA interactor in a yeast two-hybrid screen [19]. It exhibits a PIP-box that is conserved in its human relative, Tigger, and in the pogo-like Arabidopsis element Lemi1 [20]. A putative PIP-box can also be observed in the maize Ac/Ds transposon. However, the functional relevance of these motifs remains unclear.

More mechanistic insights into the significance of these interactions came from the discovery that several bacterial transposons also show interactions between their transposition machinery and the β-clamp. The first one described was the type II element Tn7 [21]. This element has two mechanisms of insertion site selection, regulated by the choice of one of two transposon encoded specificity factors, TnsD and TnsE [22]. The first one is a highly targeted insertion mechanism dependent on sequence recognition by TnsD [23]. For its part, TnsE dependent target site selection has looser sequence requirements, but shows several particularities that suggested the involvement of replication forks. Tn7 inserts via TnsE into plasmids undergoing replicative transfer [24] with a striking bias of insertion orientation that correlates with the directionality of replication. Additionally, TnsE can guide insertion into the host chromosome favoring replication termination sites and showing the same orientation bias [25] (Figure 2A). These observations suggested that TnsE could detect the presence of replication forks and direct transposition towards them. TnsE binds to substrates with recessed 3′ ends that could occur in replication forks, providing a potential explanation [26]. The mechanism for this target site selection pathway was explained when sequence conservation analysis of TnsE revealed a consensus β-clamp interaction motif [21]. In agreement with a potential role for a TnsE/β-clamp interaction in Tn7 mobility, mutation of this motif lowered transposition activity in vivo, and β-clamp overexpression increased it. A minimal in vitro transposition system with a gapped substrate to provide the recessed 3′ end enables efficient transposition, but with random position and orientation with respect to the gap in the target. However, loading the β-clamp onto the target restored the site specificity and dramatic orientation bias of the insertions. It appears that Tn7 specifically targets discontinuous DNA replication for insertion through interaction with the β-clamp [27].

Since this work was published multiple other bacterial transposons, utilizing very different insertion mechanisms, have revealed interactions between their transposition machinery and the β-clamp. The IS200/IS605 family of transposons uses a single-stranded DNA (ssDNA) "peel and paste" transposition mechanism that is profoundly influenced by replication fork dynamics [28–30]. Excision of IS608 and ISDra2, belonging to this family, is more efficient when the transposed strand is in the lagging strand template, transiently providing a ready ssDNA donor after passage of the replication fork. At the insertion side of the reaction the fork also has a strong influence, because it preferentially targets, again, the lagging strand template. As a consequence, the orientation of members of this family of transposons recapitulates the directionality of DNA replication in their hosts [30]. Notably, the IS608 TPase TnpA binds β-clamp by yeast two hybrid, and also shows affinity for fork-like structures [31], suggesting that, despite the profound differences in insertion mechanisms, IS200/IS605 and Tn7-like transposons could use common targeting strategies. These mechanisms may turn out to be very common: multiple IS families exhibit interactions between their TPases and β-clamp, also showing similar orientation biases with respect to host DNA replication [32].

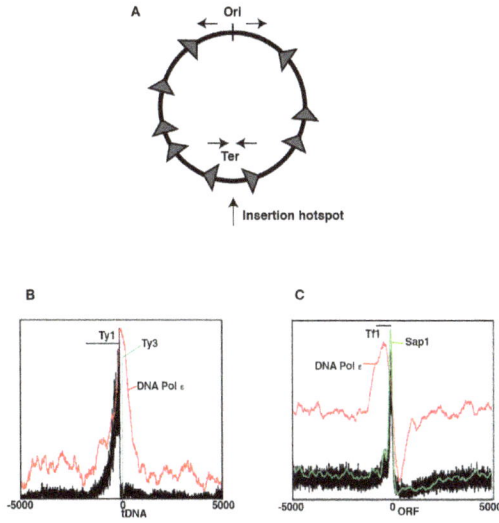

Figure 2. Fork influence on target site selection. (**A**) Insertion patterns of the Tn7 TnsE-dependent transposition into the host chromosome. Ori = origin of replication. Ter = replication termination region. Insertions are depicted as grey arrows; (**B**) Insertion patterns of Ty1 and Ty3 in type III genes. Ty1 insertions in black [33], Ty3 insertions in green [34] DNA pol ε average occupancy in red [35]; (**C**) Insertion patterns of Tf1 in type II genes. Tf1 insertion in black [36], average DNA pol ε occupancy in red [37] and average Sap1 occupancy in green [38].

The interactions between transposition machinery and sliding clamps even transcend the division between type II DNA and type I RNA-intermediate transposons: a proteomic survey of the human non-LTR retrotransposon long interspersed nuclear element-1 (LINE-1) ribonucleoprotein revealed the interaction between the endonuclease (EN)/reverse transcriptase (RT) ORF2p and PCNA, carried out via a canonical PIP-box [39]. This interacting motif is necessary for transposition activity. Interaction with PCNA was decreased in ORF2 EN and RT mutants, indicating that it is recruited in the context of the initial steps of LINE-1 transposition. In contrast with IS200/IS605 and Tn7, the mechanism of LINE-1 target-primed reverse transcription insertion does not readily provide an explanation for the involvement of PCNA, but potential roles in RT processivity or post-insertion DNA repair can be imagined without the involvement of a native DNA replication fork [39].

A common observation in the study of replication regulated transposons is the insertion preference for sites of programmed fork arrest [26,30]. Such sites are an essential part of the host replication program, because they organize the genome in domains with defined replication directionalities, usually disfavoring replication in antisense orientation over highly expressed genes [40–42]. This organization prevents head-on collisions between the advancing replisome and transcription complexes, which can result in replisome loss, leaving unreplicated regions that become fragile sites upon chromosome segregation [43]. Programmed replication fork barriers (RFB) usually require the action of sequence-specific DNA binding factors with asymmetric binding properties that impart a defined polarity to the barrier activity, blocking fork advance in one direction but allowing progression in the opposite direction [44]. The replication fork can also arrest when it encounters other types of impediments to its progression, such as G-quadruplexes, highly transcribed genes, or tightly-bound DNA binding proteins [45]. The dynamics of arrested forks is a subject of intense research because forks that stall, losing the replisome, can destabilize leading to double-strand breaks and gross chromosomal rearrangements. Unsurprisingly, arrested forks often activate DNA damage signaling pathways and engage repair mechanisms [46].

Since bacterial replication usually starts from a single origin, with two sister forks travelling around the circular chromosome, their termination and merging sites are well known. Both Tn7 and IS200/IS608 transposons exhibit preference for natural and ectopic replication termination sites [25,30] (Figure 2A). A possible explanation invokes the role of DNA replication fork structure in the insertion mechanisms of these elements: a stalled fork would exhibit the ssDNA target for a longer period, until DNA replication from a converging fork merges with it, perhaps providing an extended window of opportunity for transposition to occur. In agreement with this potential mechanism the bacterial protein that binds and protects single-stranded template DNA during replication, Ssb, is a negative regulator of IS608 transposition [31]. Several TPases show binding to unique DNA structures that could be exposed in stalled replication forks [26,31]. Interaction between the sliding clamp and transposition machinery might also be favored in stalled replisomes, because the loss of DNA polymerases could release the hydrophobic pockets in the clamp [18]. In addition, sliding clamps are involved in the signaling of DNA damage and the recruitment of repair activities, both of which could potentially modulate interactions with the transposition machinery at sites of fork arrest.

Recent work in fungal LTR retrotransposons also point to the involvement of replication fork arrest in their insertion target site selection. The high gene density in fungal genomes puts a selective pressure on these elements to evolve strategies that target insertion away from protein coding sequences, so as not to decrease host fitness. These strategies usually take the form of protein-protein interactions between the integrase (INT) and host DNA binding factors that localize at non-coding regions such as promoters and heterochromatin, providing a platform to recruit the integration complex (intasome) to "safe haven" targets [47]. Intense scrutiny has revealed the insertion preferences of the copia-like Ty1 and Ty5 as well as the gypsy-like Ty3 elements in *Saccharomyces cerevisiae*, and the gypsy-like Tf1/2 elements in *Schizosaccharomyces pombe*. These closely related elements show a variety of preferred target sites: Ty1 and Ty3 insert upstream of type III genes (*tDNA*, *5S* and *U6*; Figure 2B), Ty5 inserts in heterochromatic domains, and Tf1/2 insert in promoters of protein-coding genes. (Figure 2C) Potential INT DNA binding partners that have been identified could explain these insertion preferences. Ty1 INT interacts with the RNA Pol III subunit AC40 [48], and substituting it with a non-interacting ortholog leads to dispersal of insertions away from its usual targets in type III promoters. Ty3 can transpose in vitro into *tDNA* targets in the presence of transcription factor for polymerase III B (TFIIIB)/transcription factor for polymerase III C (TFIIIC) [49]. Ty5 INT binds to the silencing factor Sir4 [50], and the interacting domain can be transferred to a different sequence specific binding factor that can then direct insertion to ectopic binding sites [51].

The fission yeast element Tf1 element, like its close relative Tf2, shows insertions in type II protein coding gene promoters [52,53]. While interactions between Tf1 INT and host factors have been described, none fully explain this insertion specificity. The transcription factor Atf1 binds INT [54], but deletion mutants don't exhibit decreased transposition and only show a modest difference in target site preference. Together with the clear accumulation of insertions in the nucleosome-depleted regions (NDR) that are usually present in type II promoters, this led to a model whereby chromatin structure and sequence-specific DNA binding factors collaborated to determine Tf1/2 target site preferences [53].

The DNA binding factor Sap1, which also binds Tf1 INT by yeast two-hybrid analysis, is the main determinant of NDR formation in pombe genes [55]. Genome-wide analysis showed that Sap1 binding is highly predictive of Tf1 insertion [36,56]. However, Sap1 binding is not sufficient for insertion, as some very strong Sap1 binding sites are cold spots for transposition. Sap1 has an additional function required for genome integrity: in certain binding arrangements, it forms a polar RFB [57–59]. A mutation in *sap1* that abrogates this function but only mildly affects DNA binding [38] severely decreases Tf1 transposition [36]. Additionally, Sap1 binding sites constitute insertion hotspots but only if they exhibit RFB activity, and their insertion competence depends on the orientation with respect to the advancing fork. However, other programmed RFB that are independent of Sap1 are not targeted for insertion [36,53,56]. These observations indicate that Sap1 binding and RFB activity are both necessary but neither is sufficient for target site selection. Measuring intasome recruitment by chromosome conformation capture (3C) between the mature cDNA and an ectopic target site revealed that fork arrest is necessary for intasome tethering to the target. Together, these results suggest that Sap1 presence and its RFB activity collaborate to determine target site selection [36].

Unlike in the case of β-clamp interacting TPases, there is no obvious mechanism for the recognition of an arrested form by the LTR retrotransposon integrase. Could the arrested fork be the real tethering factor? Sap1-INT interaction is only detectable by yeast two-hybrid [36,56]. Weak intasome tethering can be detected at Sap1-independent RFB when in the blocking orientation with respect to fork progression, although these are not insertion targets [36,53]. The Sap1 binding and RFB requirements are separable, so a model in which the arrested fork tethers the intasome and the Sap1 interaction activates it for insertion could have merit.

Comparing the insertion preferences of fungal LTR retrotransposons may offer new insights. Type III genes, the targets for Ty1 and Ty3 insertion, are notorious RFB [60,61], and Sir4, the Ty5 tethering factor, is recruited to sites of replication fork arrest [62]. In the amoeba *Dictyostelium discoideum* several LTR and non-LTR retrotransposons also show targeting to *tDNA* genes [63]. These target site preferences could indicate an ancestral role of arrested replication forks in retrotransposon target site selection. The Tf1-like element Tj1 originating from the fission yeast *Schizosaccharomyces japonicus* can be coaxed into transposing in *S. pombe* [64]. Tj1 is present in heterochromatic regions of the *S. japonicus* centromeres, which exhibit dense clusters of *tDNA* [65]. Unlike its close cousin Tf1, the Tj1 insertion points in *S. pombe* accumulate in a small window upstream of type III genes, reminiscent of the insertion pattern observed in Ty3. In conclusion, the insertion site preference for type III and type II promoters and heterochromatin appear to be characteristic of fungal LTR retrotransposons, but the choice of one of these targets is not tied to particular families of elements, with members of Ty1/Copia, Ty3/Gypsy and Tf1/Gypsy groups showing insertion preferences as variant as all fungal LTR taken as a whole. The only commonality in all these target types is their activity as RFB.

Why are transposons fixated on the replication fork? Insertion into fork arrest sites does not impart an obvious selective advantage to the mobile element. A potential role could be to widen the potential host spectrum, increasing their chances for horizontal transfer (HT). HT is essential for the evolutionary success of transposons, because it allows them to escape vertical extinction. Despite its importance, HT is extremely poorly understood.

Since transposons rely on the cellular machinery for their vertical transmission, they may evolve specialized adaptations to the new host that ensure their persistence by increasing their copy number to avoid loss by genetic drift. Conversely, the host evolves with its transposons, defending against their destabilizing influence, and sometimes domesticating the transposon machinery, exapting it into new cellular activities [66]. This tug-of-war between the host and the transposon guides their co-evolution [67]. However, these host-specific TE adaptations may not serve after HT to a new host, and could even be detrimental. Most HT events described in eukaryotes occurred between closely related species [68], perhaps as a consequence of host specialization, but the ubiquity of some families of transposons indicates that wide leaps, even between different phyla, do occur in nature. The evolutionary success of a transposon could therefore depend not only on host-specific adaptation to ensure vertical transmission, but also on balancing generalist mechanisms that enable successful HT.

The replication fork is one potential focus point for these generalist interactions. The structure of the replicating DNA is completely universal in all cellular life forms. A transposon able to exploit this structure to facilitate its transposition would always find the same substrate no matter the host [27]. The protein factors that carry out DNA replication are also remarkably conserved, because the essential nature of this process subjects them to intense purifying selection. Interaction between sliding clamps and replication factors constitutes another universal feature DNA replication, providing transposons with a conserved point of cross-talk with the fork [21,69,70]. Convergent evolution of these interactions may explain the widespread presence of β-clamp and PCNA binding motifs in transposition machinery. A central role of sliding clamp interactions in HT was recently proposed, with supporting mechanistic evidence, in the IS1634 element from the bacterium *Acidiphilum* sp. [70]. Mutating a β-clamp binding motif present in its TPase showed that transposition efficiency is directly proportional to binding affinity, not only in its *Acidiphilum* host but also upon transfer to *E. coli*. This work also showed that *Acidiphilum* IS1634 TPase can interact with the archaeal PCNA sliding clamp in *Methanosarcina*. Conversely, an IS1634 element TPase aboriginal to *Methanosarcina* can interact with the *Acidiphilum* β-clamp. These experiments dramatically illustrate the generalist nature of interactions between sliding clamps and transposition machinery, and suggest that the similitude between the β-clamp interaction motif and the PIP-box might enable transposon HT between host species belonging to entirely different kingdoms.

The search for insertion safe havens may also benefit from an ancestral preference for arrested replication forks. Since they coordinate the direction of replication and transcription they are usually localized in intergenic regions, making them an attractive platform for new mobilizations minimizing the mutagenic potential. RFB stop fork progression through poorly understood mechanisms but they are usually associated with tight protein-DNA interactions [44,71]. Several elements that show RFB activity, such as promoters bound by transcription factors and highly compacted heterochromatin [60,62,72], would constitute safe havens in a broad variety of potential hosts. Here again, experimentally forced horizontal transfer could provide interesting information about what insertion targets are available to a transposon undergoing HT [64].

4. Influence of TEs Presence in Host DNA Replication and Homologous Recombination

Ever since their discovery, TE were observed to very strongly destabilize their surroundings. Mutations created by transposition into cellular genes or regulatory elements show high rates not just of reversion (often caused by TE excision) but also of derivation into different alleles affecting the same gene [73]. Moreover, TE can cause gross chromosomal rearrangements involving their insertion sites [74]. In the case of type II DNA TE this phenomenon is often explained by the activity of the transposition machinery, which can lead to erroneous excisions involving dispersed TE sequences. Due to the ease of generation of derived alleles, much of the early research into TE after their re-discovery in bacteria, fungi and animals concentrated in the characterization of these post-insertion rearrangements, leading to pilot models of transposition mechanisms [3].

But mobilization is not the only cause of TE-mediated rearrangements. *S. cerevisiae* mutations caused by LTR retrotransposon insertion also exhibit instability [75,76]. However, since the INT protein binds to the free cDNA ends, not the integrated element, the transposition mechanism can't explain the rearrangements. Instead, they depend on the host Homologous Recombination (HR) pathway. TE mediated rearrangements showing the hallmarks of HR are common in all organisms. Repetitive DNA is intrinsically unstable because the process of HR includes a search for homology that in repeated sequences may engage non-allelic loci, resulting in cross-over and non cross-over resolution, observable as rearrangements and gene conversions. As a result, HR of TE sequences was considered an inevitable consequence of its repetitive nature. Since the only requirement for this process is sequence homology, it also involves inactive copies, which vastly outnumber active ones.

Mobilization-dependent and HR-dependent rearrangements are now known to be major drivers of eukaryotic genome structural variation (SV) and evolution [77–79]. Examples of structural variation involving TE, with and without adaptive value to the host, are abundant in the literature. The role of fungal LTR retroelements in yeast SV has been extensively investigated, because the small genome and long history of strain domestication facilitates comparative analysis [80]. The non-autonomous type I *Alu* elements seem to be a major cause of SV, both in polymorphisms present in human populations [81] and in primate evolution [82]. Plant genomes with high transposon content exhibit extreme SV, some of which underlies important traits in commercial cultivars [83]. Finally, TE mediated rearrangements could explain some of the genomic instability observable in cancer [84], which often shows activation of TE as part of its disregulated transcriptional program [85].

The processes that lead to mobilization-dependent SV can be retraced because transposition mechanisms are relatively well understood, sometimes revealing behaviors nothing short of acrobatic [86]. But since the role of TE in HR-mediated rearrangements was considered to be passive, it has received little attention. Work in fungal LTR retrotransposons has revealed that their behavior in this process is more active than previously thought.

The recombinogenic activity of *S. cerevisiae* LTR elements was observed even before they were recognized as TE sequences. Rothstein characterized deletion and inversion mutations of the *tDNA* gene *SUP4*, locating the breakpoint regions in five Ty1 LTR (then known as delta sequences) that flanked the locus [87]. This phenomenon required the HR factor RAD52. Soon thereafter, the characterization of revertants of mutations caused by Ty1 insertions revealed that it was frequently excised through HR between the two flanking LTR [1,75,76]. This recombination explains the abundance of solo LTR that pepper eukaryotic genomes: each represents an ancient insertion that was lost through recombination, leaving a solo LTR at the insertion site. Inter-LTR recombination is therefore a very common event. In fact, it appears to be the only process that counteracts the plant genome gigantism caused by runaway LTR retrotransposon activity [88]. HR between non-allelic LTR underlies a large proportion of yeast SV [80]. The solo LTR is sufficient to mediate HR rearrangements [89], so the destabilizing influence of LTR retrotransposons could continue even after their complete extinction from the host genome.

Paradoxically, the frequency of mitotic and meiotic non-allelic HR of Ty1 sequences is low when directly compared with artificially introduced non TE repeats [89]. Some LTR are more recombinogenic than others, even in very similar contexts, suggesting that factors extrinsic to their sequence homology or repetitive nature influence this activity, and that mechanisms that prevent TE dependent HR exist. Mutation of the topoisomerase TOP3 increases the frequency of *SUP4* deletion by inter-LTR recombination [90]. TOP3 restarts stalled replication forks together with the RecQ DNA helicase slow growth suppressor 1 (SGS1). Arrested forks engage the HR machinery to restart the replisome, and mutations in TOP3 or SGS1 result in increased HR and gross chromosomal rearrangements [91]. The dependence on TOP3/SGS1 to prevent HR of LTR indicates that these elements constitute impediments to the progression of the replication fork (Figure 3A). In agreement with this model, Ty LTR exhibit accumulation of DNA polymerase ε indicative of replisome pausing, as well as DNA damage signaling by local accumulation of phosphorylated histone γ-H2A. These hallmarks of replication fork arrest are exacerbated in mutants of RRM3, a DNA helicase that aids the replication fork in overcoming obstacles to its progression [35].

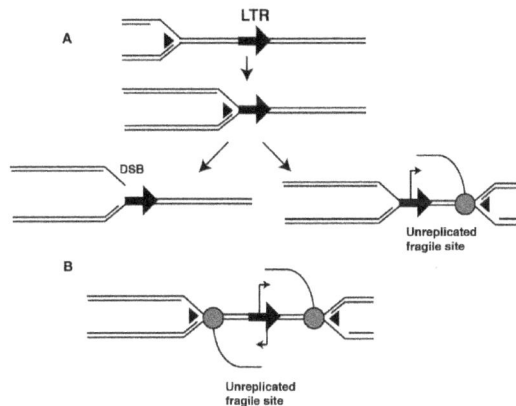

Figure 3. Fork instability at transposable elements (TE). An LTR containing replication fork barriers (RFB) can lead to replication fork stalling and double strand break (DSB) formation (left). (**A**) Active transcription of the TE can cause replisome-RNA Pol II collisions and unreplicated regions (right); (**B**) TE with actively transcribing bidirectional promoters can cause replisome-RNA Pol II collisions and unreplicated regions.

Despite their evolutionary distance with Ty elements, the *S. pombe* Tf1/2 LTR retrotransposons also exhibit this property. A genomic survey of γ-H2A localization revealed that Tf2 and solo LTR elements signaled DNA damage even during a completely undisturbed S phase [92]. Strikingly, the Tf1/2 LTR contain a conserved binding site for Sap1 (yes, the very same DNA binding factor implicated in target site selection) that exhibits polar RFB activity [38]. Sap1 is not conserved in *S. cerevisiae*, so whatever RFB activity Ty LTR have must be carried out by other mechanisms; this property could be the result of convergent evolution co-opting host factors.

Most HR at Ty elements does not occur between non-allelic copies, but instead involves gene conversion of the inserted copies by cDNA or cDNA intermediates [93,94]. A sizable proportion of mobilization events in fungal LTR retrotransposons is INT-independent, but requires HR machinery. In the case of *S. pombe* Tf2 this pathway constitutes the majority (~70%) of mobilization events observed upon overexpression of the transposon [95]. Screens for regulators of Ty mobility seldom distinguish between mobilization by insertion and HR mediated gene conversion events, so some negative regulators of mobilization could be in fact repressors of Ty mediated HR. As an example, the PCNA unloader ELG1 was independently identified as a negative regulator both of inter-LTR recombination [96] and of Ty1 mobility [97], and multiple host factors that repress mobility have known functions to repress HR. However, it is difficult to separate the contribution of LTR-initiated HR from the effect of DNA damage prevention, checkpoint, signaling and repair pathways on Ty cDNA formation and mobility. For example, mutation of SGS1 or RRM3 increases Ty1 mobility dependent on RAD52, but rather than stimulating cDNA mediated gene conversion the increase is due to the formation of cDNA multimers [98,99], which are the main mediators of mobility when INT activity is prevented [100]. The dissection of this phenomenon will require specifically designed models that address these multiple pathways.

Transcription also plays an important role in this process that is independent from cDNA generation. Inducing the transcription of a Ty1 copy via a regulated promoter increases its competence as a recipient of cDNA mediated gene conversion by up to an order of magnitude [101]. Similarly, mutations that activate transcription of Tf1/2 in *S. pombe* increase mobility by HR [102]. Tf1/2 elements are silenced by three partially redundant domesticated Pogo/Tigger TPase-like factors collectively known as centromere protein B (CENP-B). Besides Tf1/2 increased transcription, mutations in these factors also cause a dramatic loss of genome integrity and recruitment of HR factors to LTR [38]. Mutations of Sap1 abrogating RFB activity suppress the loss of genome integrity, indicating that forks arrested by Sap1 at LTR become destabilized in CENP-B mutants. As a result, CENP-B mutants require an intact HR pathway for viability. While INT-mediated mobility is not affected, HR-dependent mobility increases dramatically in a CENP-B mutant [102]. Conversely, mutation of *sap1* removing RFB activity practically eliminates INT-independent transposition by HR [36]. These observations suggest that fork arrest and transcription at the recipient elements have a synergistic effect on HR-dependent mobility.

Transcription poses a formidable obstacle to replication fork progression [43]. The presence of programmed RFB at LTR could exacerbate replication-transcription conflicts (Figure 3), leading to the genome-wide proliferation of arrested forks and unreplicated regions that engage HR to resume replication and prevent instability. Increased HR, if directed at the offending repetitive elements, could cause gross chromosomal rearrangements. This model would explain the role of heterochromatin in maintenance of genome integrity [103]. Loss of silencing of TE and other forms of repetitive elements leads to widespread replication-transcription conflicts that cause DNA damage localized at heterochromatic DNA, and rearrangements through non-allelic or improperly resolved HR. Since this source of genome instability does not require transposition mechanisms, non-autonomous and even highly mutated copies of TE could participate. This phenomenon has been observed in multiple model organisms, affecting centromeric and rDNA repeats as well as TE [38,72,104–107].

What function could RFB activity bring to these elements? A possible explanation invokes HR-mediated mobilization. An element able to exploit this process may paradoxically stabilize its presence in the host genome, perhaps counteracting inter-LTR recombination [38]. Such a mechanism would enable a transposon colony to use the cDNA pool as a community resource and a communication tool, enforcing sequence consensus or spreading variants with favorable characteristics [108,109]. Alternatively, if the RFB contained in the LTR mediate target site selection they could aid genome colonization by dispersing insertion hotspots to new safe havens. The LTR of Ty1 and Tf1/2 elements show this activity [36,110], and if extensible to other elements it could explain the tendency of transposons to accumulate as clusters and nested insertions.

Regardless of its role in TE biology, the consequences for the host genome can be quite dramatic. The proliferation of RFB could change the replication program and increase genome plasticity, particularly under conditions of active TE transcription. Since many TE are transcriptionally activated by cellular stress, TE-mediated HR could represent an additional layer of the long-proposed role of transposons in host adaptability. Gene amplification is a common mechanism for adaptation to stress. Some TE, such as the Tf1 element, show a preference for insertion in promoters of stress-regulated genes, and could therefore poise them for amplification by HR. This activity has been observed in a case of Histone gene amplification mediated by Ty1 [111] which can be induced by treatment with hydroxyurea, a drug that stalls replication, and by mutation of factors required for fork progression [112]. Similarly, experimental evolution of yeast grown under limiting glucose yields adaptive rearrangements, such as amplification of hexose transporters, through non-allelic HR between transposon sequences [113].

It is not known whether other TE present RFB like fungal LTR elements. Inverted repeats of the primate short interspersed nuclear element (SINE) *Alu* form hairpins that arrest replication forks in bacteria, yeast and mammalian cells [114]. *Alu* elements constitute the majority of inverted repeats in the human genome and could therefore influence genome plasticity via their interaction with replication and HR. The non-LTR retrotransposon LINE-1 are the most abundant autonomous TE in humans, and their role in cancer progression is the subject of much debate because multiple cancers exhibit LINE-1 transcription activation and mobilization. LINE-1 contain bidirectional promoters that, if activated, could arrest replication forks converging on the transcribed element resulting in fragile sites (Figure 3B). Oncogenic transformation is often accompanied by increased endogenous replication stress and DNA damage [84], and the resulting genomic instability that drives cancer progression could have a TE component. Increased activity of the LINE-1 transposition machinery is a likely culprit [115,116], but considering the high TE content of the human genome, and the genome integrity phenotypes of heterochromatin mutations observed in model organisms, loss of seamless repetitive element replication might also be a significant contributor [103].

Acknowledgments: I am thankful to the members of the Zarategui lab and to Henry Levin for fruitful discussions on the topic. Research in the Zarategui laboratory is supported by NIGMS/NIH 1R01GM105831 and a Busch Biomedical Research Grant.

Conflicts of Interest: The author declares no conflict of interest.

References

1. Winston, F.; Chaleff, D.T.; Valent, B.; Fink, G.R. Mutations affecting Ty-mediated expression of the HIS4 gene of Saccharomyces cerevisiae. *Genetics* **1984**, *107*, 179–197. [PubMed]
2. Slotkin, R.K.; Martienssen, R. Transposable elements and the epigenetic regulation of the genome. *Nat. Rev. Genet.* **2007**, *8*, 272–285. [CrossRef] [PubMed]
3. Shapiro, J.A. Molecular model for the transposition and replication of bacteriophage Mu and other transposable elements. *Proc. Natl. Acad. Sci. USA* **1979**, *76*, 1933–1937. [CrossRef] [PubMed]
4. Thomas, J.; Pritham, E.J. Helitrons, the Eukaryotic Rolling-circle Transposable Elements. *Microbiol. Spectr.* **2015**, *3*. [CrossRef] [PubMed]
5. Jang, S.; Harshey, R.M. Repair of transposable phage Mu DNA insertions begins only when the *E. coli* replisome collides with the transpososome. *Mol. Microbiol.* **2015**, *97*, 746–758. [CrossRef] [PubMed]
6. Nakai, H.; Doseeva, V.; Jones, J.M. Handoff from recombinase to replisome: Insights from transposition. *Proc. Natl. Acad. Sci. USA* **2001**, *98*, 8247–8254. [CrossRef] [PubMed]

7. Greenblatt, I.M.; Brink, R.A. Twin Mutations in Medium Variegated Pericarp Maize. *Genetics* **1962**, *47*, 489–501. [PubMed]
8. Greenblatt, I.M. A chromosome replication pattern deduced from pericarp phenotypes resulting from movements of the transposable element, modulator, in maize. *Genetics* **1984**, *108*, 471–485. [PubMed]
9. Chen, J.; Greenblatt, I.M.; Dellaporta, S.L. Transposition of Ac from the P locus of maize into unreplicated chromosomal sites. *Genetics* **1987**, *117*, 109–116. [PubMed]
10. Ros, F.; Kunze, R. Regulation of activator/dissociation transposition by replication and DNA methylation. *Genetics* **2001**, *157*, 1723–1733.
11. Kunze, R.; Starlinger, P. The putative transposase of transposable element Ac from Zea mays L. interacts with subterminal sequences of Ac. *EMBO J.* **1989**, *8*, 3177–3185. [PubMed]
12. Wang, L.; Heinlein, M.; Kunze, R. Methylation pattern of Activator transposase binding sites in maize endosperm. *Plant Cell* **1996**, *8*, 747–758. [CrossRef] [PubMed]
13. Roberts, D.; Hoopes, B.C.; McClure, W.R.; Kleckner, N. IS10 transposition is regulated by DNA adenine methylation. *Cell* **1985**, *43*, 117–130. [CrossRef]
14. Claeys Bouuaert, C.; Liu, D.; Chalmers, R. A simple topological filter in a eukaryotic transposon as a mechanism to suppress genome instability. *Mol. Cell. Biol.* **2010**, *31*, 3925–3932. [CrossRef] [PubMed]
15. Claeys Bouuaert, C.; Chalmers, R. Hsmar1 transposition is sensitive to the topology of the transposon donor and the target. *PLoS ONE* **2013**, *8*, e53690. [CrossRef] [PubMed]
16. Saredi, G.; Huang, H.; Hammond, C.M.; Alabert, C.; Bekker-Jensen, S.; Forne, I.; Reverón-Gómez, N.; Foster, B.M.; Mlejnkova, L.; Bartke, T.; et al. H4K20me0 marks post-replicative chromatin and recruits the TONSL–MMS22L DNA repair complex. *Nature* **2016**, *534*, 714–718. [CrossRef] [PubMed]
17. Hatanaka, Y.; Inoue, K.; Oikawa, M.; Kamimura, S.; Ogonuki, N.; Kodama, E.N.; Ohkawa, Y.; Tsukada, Y.-I.; Ogura, A. Histone chaperone CAF-1 mediates repressive histone modifications to protect preimplantation mouse embryos from endogenous retrotransposons. *Proc. Natl. Acad. Sci. USA* **2015**, *112*, 14641–14646. [CrossRef] [PubMed]
18. Bloom, L.B. Loading clamps for DNA replication and repair. *DNA Repair (Amst.)* **2009**, *8*, 570–578. [CrossRef] [PubMed]
19. Warbrick, E.; Heatherington, W.; Lane, D.P.; Glover, D.M. PCNA binding proteins in Drosophila melanogaster: The analysis of a conserved PCNA binding domain. *Nucleic Acids Res.* **1998**, *26*, 3925–3932. [CrossRef] [PubMed]
20. Warbrick, E. The puzzle of PCNA's many partners. *Bioessays* **2000**, *22*, 997–1006. [CrossRef]
21. Parks, A.R.; Li, Z.; Shi, Q.; Owens, R.M.; Jin, M.M.; Peters, J.E. Transposition into replicating DNA occurs through interaction with the processivity factor. *Cell* **2009**, *138*, 685–695. [CrossRef] [PubMed]
22. Waddell, C.S.; Craig, N.L. Tn7 transposition: Two transposition pathways directed by five Tn7-encoded genes. *Genes Dev.* **1988**, *2*, 137–149. [CrossRef] [PubMed]
23. Waddell, C.S.; Craig, N.L. Tn7 transposition: Recognition of the attTn7 target sequence. *Proc. Natl. Acad. Sci. USA* **1989**, *86*, 3958–3962. [CrossRef] [PubMed]
24. Wolkow, C.A.; DeBoy, R.T.; Craig, N.L. Conjugating plasmids are preferred targets for Tn7. *Genes Dev.* **1996**, *10*, 2145–2157. [CrossRef] [PubMed]
25. Peters, J.E.; Craig, N.L. Tn7 transposes proximal to DNA double-strand breaks and into regions where chromosomal DNA replication terminates. *Mol. Cell* **2000**, *6*, 573–582. [CrossRef]
26. Peters, J.E.; Craig, N.L. Tn7 recognizes transposition target structures associated with DNA replication using the DNA-binding protein TnsE. *Genes Dev.* **2001**, *15*, 737–747. [CrossRef] [PubMed]
27. Fricker, A.D.; Peters, J.E. Vulnerabilities on the Lagging-Strand Template: Opportunities for Mobile Elements. *Annu. Rev. Genet.* **2014**, *48*, 167–186. [CrossRef] [PubMed]
28. Ton-Hoang, B.; Guynet, C.; Ronning, D.R.; Cointin-Marty, B.; Dyda, F.; Chandler, M. Transposition of ISHp608, member of an unusual family of bacterial insertion sequences. *EMBO J.* **2005**, *24*, 3325–3338. [CrossRef] [PubMed]
29. Guynet, C.; Hickman, A.B.; Barabas, O.; Dyda, F.; Chandler, M.; Ton-Hoang, B. In vitro reconstitution of a single-stranded transposition mechanism of IS608. *Mol. Cell* **2008**, *29*, 302–312. [CrossRef] [PubMed]
30. Ton-Hoang, B.; Pasternak, C.; Siguier, P.; Guynet, C.; Hickman, A.B.; Dyda, F.; Sommer, S.; Chandler, M. Single-stranded DNA transposition is coupled to host replication. *Cell* **2010**, *142*, 398–408. [CrossRef] [PubMed]

31. Lavatine, L.; He, S.; Caumont-Sarcos, A.; Guynet, C.; Marty, B.; Chandler, M.; Ton-Hoang, B. Single strand transposition at the host replication fork. *Nucleic Acids Res.* **2016**, *44*, 7866–7883. [CrossRef] [PubMed]

32. Gómez, M.J.; Díaz-Maldonado, H.; González-Tortuero, E.; López de Saro, F.J. Chromosomal replication dynamics and interaction with the β sliding clamp determine orientation of bacterial transposable elements. *Genome Biol. Evol.* **2014**, *6*, 727–740. [CrossRef] [PubMed]

33. Mularoni, L.; Zhou, Y.; Bowen, T.; Gangadharan, S.; Wheelan, S.J.; Boeke, J.D. Retrotransposon Ty1 integration targets specifically positioned asymmetric nucleosomal DNA segments in tRNA hotspots. *Genome Res.* **2012**, *22*, 693–703. [CrossRef] [PubMed]

34. Qi, X.; Daily, K.; Nguyen, K.; Wang, H.; Mayhew, D.; Rigor, P.; Forouzan, S.; Johnston, M.; Mitra, R.D.; Baldi, P.; et al. Retrotransposon profiling of RNA polymerase III initiation sites. *Genome Res.* **2012**, *22*, 681–692. [CrossRef] [PubMed]

35. Szilard, R.K.; Jacques, P.-E.; Laramée, L.; Cheng, B.; Galicia, S.; Bataille, A.R.; Yeung, M.; Mendez, M.; Bergeron, M.; Robert, F.; et al. Systematic identification of fragile sites via genome-wide location analysis of gamma-H2AX. *Nat. Struct. Mol. Biol.* **2010**, *17*, 299–305. [CrossRef] [PubMed]

36. Jacobs, J.Z.; Rosado-Lugo, J.D.; Cranz-Mileva, S.; Ciccaglione, K.M.; Tournier, V.; Zaratiegui, M. Arrested replication forks guide retrotransposon integration. *Science* **2015**, *349*, 1549–1553. [CrossRef] [PubMed]

37. Sabouri, N.; Capra, J.A.; Zakian, V.A. The essential Schizosaccharomyces pombe Pfh1 DNA helicase promotes fork movement past G-quadruplex motifs to prevent DNA damage. *BMC Biol.* **2014**, *12*, 101. [CrossRef] [PubMed]

38. Zaratiegui, M.; Vaughn, M.W.; Irvine, D.V.; Goto, D.; Watt, S.; Bähler, J.; Arcangioli, B.; Martienssen, R.A. CENP-B preserves genome integrity at replication forks paused by retrotransposon LTR. *Nature* **2011**, *469*, 112–115. [CrossRef] [PubMed]

39. Taylor, M.S.; LaCava, J.; Mita, P.; Molloy, K.R.; Huang, C.R. L.; Li, D.; Adney, E.M.; Jiang, H.; Burns, K.H.; Chait, B.T.; et al. Affinity proteomics reveals human host factors implicated in discrete stages of LINE-1 retrotransposition. *Cell* **2013**, *155*, 1034–1048. [CrossRef] [PubMed]

40. Bermejo, R.; Lai, M.S.; Foiani, M. Preventing replication stress to maintain genome stability: Resolving conflicts between replication and transcription. *Mol. Cell* **2012**, *45*, 710–718. [CrossRef] [PubMed]

41. Lambert, S.; Carr, A.M. Replication stress and genome rearrangements: Lessons from yeast models. *Curr. Opin. Genet. Dev.* **2013**, *23*, 132–139. [CrossRef] [PubMed]

42. Labib, K.; Hodgson, B. Replication fork barriers: Pausing for a break or stalling for time? *EMBO Rep.* **2007**, *8*, 346–353. [CrossRef] [PubMed]

43. García-Muse, T.; Aguilera, A. Transcription-replication conflicts: How they occur and how they are resolved. *Nat. Rev. Mol. Cell Biol.* **2016**, *17*, 553–563. [CrossRef] [PubMed]

44. Kaplan, D.L.; Bastia, D. Mechanisms of polar arrest of a replication fork. *Mol. Microbiol.* **2009**, *72*, 279–285. [CrossRef] [PubMed]

45. Sabouri, N. The functions of the multi-tasking Pfh1(Pif1) helicase. *Curr. Genet.* **2017**. [CrossRef] [PubMed]

46. Lambert, S.; Froget, B.; Carr, A.M. Arrested replication fork processing: Interplay between checkpoints and recombination. *DNA Repair (Amst.)* **2007**, *6*, 1042–1061. [CrossRef] [PubMed]

47. Bushman, F.D. Targeting survival: Integration site selection by retroviruses and LTR-retrotransposons. *Cell* **2003**, *115*, 135–138. [CrossRef]

48. Bridier-Nahmias, A.; Tchalikian-Cosson, A.; Baller, J.A.; Menouni, R.; Fayol, H.; Flores, A.; Saïb, A.; Werner, M.; Voytas, D.F.; Lesage, P. An RNA polymerase III subunit determines sites of retrotransposon integration. *Science* **2015**, *348*, 585–588. [CrossRef] [PubMed]

49. Kirchner, J.; Connolly, C.M.; Sandmeyer, S.B. Requirement of RNA polymerase III transcription factors for in vitro position-specific integration of a retrovirulike element. *Science* **1995**, *267*, 1488–1491. [CrossRef] [PubMed]

50. Xie, W.; Gai, X.; Zhu, Y.; Zappulla, D.C.; Sternglanz, R.; Voytas, D.F. Targeting of the yeast Ty5 retrotransposon to silent chromatin is mediated by interactions between integrase and Sir4p. *Mol. Cell. Biol.* **2001**, *21*, 6606–6614. [CrossRef] [PubMed]

51. Zhu, Y.; Dai, J.; Fuerst, P.G.; Voytas, D.F. Controlling integration specificity of a yeast retrotransposon. *Proc. Natl. Acad. Sci. USA* **2003**, *100*, 5891–5895. [CrossRef] [PubMed]

52. Bowen, N.J.; Jordan, I.K.; Epstein, J.A.; Wood, V.; Levin, H.L. Retrotransposons and their recognition of pol II promoters: A comprehensive survey of the transposable elements from the complete genome sequence of Schizosaccharomyces pombe. *Genome Res.* **2003**, *13*, 1984–1997. [CrossRef] [PubMed]

53. Guo, Y.; Levin, H.L. High-throughput sequencing of retrotransposon integration provides a saturated profile of target activity in Schizosaccharomyces pombe. *Genome Res.* **2010**, *20*, 239–248. [CrossRef] [PubMed]

54. Majumdar, A.; Chatterjee, A.G.; Ripmaster, T.L.; Levin, H.L. Determinants that specify the integration pattern of retrotransposon Tf1 in the fbp1 promoter of Schizosaccharomyces pombe. *J. Virol.* **2011**, *85*, 519–529. [CrossRef] [PubMed]

55. Tsankov, A.; Yanagisawa, Y.; Rhind, N.; Regev, A.; Rando, O.J. Evolutionary divergence of intrinsic and trans-regulated nucleosome positioning sequences reveals plastic rules for chromatin organization. *Genome Res.* **2011**, *21*, 1851–1862. [CrossRef] [PubMed]

56. Hickey, A.; Esnault, C.; Majumdar, A.; Chatterjee, A.G.; Iben, J.R.; McQueen, P.G.; Yang, A.X.; Mizuguchi, T.; Grewal, S.I.S.; Levin, H.L. Single-Nucleotide-Specific Targeting of the Tf1 Retrotransposon Promoted by the DNA-Binding Protein Sap1 of Schizosaccharomyces pombe. *Genetics* **2015**, *201*, 905–924. [CrossRef] [PubMed]

57. Mejía-Ramírez, E.; Sánchez-Gorostiaga, A.; Krimer, D.B.; Schvartzman, J.B.; Hernández, P. The mating type switch-activating protein Sap1 Is required for replication fork arrest at the rRNA genes of fission yeast. *Mol. Cell. Biol.* **2005**, *25*, 8755–8761. [CrossRef] [PubMed]

58. Krings, G.; Bastia, D. Sap1p binds to Ter1 at the ribosomal DNA of Schizosaccharomyces pombe and causes polar replication fork arrest. *J. Biol. Chem.* **2005**, *280*, 39135–39142. [CrossRef] [PubMed]

59. Krings, G.; Bastia, D. Molecular architecture of a eukaryotic DNA replication terminus-terminator protein complex. *Mol. Cell. Biol.* **2006**, *26*, 8061–8074. [CrossRef] [PubMed]

60. Deshpande, A.M.; Newlon, C.S. DNA replication fork pause sites dependent on transcription. *Science* **1996**, *272*, 1030–1033. [CrossRef] [PubMed]

61. Sabouri, N.; McDonald, K.R.; Webb, C.J.; Cristea, I.M.; Zakian, V.A. DNA replication through hard-to-replicate sites, including both highly transcribed RNA Pol II and Pol III genes, requires the S. pombe Pfh1 helicase. *Genes Dev.* **2012**, *26*, 581–593. [CrossRef] [PubMed]

62. Dubarry, M.; Loiodice, I.; Chen, C.L.; Thermes, C.; Taddei, A. Tight protein-DNA interactions favor gene silencing. *Genes Dev.* **2011**, *25*, 1365–1370. [CrossRef] [PubMed]

63. Spaller, T.; Kling, E.; Glöckner, G.; Hillmann, F.; Winckler, T. Convergent evolution of tRNA gene targeting preferences in compact genomes. *Mob. DNA* **2016**, *7*, 17. [CrossRef] [PubMed]

64. Guo, Y.; Singh, P.K.; Levin, H.L. A long terminal repeat retrotransposon of Schizosaccharomyces japonicus integrates upstream of RNA pol III transcribed genes. *Mob. DNA* **2015**, *6*, 19. [CrossRef] [PubMed]

65. Rhind, N.; Chen, Z.; Yassour, M.; Thompson, D.A.; Haas, B.J.; Habib, N.; Wapinski, I.; Roy, S.; Lin, M.F.; Heiman, D.I.; et al. Comparative Functional Genomics of the Fission Yeasts. *Science* **2011**, *332*, 930–936. [CrossRef] [PubMed]

66. Le Rouzic, A.; Boutin, T.S.; Capy, P. Long-term evolution of transposable elements. *Proc. Natl. Acad. Sci. USA* **2007**, *104*, 19375–19380. [CrossRef] [PubMed]

67. McLaughlin, R.N.; Malik, H.S. Genetic conflicts: The usual suspects and beyond. *J. Exp. Biol.* **2017**, *220*, 6–17. [CrossRef] [PubMed]

68. Schaack, S.; Gilbert, C.; Feschotte, C. Promiscuous DNA: Horizontal transfer of transposable elements and why it matters for eukaryotic evolution. *Trends Ecol. Evol. (Amst.)* **2010**, *25*, 537–546. [CrossRef] [PubMed]

69. Peters, J.E.; Craig, N.L. Tn7: Smarter than we thought. *Nat. Rev. Mol. Cell Biol.* **2001**, *2*, 806–814. [CrossRef] [PubMed]

70. Díaz-Maldonado, H.; Gómez, M.J.; Moreno-Paz, M.; San Martín-Úriz, P.; Amils, R.; Parro, V.; López de Saro, F.J. Transposase interaction with the β sliding clamp: Effects on insertion sequence proliferation and transposition rate. *Sci. Rep.* **2015**, *5*, 13329. [CrossRef] [PubMed]

71. Bastia, D.; Zaman, S. Mechanism and physiological significance of programmed replication termination. *Semin. Cell Dev. Biol.* **2014**, *30*, 165–173. [CrossRef] [PubMed]

72. Zaratiegui, M.; Castel, S.E.; Irvine, D.V.; Kloc, A.; Ren, J.; Li, F.; de Castro, E.; Marín, L.; Chang, A.-Y.; Goto, D.; et al. RNAi promotes heterochromatic silencing through replication-coupled release of RNA Pol II. *Nature* **2011**, *479*, 135–138. [CrossRef] [PubMed]

73. McClintock, B. Induction of Instability at Selected Loci in Maize. *Genetics* **1953**, *38*, 579–599. [PubMed]

74. McClintock, B. The origin and behavior of mutable loci in maize. *Proc. Natl. Acad. Sci. USA* **1950**, *36*, 344–355. [CrossRef] [PubMed]

75. Roeder, G.S.; Fink, G.R. DNA rearrangements associated with a transposable element in yeast. *Cell* **1980**, *21*, 239–249. [CrossRef]

76. Chaleff, D.T.; Fink, G.R. Genetic events associated with an insertion mutation in yeast. *Cell* **1980**, *21*, 227–237. [CrossRef]

77. Eichler, E.E.; Sankoff, D. Structural dynamics of eukaryotic chromosome evolution. *Science* **2003**, *301*, 793–797. [CrossRef] [PubMed]

78. Kazazian, H.H. Mobile elements: Drivers of genome evolution. *Science* **2004**, *303*, 1626–1632. [CrossRef] [PubMed]

79. Feschotte, C.; Pritham, E.J. DNA transposons and the evolution of eukaryotic genomes. *Annu. Rev. Genet.* **2007**, *41*, 331–368. [CrossRef] [PubMed]

80. Garfinkel, D.J. Genome evolution mediated by Ty elements in Saccharomyces. *Cytogenet. Genome Res.* **2005**, *110*, 63–69. [CrossRef] [PubMed]

81. Batzer, M.A.; Deininger, P.L. Alu repeats and human genomic diversity. *Nat. Rev. Genet.* **2002**, *3*, 370–379. [CrossRef] [PubMed]

82. Johnson, M.E.; National Institute of Health Intramural Sequencing Center Comparative Sequencing Program; Cheng, Z.; Morrison, V.A.; Scherer, S.; Ventura, M.; Gibbs, R.A.; Green, E.D.; Eichler, E.E. Recurrent duplication-driven transposition of DNA during hominoid evolution. *Proc. Natl. Acad. Sci. USA* **2006**, *103*, 17626–17631. [PubMed]

83. Meyer, R.S.; Purugganan, M.D. Evolution of crop species: Genetics of domestication and diversification. *Nat. Rev. Genet.* **2013**, *14*, 840–852. [CrossRef] [PubMed]

84. Hanahan, D.; Weinberg, R.A. Hallmarks of cancer: The next generation. *Cell* **2011**, *144*, 646–674. [CrossRef] [PubMed]

85. Lee, E.; Iskow, R.; Yang, L.; Gokcumen, O.; Haseley, P.; Luquette, L.J.; Lohr, J.G.; Harris, C.C.; Ding, L.; Wilson, R.K.; et al. Cancer Genome Atlas Research Network Landscape of somatic retrotransposition in human cancers. *Science* **2012**, *337*, 967–971. [CrossRef] [PubMed]

86. Zhang, J.; Zuo, T.; Peterson, T. Generation of tandem direct duplications by reversed-ends transposition of maize ac elements. *PLoS Genet.* **2013**, *9*, e1003691. [CrossRef] [PubMed]

87. Rothstein, R. Deletions of a tyrosine tRNA gene in S. cerevisiae. *Cell* **1979**, *17*, 185–190. [CrossRef]

88. Vitte, C.; Panaud, O. LTR retrotransposons and flowering plant genome size: Emergence of the increase/decrease model. *Cytogenet. Genome Res.* **2005**, *110*, 91–107. [CrossRef] [PubMed]

89. Kupiec, M.; Petes, T.D. Allelic and ectopic recombination between Ty elements in yeast. *Genetics* **1988**, *119*, 549–559. [PubMed]

90. Wallis, J.W.; Chrebet, G.; Brodsky, G.; Rolfe, M.; Rothstein, R. A hyper-recombination mutation in S. cerevisiae identifies a novel eukaryotic topoisomerase. *Cell* **1989**, *58*, 409–419. [CrossRef]

91. Myung, K.; Datta, A.; Chen, C.; Kolodner, R.D. SGS1, the Saccharomyces cerevisiae homologue of BLM and WRN, suppresses genome instability and homeologous recombination. *Nat. Genet.* **2001**, *27*, 113–116. [PubMed]

92. Rozenzhak, S.; Mejía-Ramírez, E.; Williams, J.S.; Schaffer, L.; Hammond, J.A.; Head, S.R.; Russell, P. Rad3 decorates critical chromosomal domains with gammaH2A to protect genome integrity during S-Phase in fission yeast. *PLoS Genet.* **2010**, *6*, e1001032. [CrossRef] [PubMed]

93. Melamed, C.; Nevo, Y.; Kupiec, M. Involvement of cDNA in homologous recombination between Ty elements in Saccharomyces cerevisiae. *Mol. Cell. Biol.* **1992**, *12*, 1613–1620. [CrossRef] [PubMed]

94. Nevo-Caspi, Y.; Kupiec, M. cDNA-mediated Ty recombination can take place in the absence of plus-strand cDNA synthesis, but not in the absence of the integrase protein. *Curr. Genet.* **1997**, *32*, 32–40. [CrossRef] [PubMed]

95. Hoff, E.F.; Levin, H.L.; Boeke, J.D. Schizosaccharomyces pombe retrotransposon Tf2 mobilizes primarily through homologous cDNA recombination. *Mol. Cell. Biol.* **1998**, *18*, 6839–6852. [CrossRef] [PubMed]

96. Ben-Aroya, S.; Koren, A.; Liefshitz, B.; Steinlauf, R.; Kupiec, M. ELG1, a yeast gene required for genome stability, forms a complex related to replication factor C. *Proc. Natl. Acad. Sci. USA* **2003**, *100*, 9906–9911. [CrossRef] [PubMed]

97. Scholes, D.T.; Banerjee, M.; Bowen, B.; Curcio, M.J. Multiple regulators of Ty1 transposition in Saccharomyces cerevisiae have conserved roles in genome maintenance. *Genetics* **2001**, *159*, 1449–1465. [PubMed]

98. Bryk, M.; Banerjee, M.; Conte, D.; Curcio, M.J. The Sgs1 helicase of Saccharomyces cerevisiae inhibits retrotransposition of Ty1 multimeric arrays. *Mol. Cell. Biol.* **2001**, *21*, 5374–5388. [CrossRef] [PubMed]

99. Stamenova, R.; Maxwell, P.H.; Kenny, A.E.; Curcio, M.J. Rrm3 protects the Saccharomyces cerevisiae genome from instability at nascent sites of retrotransposition. *Genetics* **2009**, *182*, 711–723. [CrossRef] [PubMed]

100. Sharon, G.; Burkett, T.J.; Garfinkel, D.J. Efficient homologous recombination of Ty1 element cDNA when integration is blocked. *Mol. Cell. Biol.* **1994**, *14*, 6540–6551. [CrossRef] [PubMed]

101. Nevo-Caspi, Y.; Kupiec, M. Transcriptional induction of Ty recombination in yeast. *Proc. Natl. Acad. Sci. USA* **1994**, *91*, 12711–12715. [CrossRef] [PubMed]

102. Cam, H.P.; Noma, K.-I.; Ebina, H.; Levin, H.L.; Grewal, S.I.S. Host genome surveillance for retrotransposons by transposon-derived proteins. *Nature* **2008**, *451*, 431–436. [CrossRef] [PubMed]

103. Nikolov, I.; Taddei, A. Linking replication stress with heterochromatin formation. *Chromosoma* **2015**, 1–11. [CrossRef] [PubMed]

104. Peng, J.C.; Karpen, G.H. H3K9 methylation and RNA interference regulate nucleolar organization and repeated DNA stability. *Nat. Cell. Biol.* **2006**, *9*, 25–35. [CrossRef] [PubMed]

105. Peng, J.C.; Karpen, G.H. Heterochromatic genome stability requires regulators of histone H3 K9 methylation. *PLoS Genet.* **2009**, *5*, e1000435. [CrossRef] [PubMed]

106. Castel, S.E.; Ren, J.; Bhattacharjee, S.; Chang, A.-Y.; Sánchez, M.; Valbuena, A.; Antequera, F.; Martienssen, R.A. Dicer promotes transcription termination at sites of replication stress to maintain genome stability. *Cell* **2014**, *159*, 572–583. [CrossRef] [PubMed]

107. Zeller, P.; Padeken, J.; van Schendel, R.; Kalck, V.; Tijsterman, M.; Gasser, S.M. Histone H3K9 methylation is dispensable for Caenorhabditis elegans development but suppresses RNA:DNA hybrid-associated repeat instability. *Nat. Genet.* **2016**, *48*, 1385–1395. [CrossRef] [PubMed]

108. Roy-Engel, A.M.; Carroll, M.L.; El-Sawy, M.; Salem, A.-H.; Garber, R.K.; Nguyen, S.V.; Deininger, P.L.; Batzer, M.A. Non-traditional Alu evolution and primate genomic diversity. *J. Mol. Biol.* **2002**, *316*, 1033–1040. [CrossRef] [PubMed]

109. Ellison, C.E.; Bachtrog, D. Non-allelic gene conversion enables rapid evolutionary change at multiple regulatory sites encoded by transposable elements. *eLife* **2015**, *4*. [CrossRef] [PubMed]

110. Ji, H.; Moore, D.P.; Blomberg, M.A.; Braiterman, L.T.; Voytas, D.F.; Natsoulis, G.; Boeke, J.D. Hotspots for unselected Ty1 transposition events on yeast chromosome III are near tRNA genes and LTR sequences. *Cell* **1993**, *73*, 1007–1018. [CrossRef]

111. Libuda, D.E.; Winston, F. Amplification of histone genes by circular chromosome formation in Saccharomyces cerevisiae. *Nature* **2006**, *443*, 1003–1007. [CrossRef] [PubMed]

112. Libuda, D.E.; Winston, F. Alterations in DNA replication and histone levels promote histone gene amplification in Saccharomyces cerevisiae. *Genetics* **2010**, *184*, 985–997. [CrossRef] [PubMed]

113. Dunham, M.J.; Badrane, H.; Ferea, T.; Adams, J.; Brown, P.O.; Rosenzweig, F.; Botstein, D. Characteristic genome rearrangements in experimental evolution of Saccharomyces cerevisiae. *Proc. Natl. Acad. Sci. USA* **2002**, *99*, 16144–16149. [CrossRef] [PubMed]

114. Voineagu, I.; Narayanan, V.; Lobachev, K.S.; Mirkin, S.M. Replication stalling at unstable inverted repeats: Interplay between DNA hairpins and fork stabilizing proteins. *Proc. Natl. Acad. Sci. USA* **2008**, *105*, 9936–9941. [CrossRef] [PubMed]

115. Wallace, N.A.; Belancio, V.P.; Deininger, P.L. L1 mobile element expression causes multiple types of toxicity. *Gene* **2008**, *419*, 75–81. [CrossRef] [PubMed]

116. Gilbert, N.; Lutz-Prigge, S.; Moran, J.V. Genomic deletions created upon LINE-1 retrotransposition. *Cell* **2002**, *110*, 315–325. [CrossRef]

viruses

MDPI

Review

Mechanisms of LTR-Retroelement Transposition: Lessons from *Drosophila melanogaster*

Lidia N. Nefedova and Alexander I. Kim *

Department of Biology, Moscow State University, Moscow 119992, Russia; lidia_nefedova@mail.ru
* Correspondence: aikim57@mail.ru; Tel.: +7-495-939-4253

Academic Editors: David J. Garfinkel and Katarzyna J. Purzycka
Received: 31 January 2017; Accepted: 10 April 2017; Published: 16 April 2017

Abstract: Long terminal repeat (LTR) retrotransposons occupy a special place among all mobile genetic element families. The structure of LTR retrotransposons that have three open reading frames is identical to DNA forms of retroviruses that are integrated into the host genome. Several lines of evidence suggest that LTR retrotransposons share a common ancestry with retroviruses and thus are highly relevant to understanding mechanisms of transposition. *Drosophila melanogaster* is an exceptionally convenient model for studying the mechanisms of retrotransposon movement because many such elements in its genome are transpositionally active. Moreover, two LTR-retrotransposons of *D. melanogaster*, *gypsy* and *ZAM*, have been found to have infectious properties and have been classified as errantiviruses. Despite numerous studies focusing on retroviral integration process, there is still no clear understanding of integration specificity in a target site. Most LTR retrotransposons non-specifically integrate into a target site. Site-specificity of integration at vertebrate retroviruses is rather relative. At the same time, sequence-specific integration is the exclusive property of errantiviruses and their derivatives with two open reading frames. The possible basis for the errantivirus integration specificity is discussed in the present review.

Keywords: *Drosophila*; LTR-retrotransposon; errantivirus; retrovirus; transposition

1. Introduction

Unlike the human genome in which retrotransposons occupy more than 65% of the genome [1], in *Drosophila melanogaster*, retrotransposons accounts for only approximately 5% of the genome [2]. Nevertheless, *D. melanogaster* is an exceptionally convenient model for studying the mechanisms of retrotransposon movement because many of the retrotransposons in its genome are transpositionally active (in contrast to human retrotransposons) [3].

Retrotransposons that have long terminal repeat (LTR retrotransposons) occupy a special place among *D. melanogaster* retrotransposons. LTR retrotransposons have a varying structure and can contain different open reading frames (ORFs): from one to three. Most complex forms are retrotransposons that contain three ORFs: ORF1 (corresponding to the *gag* gene of retroviruses) encodes capsid proteins, ORF2 (*pol*) encodes protease, reverse transcriptase, RNase H, and integrase, and ORF3 (*env*) encodes a product that is responsible for cell receptor recognition and virus penetration into the cell. Thus, the structure of LTR retrotransposons that have three ORFs is identical to DNA forms of vertebrate retroviruses that are integrated into the host genome. Not by chance, two LTR retrotransposons of *D. melanogaster*, *gypsy* and *ZAM*, have been found to have infectious properties and have been classified as endogenous retroviruses [4,5].

According to the classification of LTR retrotransposons, which is based on a comparative analysis of the conserved domain of reverse transcriptases, there are three groups of LTR retrotransposons that correspond to the individual phylogenetic clades: Gypsy, Copia, and BEL [6]. The *D. melanogaster* genome contains 36 families of LTR retrotransposons [7]. The BEL and Copia groups are represented by five and four families, respectively, and the Gypsy group is represented by 27 families of retrotransposons. While the LTR retrotransposons of the Copia and BEL groups have one ORF, the Gypsy group is heterogeneous in composition and is represented by LTR retrotransposons with one, two, or three ORFs (Figure 1). The observed diversity of the Gypsy group of LTR retrotransposons obviously shows the recent origin of the currently existing families and transposition activity of the retrotransposons in this group. A high level of polymorphism is observed not only between families of the group but also within some families of the group, e.g., within the family of *gypsy* [7]. There are 12 subfamilies in *gypsy* (*gypsy1–gypsy12*), which are polymorphic in sequence and length (the difference can be up to 1000 base pair; bp). The presence of such a large number of subfamilies indicates that *gypsy* has the highest rate of diversification in the *D. melanogaster* genome.

Some *D. melanogaster* LTR retrotransposons are included in the international classification of viruses (ICTV) [8], in the *Metaviridae* family, which includes 3 genera: *Errantivirus* (includes six families of the Gypsy group of LTR retrotransposons with three ORFs: *gypsy, ZAM, Idefix, Tirant*, 297, and 17.6), *Metavirus* (includes five families of the Gypsy group of LTR retrotransposons with one and two ORFs: 412, *mdg1, mdg3, blastopia*, and *micropia*), and *Semotivirus* (includes two families of the BEL group of LTR retrotransposons: *roo* and 3S18). The mobile element *copia* was assigned by the ICTV to the *Hemivirus* genus of the *Pseudoviridae* family.

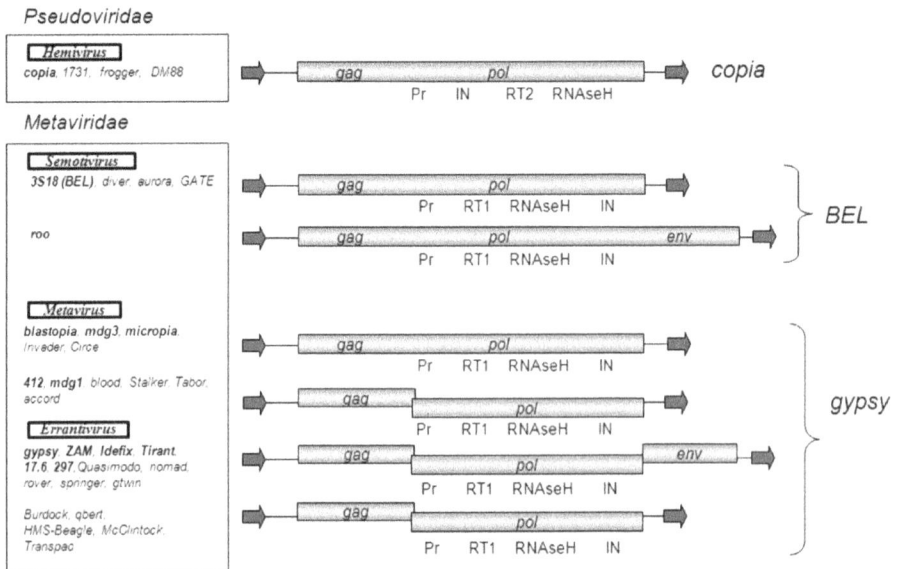

Figure 1. Structural organization and classification of *Drosophila melanogaster* long terminal repeat (LTR) retrotransposons. As shown: open reading frames (*gag, pol, env*) and the Pol domains (Pr, protease; RT1 and RT2, reverse transcriptase types 1 and 2; IN, integrase). The arrows indicate LTRs. LTR retrotransposons are distributed in groups and genera according to the phylogenetic analysis conducted in [9]. The LTR retrotransposon families introduced by the International Committee on Taxonomy of Viruses (ICTV) in *Metaviridae* and *Pseudoviridae* are highlighted in bold.

According to the phylogenetic analysis, which is based on a comparative analysis of *gag* and *pol* ORFs, all Gypsy group LTR retrotransposons can be classified either as genus *Metavirus* or *Errantivirus*. *Metavirus* contains LTR retrotransposons that are divided into two separate subgroups: blastopia (with one ORF) and 412 (with two ORFs) (Figure 1). The LTR retrotransposons with two ORFs, *McClintock*, *qbert*, *accord*, *Burdock*, *HMS-Beagle*, and *Transpac*, are derived from errantiviruses and have lost their infectious properties [9]. Therefore, these LTR retrotransposons that have two ORFs should be classified to the *Errantivirus* genus.

Retroviruses of vertebrates belong to the *Retroviridae* family, which is divided into two subfamilies (*Orthoretrovirinae* and *Spumaretrovirinae*) that include six genera (*Alpha-*, *Beta-*, *Delta-*, *Gamma-*, *Epsilonretrovirus* and *Lentivirus*), and one genus (*Spumavirus*), respectively [8]. According to the phylogenetic analysis of Gag and Pol sequences, the *Retroviridae* family can be divided into three classes [10]. Class 1 includes *Gamma-* and *Epsilonretrovirus*, class 2 includes *Lentivirus*, *Alpha-*, *Beta-*, and *Deltaretrovirus*, and class 3 includes *Spumaretrovirus* and endogenous retroviral (ERV) elements. For a number of structural features, the same phylogenetic analysis shows that *D.melanogaster* retrotransposons of the 412 subgroup of the *Metavirus* genus are similar to class 1 retroviruses. This also shows that retrotransposons in the genus of the blastopia subgroup of *Metavirus* are similar to class 2 retroviruses, and that errantiviruses are similar to class 3 retroviruses [11]. Thus, vertebrate retroviruses and LTR retrotransposons/retroviruses of *Drosophila* have a common evolutionary history and should be considered in parallel. Because the *D. melanogaster* genome has a large variety of LTR retrotransposons and retroviruses, we can use this organism as a model to generally analyze the evolutionary mechanisms of retroelement transposition in eukaryotes.

2. Errantiviruses Specifically Integrate into the Target DNA

There is still no clear understanding of the specificity of retroviral integration within the target site. The efficiency of integration primarily depends on the efficiency of the integrase enzyme. The interaction of integrase with tethering factors is the basis of integration targeting, at least for murine leukemia virus (MLV) and human immunodeficiency virus (HIV) retroviruses and yeast retroelements Ty1, Ty3 and Ty5 [12–16]. On the other hand, the target structure can contribute to the targeted integration. It is believed that the choice of the target DNA can affect a variety of factors, including the transcriptional status of DNA, methylation, association of DNA with histones and other DNA-binding proteins, DNA bending, etc. [17–20]. Furthermore, the strict specificity of integration is not characteristic of vertebrate retroviruses [21–24]. During the analysis of the retroviral integration sites of HIV-1, a "weak" target site consensus, GT(A/T)AC, was found [23]; it is similar to the target of the *D. melanogaster* Copia group LTR retrotransposons (Figure 2).

It was found that errantiviruses and their derivatives that have two ORFs exhibit a specificity of choice for the target DNA. These LTR retrotransposons can be divided into three subgroups, *gypsy*, *ZAM*, and *Idefix*, the representatives of which have a different specificity for the target [25]. In all three cases, the more frequent target is a palindromic (or imperfect palindromic) sequence: TATA, CGCG, or ATAT (Figure 2). More recent studies, using population genomic resequencing data from hundreds of strains of *D. melanogaster* as well as computational analyses, reveal the same specific target site preferences of *D.melanogaster* retroelements [26–28].

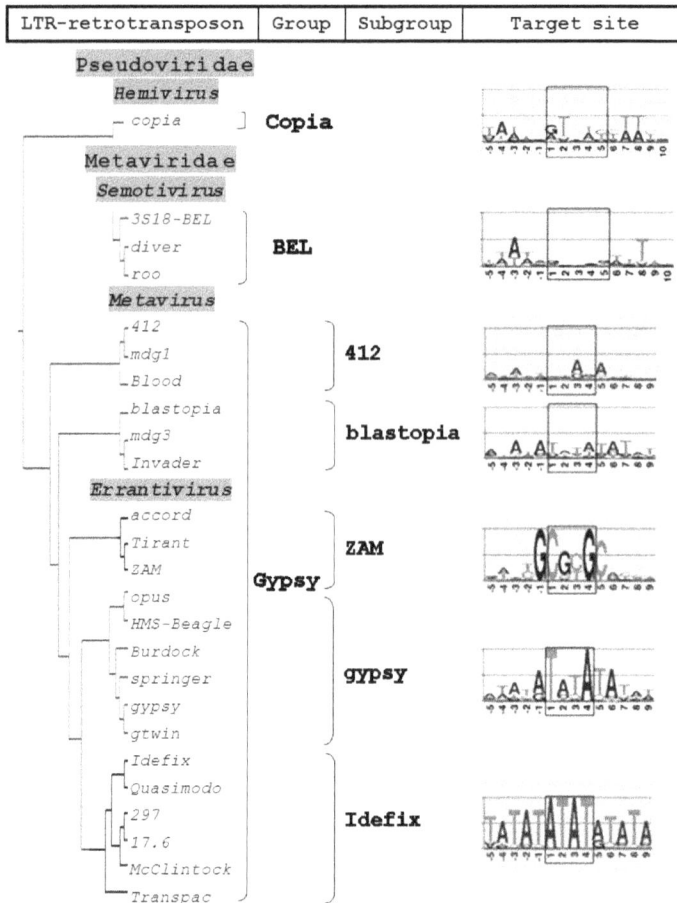

Figure 2. Integration sites of *D. melanogaster* LTR retrotransposons. A phylogenetic tree construction is based on a comparison of the amino acid sequences of the integrases of *D. melanogaster* LTR retrotransposons [25]. Visualization of the target site duplication was made using WEBLOGO (version 3) [29].

The specificity of integration is an exclusive property of errantiviruses and derived LTR retrotransposons that have two ORFs. The fact that the sequence motifs at errantivirus target sites are always palindromes is quite remarkable. Recent results indicate that vertebrate retrovirus integration sites contain a shared non-palindromic motif [30]. The shared motif is 5′-T(N1/2) [C(N0/1)T | (W1/2)C]CW-3′, where the square brackets represent the duplicated region, W denotes A or T, and | represents the axis of symmetry.

3. Repeats in the 5′-UTR Can Direct Heterochromatic Localization of Errantiviruses

It is noteworthy that ZAM subgroup errantiviruses integrate preferentially into GC-rich repeats. According to data in FlyBase [31], *ZAM* errantivirus insertions were found only in the constitutive heterochromatin, and *Tirant* insertions were found only in the euchromatin and facultative heterochromatin in the reference *Drosophila* genome. The other LTR retrotransposons were found both in euchromatin and heterochromatin. Of note, both *ZAM* and *Tirant* have tandem repeats in the 5′-untranslated region (5′-UTR): the number of repeats in *ZAM* 5′-UTR is 2.3 repeats (each one is 307 bp in length), and the number of repeats in *Tirant* 5′-UTR varies from two to six (each one is 102 bp). The role of the repeats in the *Tirant* 5′-UTR is still unclear. Previously, it was shown that repetitive sequences in the 5′-UTR *ZAM* errantivirus that are phylogenetically similar to *Tirant* interact with the heterochromatin protein, HP1a, which probably directs its heterochromatic localization [32].

Earlier in the *Drosophila simulans* genome, two subfamilies of *Tirant* were found: C-euchromatic (found both in *D. simulans* and *D. melanogaster*) and S-heterochromatic (found only in *D. simulans*) [33,34]. Localization of each subfamily in a certain type of chromatin was determined via association with modified histones, H3K9me2, H3K4me2, and H3K27me3, which are epigenetic markers of constitutive heterochromatin, facultative heterochromatin, and euchromatin, respectively. *Tirant* primarily associates with facultative heterochromatin [34]. By analyzing the heterochromatin component of the sequenced genome, we discovered a new heterochromatin subfamily of *Tirant* that consists of four copies and named this subfamily *Tirant_het*. The *Tirant_het* subfamily is not the same as the S-subfamily and represents an older, now non-functional, individually evolving heterochromatic branch of the ZAM subgroup of the Gypsy group. It contains two repeat modules in the 5′-UTR. The sequence similarity of the Pol sequences of *Tirant* and *Tirant_het* is approximately 80%, and the similarity of the repeat modules in 5′-UTR is 85% (Figure 3).

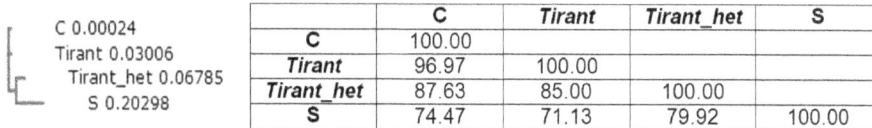

C 0.00024
Tirant 0.03006
Tirant_het 0.06785
S 0.20298

	C	Tirant	Tirant_het	S
C	100.00			
Tirant	96.97	100.00		
Tirant_het	87.63	85.00	100.00	
S	74.47	71.13	79.92	100.00

Figure 3. Phylogenetic tree of the 5′-untranlsated region (5′-UTR) repeat module (102 base pair; bp) in *Tirant* of *D. simulans* (subfamilies S and C) and *D. melanogaster* (subfamilies *Tirant* and *Tirant_het*) and sequence identity matrix (%).

Thus, along with the specificity for a nucleotide integration target, the ZAM subgroup elements have specificity for integration into the euchromatin/heterochromatin that correlates with the structure of the regulatory region in 5′-UTR. Tandem repeats in the 5′-UTR of *Tirant* errantivirus seem to have been captured in the host genome. Possibly, targeted integration into the active chromatin allows the retrotransposon to escape from host defenses. Many viruses clearly have acquired accessory genes and regulatory sequences from their hosts. In particular, lentiviruses contain accessory genes that antagonize or circumvent host restriction factors [35].

4. Specific Terminal Nucleotides of Errantivirus Long Terminal Repeats Are Involved in the Interaction with Integrase

The integration process can be divided into the following steps: (1) processing of the LTR ends; (2) recognition and cutting of target DNA in the host genome; and (3) integration of the LTR sequences into the target DNA [36]. All three steps are catalyzed by a retroviral integrase. Integrase is a part of the preintegration complex that recognizes the nucleotide sequences at the ends of the LTRs and prepares them for integration by removing the TG dinucleotide at the 3′-terminus of each chain (reaction of the 3′-end processing). The integration scheme is represented in Figure 4.

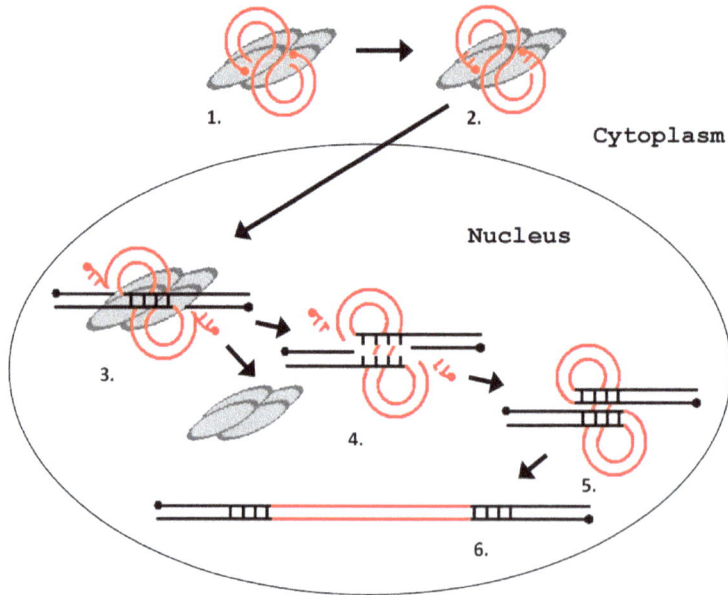

Figure 4. Schematic representation of the process of integration of a retrovirus (LTR retrotransposon) into the host genome. **(1)** interaction of integrase with a blunt-ended DNA substrate (other proteins are not shown); **(2)** removal of two terminal nucleotides from the 3′-ends of the DNA substrate (3′-end processing); **(3)** cleavage of the integration site; **(4)** removal of unpaired nucleotides at the 5′ ends of the DNA substrate; **(5)** filling-in of the gaps of the target DNA and ligation of discontinuities; and **(6)** repeating of the provirus integration site. The gray ovals represent integrase monomers. The red lines represent viral DNA, and the black lines represent chromosomal DNA. The dots indicate 5′-ends of the DNA.

All LTR-retroelements (LTR retrotransposons and retroviruses) obligatorily have inverted dinucleotides at the ends. Vertebrate proviruses have conserved 5′-TG/CA-3′ dinucleotides at the ends. It is believed that they are specific recognition sites for integrase and are a signal for 3′-end processing [37]. The protruding ends formed after processing comprise two terminal CA nucleotides that interact with the integrase. It has been shown that 12–15 subterminal nucleotides, in addition to the CA dinucleotide, can be employed in conjunction with the integrase [38]. The occurrence of 5′-TG/CA-3′ dinucleotides at the retrotransposon ends can be explained by the fact that TG, CA, and TA dinucleotides are the most deformable links in a DNA structure and are capable of local bending of the double helix due to the low energies of stacking interactions. Therefore, these three dinucleotides are often recognition sites for proteins that are involved in recombination, replication, and insertional events [39]. However, one exception is the integrase of *Drosophila* errantiviruses. According to an analysis of the terminal sequences, *D. melanogaster* LTR retrotransposons can be subdivided into two groups (Figure 5). The dinucleotides, TG/CA, are present at the ends of LTR retrotransposons of the BEL, Copia, and Gypsy groups of the *Metavirus* genus. However, errantiviruses (subgroups of *gypsy*, *ZAM*, and *Idefix*) and their derivatives have AGT/AnT trinucleotides at the ends, where "n" is usually A or C. In five of the eleven errantiviruses, all three terminal nucleotides are completely complementary; moreover, the errantiviruses, *Tirant*, *opus*, and *ZAM*, have five, six, and seven completely complementary nucleotides at the ends, respectively [25]. It is unclear how many and which nucleotides of the errantivirus ends are involved in the integrase and processing interactions.

Genus	Group/subgroup	5'-LTR	3'-LTR
Errantivirus	Gypsy, ZAM, Idefix	AGT.A....A.	...T.A.T
Hemivirus Metavirus Semotivirus	BEL, Copia, 412, blastopia	TGT.....A....	...TACA

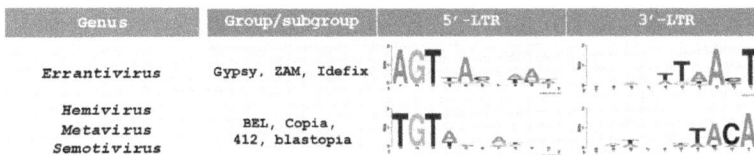

Figure 5. Multiple alignment of the 5'- and 3'-terminal sequences of the *D. melanogaster* LTR retrotransposons. The LTR retrotransposons of *D. melanogaster* can be divided into two groups depending on the composition of the end sequences [25]. Visualization performed using WEBLOGO [29].

5. LTR Retrotransposons of the *Metavirus* Genus Have a Chromodomain in the Integrase Structure

For obvious reasons, the most studied retroviral integrase is HIV-1 integrase. However, despite numerous attempts to establish an accurate pattern of DNA–protein interactions, the exact interaction mechanism of integrase with the target DNA is poorly understood [40]. Even the spatial structure of integrase is still uncertain. There are three domains in the integrase structure: N-terminal, central catalytic, and C-terminal [41]. Specific binding is obviously carried out by the most conservative central domain [42]. The role of the N-terminal domain during the process of integration is the least clear. This region contains a His-His-Cys-Cys motif, which is characteristic for the majority of retroviral integrases [43]. This domain appears to be involved in protein dimerization; its role in the binding with DNA is not significant. A mutant enzyme in which the N-terminal domain or HHCC-motif is absent loses the ability to carry out 3'-end processing and strand transfer [44]. The C-terminal is believed to participate in nonspecific binding of DNA [45].

In some LTR retrotransposons of the Gypsy group, including many chromoviruses of plants, algae, and fungi (but not yeast), the chromodomain is localized in the C-terminal domain of integrase and plays an important role in the interaction with the LTRs [10]. This domain is characterized by a conserved GPY/F motif. It is believed that this domain facilitates the interaction of integrase with chromatin. Chromodomains are found in integrases of vertebrate retroviruses of class 1, i.e., gamma- and epsilonretroviruses, including MLV. Of note, the GPY/F motif is present in *D. melanogaster* LTR retrotransposons of the *Metavirus* genus in the representatives of the two subgroups, 412 and blastopia [25]. Errantivirus integrases do not have a GPY/F motif.

6. LTR Retrotransposons of the *Metavirus* Genus Are Able to Transfer Horizontally

The main difference between retroviruses and LTR retrotransposons is the presence of the *env* gene, which is responsible for infectivity. It is believed that *Drosophila* LTR retrotransposons of the Gypsy group initially had two ORFs. Then, they acquired the *env* gene from baculoviruses and, therefore, their infective properties [46]. However, in *Drosophila*, besides errantiviruses, an additional LTR retrotransposon has the *env* gene, *roo*. It is the LTR retrotransposon of the BEL group with one ORF (Figure 1). Meanwhile, the *env* genes of errantiviruses and the *roo* LTR retrotransposon are homologous; therefore, they have a common origin. It is shown that the acquisition of the *env* gene by the *roo* LTR retrotransposon occurred after the separation of Drosophilidae into a separate evolutionary branch of insects. Thus, errantiviruses may be the source of the *env* gene used by the *roo* element [46]. LTR retrotransposons of the *Drosophila* Copia group do not have the *env* gene. However, this does not mean that the appearance of *env* as part of the retroelement is impossible. Thus far, the only case of LTR retrotransposon of the Copia group with the *env* gene is the *SIRE* retroelement, which has been described only in soybeans [47].

Of note, LTR retrotransposons of the 412 subgroup have substantially identical copies in the genomes of different species of *Drosophila* and very close homologs (the identity in amino sequences of reverse transcriptase is more than 90%) in a very distant species (melanogaster, willistoni, virilis, and replete groups) [9]. This implies that LTR retrotransposons of the 412 subgroup can horizontally transfer without their own *env* gene. The question of how these elements move between species remains open: either they do not need the *env* gene function for infection, or the elements of this subgroup use a foreign envelope protein to move. The most likely possibility is a transmission of the retrotransposons through pseudotyping with envelope glycoproteins derived from errantiviruses. The presence of close homologs of 412 in the genomes of different species of *Drosophila* is correlated with the presence of close homologs of *gypsy* and *springer* errantiviruses in the genomes of the same species [9]. This does not preclude that 412 LTR retrotransposon uses the errantivirus *env* gene function for movement.

7. Consequences of the Retroelement Transposition

For a long time, it was believed that mobile elements are genomic parasites and nature removes them from participation in the functioning of the genome via the heterochromatization of sites where they are localized. However, recent molecular studies have shown that mobile element sequences, including retroelements, may acquire functional significance for the host genome during the course of evolution. The DNA sequence of any retroelement (retrotransposon or retrovirus) incorporated in the gene eventually accumulates mutations and degrades. Meanwhile, certain genes or regulatory sequences from the retroelement can be stored and undergo domestication and/or exaptation (change of function). As a rule, retroelement gene function is adapted to benefit the host genome. Thus, domestication of heterologous genes, including genes of retroelements, is one of the mechanisms of gene origin. The domestication of *gag* and *env* genes deserves special attention. Obviously, their functions can be adapted to protect the host genome from a retroviral infection via competition with homologous viral gene products. Some examples, known as the mammalian homologs of *gag* and *env* genes, participate not only in protection against viral infection but also in the control of cell division, apoptosis, placenta functioning, and other biological processes [48–50]. Therefore, "the scope" of domesticated capsid and envelope proteins could be much wider than previously thought and requires further study. *D. melanogaster* could be a good model for such research because its genome contains both *gag* and *env* homologs. It has been shown that both genes are under strong selection [51,52]. Currently, their functions are being actively studied, and it is possible that both genes are involved in the defense against viral infections.

8. Conclusions

The interaction of integrase with a target DNA sequence is a process dependent on the "complementarity" of DNA-binding domain of enzyme and DNA region that it connects. Mostly, three factors can influence on integration process: host chromatin status; genomic features such as histone modifications and transcription factor binding sites; and primary sequence of a target DNA. Specificity of vertebrate retrovirus integration into a target site is rather relative. The search for retroviruses and LTR-retrotransposons specifically integrating into a target is of great interest for the studies concerning the use of a site-directed mutagenesis.

Errantiviruses specifically integrate into the target DNA. In addition, tandem repeats in the 5'-UTR of *Tirant* errantivirus seem to direct its euchromatic localization. The integration specificity correlates with the structural features of the target DNA and the distinctive sequence of errantivirus LTR terminal nucleotides. The end sequences of LTRs in "nonspecific" LTR-retrotransposons of *Drosophila* (GT/CA dinucleotides) are, like in vertebrate retroviruses, highly conservative. It is believed that these dinucleotides have a low energy of stacking interaction and are, therefore, the most deformable links in the DNA structure, which are capable of forming a local bending to promote integration. In some "nonspecific" LTR retrotransposons of the Gypsy group (subgroups 412 and blastopia), chromodomain, which is localized in the C-terminal domain of integrase, probably plays an important role in the interaction with LTRs.

LTR retrotransposons of *D. melanogaster*, especially representatives of the Gypsy group, clearly demonstrate the possibility of mobile element evolution, which is based not only on their high rate of diversification but also on the ability to acquire individual modules or genes. The molecular rearrangement, transposition, recombination, and horizontal transfer, coupled with the selection of viable and adaptive variants of newly formed retrotransposons, play a key role in the evolution of retrotransposons and retroviruses. As a result of these changes, some retrotransposons or retroviruses acquire specific opportunities to integrate into actively transcribed regions of the genome, which is important for their future activity. The lack of molecular barriers for recombination between genes (or their fragments) can lead to multidirectional pathways of retroelement evolution followed by diversification of mechanisms of retroelement integration.

Acknowledgments: This work was supported by the Russian Foundation for Basic Research (RFBR) (Grant 14-04-01450).

Author Contributions: L.N.N. was responsible for writing the manuscript and creating the figures. L.N.N. and A.I.K. were responsible for organizing the contents. A.I.K. was involved in critical reading of the manuscript. Both authors read and approved the final manuscript.

Conflicts of Interest: The authors declare no conflict of interest. The founding sponsors had no role in the design of the study; in the collection, analyses, or interpretation of data; in the writing of the manuscript, and in the decision to publish the results.

References

1. De Koning, A.P.; Gu, W.; Castoe, T.A.; Batzer, M.A.; Pollock, D.D. Repetitive elements may comprise over two-thirds of the human genome. *PLoS Genet.* **2011**, *7*, e1002384. [CrossRef] [PubMed]
2. Bergman, C.M.; Quesneville, H. Discovering and detecting transposable elements in genome sequences. *Brief. Bioinform.* **2007**, *8*, 382–392. [CrossRef] [PubMed]
3. Petrov, D.A.; Fiston-Lavier, A.S.; Lipatov, M.; Lenkov, K.; González, J. Population genomics of transposable elements in *Drosophila melanogaster*. *Mol. Biol. Evol.* **2011**, *28*, 633–644. [CrossRef] [PubMed]
4. Kim, A.; Terzian, C.; Santamaria, P.; Pélisson, A.; Prud'homme, N.; Bucheton, A. Retroviruses in vertebrates: The *gypsy* retrotransposon is apparently an infectious retrovirus of *Drosophila melanogaster*. *Proc. Natl. Acad. Sci. USA* **1994**, *91*, 1285–1289. [CrossRef] [PubMed]
5. Leblanc, P.; Desset, S.; Giorgi, F.; Taddei, A.R.; Fausto, A.M.; Mazzini, M.; Dastugue, B.; Vaury, C. Life cycle of an endogenous retrovirus, *ZAM*, in *Drosophila melanogaster*. *J. Virol.* **2000**, *74*, 10658–10669. [CrossRef] [PubMed]
6. Bowen, N.J.; McDonald, J.F. *Drosophila* euchromatic LTR retrotransposons are much younger than the host species in which they reside. *Genome Res.* **2001**, *11*, 1527–1540. [CrossRef] [PubMed]
7. Kaminker, J.S.; Bergman, C.M.; Kronmiller, B.; Carlson, J.; Svirskas, R.; Patel, S.; Frise, E.; Wheeler, D.A.; Lewis, S.E.; Rubin, G.M.; et al. The transposable elements of the *Drosophila melanogaster* euchromatin: A genomics perspective. *Genome Biol.* **2002**, *3*. [CrossRef]
8. King, A.M.Q.; Adams, M.J.; Carsten, E.B.; Lefkowitz, E. *Virus Taxonomy, Ninth Report of the International Committee for the Taxonomy of Viruses*, 1st ed.; Elsevier: San Diego, CA, USA, 2012; 1338p.
9. Nefedova, L.N.; Kim, A.I. Molecular phylogeny and systematics of *Drosophila* retrotransposons and retroviruses. *Mol. Biol.* **2009**, *43*, 747–756. [CrossRef]
10. Llorens, C.; Muñoz-Pomer, A.; Bernad, L.; Botella, H.; Moya, A. Network dynamics of eukaryotic LTR retroelements beyond phylogenetic trees. *Biol. Direct.* **2009**, *4*, 41. [CrossRef] [PubMed]
11. Llorens, C.; Fares, M.A.; Moya, A. Relationships of gag-pol diversity between *Ty3/Gypsy* and *Retroviridae* LTR retroelements and the three kings hypothesis. *BMC Evol. Biol.* **2008**, *8*, 276. [CrossRef] [PubMed]
12. Xie, W.; Gai, X.; Zhu, Y.; Zappulla, D.C.; Sternglanz, R.; Voytas, D.F. Targeting of the yeast *Ty5* retrotransposon to silent chromatin is mediated by interactions between integrase and Sir4p. *Mol. Cell. Biol.* **2001**, *21*, 6606–6614. [CrossRef] [PubMed]
13. Aye, M.; Dildine, S.L.; Claypool, J.A.; Jourdain, S.; Sandmeyer, S.B. A truncation mutant of the 95-kilodalton subunit of transcription factor IIIC reveals asymmetry in Ty3 integration. *Mol. Cell. Biol.* **2001**, *21*, 7839–7851. [CrossRef] [PubMed]

14. Cheung, S.; Ma, L.; Chan, P.H.; Hu, H.L.; Mayor, T.; Chen, H.T.; Measday, V. Ty1 integrase interacts with RNA polymerase III-specific subcomplexes to promote insertion of Ty1 elements upstream of polymerase (Pol)III-transcribed genes. *J. Biol. Chem.* **2016**, *291*, 6396–6411. [CrossRef] [PubMed]

15. Shun, M.-C.; Raghavendra, N.K.; Vandegraaff, N.; Daigle, J.E.; Hughes, S.; Kellam, P.; Cherepanov, P.; Engelman, A. LEDGF/p75 functions downstream from preintegration complex formation to effect genespecific HIV-1 integration. *Genes Dev.* **2007**, *21*, 1767–1778. [CrossRef] [PubMed]

16. Gupta, S.S.; Maetzig, T.; Maertens, G.N.; Sharif, A.; Rothe, M.; Weidner-Glunde, M.; Galla, M.; Schambach, A.; Cherepanov, P.; Schulz, T.F. Bromo-and extraterminal domain chromatin regulators serve as cofactors for murine leukemia virus integration. *J. Virol.* **2013**, *87*, 12721–12736. [CrossRef] [PubMed]

17. Muller, H.P.; Varmus, H.E. DNA bending creates favored sites for retroviral integration, an explanation for preferred insertion sites in nucleosomes. *EMBO J.* **1994**, *13*, 4704–4714. [PubMed]

18. Kitamura, Y.; Lee, Y.M.H.; Coffin, J.M. Nonrandom integration of retroviral DNA in vitro, Effect of CpG methylation. *Proc. Natl. Acad. Sci. USA* **1992**, *89*, 5532–5536. [CrossRef] [PubMed]

19. Pryciak, P.M.; Varmus, H.E. Nucleosomes, DNA-binding proteins, and DNA sequence modulate retroviral integration target site selection. *Cell* **1992**, *69*, 769–780. [CrossRef]

20. Pruss, D.; Reeves, R.; Bushman, F.D.; Wolffe, A.P. The influence of DNA and nucleosome structure on integration events directed by HIV integrase. *J. Biol. Chem.* **1994**, *269*, 25031–25041. [PubMed]

21. Mitchell, R.S.; Beitzel, B.F.; Schroder, A.R.; Shinn, P.; Chen, H.; Berry, C.C.; Ecker, J.R.; Bushman, F.D. Retroviral DNA integration, ASLV, HIV, and MLV show distinct target site preferences. *PLoS Biol.* **2004**, *2*, e234. [CrossRef] [PubMed]

22. Derse, D.; Crise, C.; Li, Y.; Princler, G.; Lum, N.; Stewart, C.; McGrath, C.F.; Hughes, S.H.; Munroe, D.J.; Wu, X. Human T-Cell Leukemia Virus Type 1 Integration Target Sites in the Human Genome: Comparison with Those of Other Retroviruses. *J. Virol.* **2007**, *81*, 6731–6741. [CrossRef] [PubMed]

23. Wu, X.; Li, Y.; Crise, B.; Burgess, S.M.; Munroe, D.J. Weak palindromic consensus sequences are a common feature found at the integration target sites of many retroviruses. *J. Virol.* **2005**, *79*, 5211–5214. [CrossRef] [PubMed]

24. Jin, Y.F.; Ishibashi, T.; Nomoto, A.; Masuda, M. Isolation and analysis of retroviral integration targets by solo long terminal repeat inverse PCR. *J. Virol.* **2002**, *76*, 5540–5547. [CrossRef] [PubMed]

25. Nefedova, L.N.; Mannanova, M.M.; Kim, A.I. Integration specificity of LTR-retrotransposons and retroviruses in the *Drosophila melanogaster* genome. *Virus Genes* **2011**, *42*, 297–306. [CrossRef] [PubMed]

26. Linheiro, R.S.; Bergman, C.M. Whole genome resequencing reveals natural target site preferences of transposable elements in *Drosophila melanogaster*. *PLoS ONE* **2012**, *7*, e30008. [CrossRef] [PubMed]

27. Zhuang, J.; Wang, J.; Theurkauf, W.; Weng, Z. TEMP: A computational method for analyzing transposable element polymorphism in populations. *Nucleic Acids Res.* **2014**, *42*, 6826–6838. [CrossRef] [PubMed]

28. Fiston-Lavier, A.S.; Barrón, M.G.; Petrov, D.A.; González, J. T-lex2: Genotyping, frequency estimation and re-annotation of transposable elements using single or pooled next-generation sequencing data. *Nucleic Acids Res.* **2015**, *43*, e22. [CrossRef] [PubMed]

29. Crooks, C.E.; Hon, G.; Chandonia, J.-M.; Brenner, S.E. WebLogo: A sequence logo generator. *Genome Res.* **2004**, *14*, 1188–1190. [CrossRef] [PubMed]

30. Kirk, P.D.; Huvet, M.; Melamed, A.; Maertens, G.N.; Bangham, C.R. Retroviruses integrate into a shared, non-palindromic DNA motif. *Nat. Microbiol.* **2016**, *2*, 16212. [CrossRef] [PubMed]

31. Gramates, L.S.; Marygold, S.J.; dos Santos, G.; Urbano, J.-M.; Antonazzo, G.; Matthews, B.B.; Rey, A.J.; Tabone, C.J.; Crosby, M.A.; Emmert, D.B.; et al. FlyBase at 25: Looking to the future. *Nucleic Acids Res.* **2017**, *5*(D1), D663–D671. [CrossRef] [PubMed]

32. Minervini, C.; Marsano, R.; Casieri, P.; Fanti, L.; Caizzi, R.; Pimpinelli, S.; Mariano, R.; Luigi, V. Heterochromatin protein 1 interacts with 5′UTR of transposable element *ZAM* in a sequence-specific fashion. *Gene* **2007**, *393*, 1–10. [CrossRef] [PubMed]

33. Fablet, M.; McDonald, J.F.; Biemont, C.; Vieira, C. Ongoing loss of the *tirant* transposable element in natural populations of *Drosophila simulans*. *Gene* **2006**, *375*, 54–62. [CrossRef] [PubMed]

34. Fablet, M.; Lerat, E.; Rebollo, R.; Horard, B.; Burlet, N.; Martinez, S.; Brasset, E.; Gilson, E.; Vaury, C.; Vieira, C. Genomic environment influences the dynamics of the *tirant* LTR retrotransposon in *Drosophila*. *FASEB J* **2009**, *23*, 1482–1489. [CrossRef] [PubMed]

35. McCarthy, K.R.; Johnson, W.E. Plastic proteins and monkey blocks, how lentiviruses evolved to replicate in the presence of primate restriction factors. *PLoS Pathog.* **2014**, *10*, e1004017. [CrossRef] [PubMed]
36. Gao, K.; Butler, S.L.; Bushman, F. Human immunodeficiency virus type 1 integrase, arrangement of protein domains in active cDNA complexes. *EMBO J.* **2001**, *20*, 3565–3576. [CrossRef] [PubMed]
37. Brown, H.E.; Chen, H.; Engelman, A. Structure-based mutagenesis of the human immunodeficiency virus type 1 DNA attachment site, effects on integration and cDNA synthesis. *J. Virol.* **1999**, *73*, 9011–9020. [PubMed]
38. LaFemina, R.L.; Callahan, P.L.; Cordingley, M.G. Substrate specificity of recombinant human immunodeficiency virus integrase protein. *J. Virol.* **1991**, *65*, 5624–5630. [PubMed]
39. Mashkova, T.D.; Oparina, N.Y.; Lacroix, M.H.; Fedorova, L.I.; Tumeneva, G.; Zinovieva, O.L.; Kisselev, L.L. Structural rearrangements and insertions of dispersed elements in pericentromeric alpha satellites occur preferably at kinkable DNA sites. *J. Mol. Biol.* **2001**, *305*, 33–48. [CrossRef] [PubMed]
40. Delelis, O.; Carayon, K.; Saib, A.; Deprez, E.; Mouscadet, J.F. Integrase and integration, biochemical activities of HIV-1 integrase. *Retrovirology* **2008**, *5*, 114. [CrossRef] [PubMed]
41. Dyda, F.; Hickman, A.B.; Jenkins, T.M.; Engelman, A.; Craigie, R.; Davies, D.R. Crystal structure of the catalytic domain of HIV-1 integrase, similarity to other polynucleotidyl transferases. *Science* **1994**, *266*, 1981–1986. [CrossRef] [PubMed]
42. Katzman, M.; Sudol, M. Mapping viral DNA specificity to the central region of integrase by using functional human immunodeficiency virus type 1/Visna virus chimeric proteins. *J. Virol.* **1998**, *72*, 1744–1753.
43. Appa, R.S.; Shin, C.G.; Lee, P.; Chow, S.A. Role of the nonspecific DNA-binding region and alpha helices within the core domain of retroviral integrase in selecting target DNA sites for integration. *J. Biol. Chem.* **2001**, *276*, 45848–45855. [CrossRef] [PubMed]
44. Zheng, R.; Jenkins, T.M.; Craigie, R. Zinc folds the N-terminal domain of HIV-1 integrase, promotes multimerization, and enhances catalytic activity. *Proc. Natl. Acad. Sci. USA* **1996**, *93*, 13659–13664. [CrossRef] [PubMed]
45. Woerner, A.M.; Marcus-Sekura, C.J. Characterization of a DNA binding domain in the C-terminus of HIV-1 integrase by deletion mutagenesis. *Nucleic Acids Res.* **1993**, *21*, 3507–3511. [CrossRef] [PubMed]
46. Malik, H.S.; Henikoff, S.; Eickbush, T.H. Poised for contagion, evolutionary origins of the infectious abilities of invertebrate retroviruses. *Genome Res.* **2000**, *10*, 1307–1318. [CrossRef] [PubMed]
47. Pearce, S.R. SIRE-1, a putative plant retrovirus is closely related to a legume *Ty1-copia* retrotransposon family. *Cell. Mol. Biol. Lett.* **2007**, *12*, 120–126. [CrossRef] [PubMed]
48. Brandt, J.; Veith, A.M.; Volff, J.N. A family of neofunctionalized *Ty3/gypsy* retrotransposon genes in mammalian genomes. *Cytogenet. Genome Res.* **2005**, *110*, 307–317. [CrossRef] [PubMed]
49. Casacuberta, E.; González, J. The impact of transposable elements in environmental adaptation. *J. Mol. Ecol.* **2013**, *22*, 1503–1517. [CrossRef] [PubMed]
50. Elbarbary, R.A.; Lucas, B.A.; Maquat, L.E. Retrotransposons as regulators of gene expression. *Science* **2016**, *351*, aac7247. [CrossRef] [PubMed]
51. Nefedova, L.; Kuzmin, I.; Makhnovskii, P.; Kim, A. Domesticated retroviral *gag* gene in *Drosophila*, new functions for an old gene. *Virology* **2014**, *450–451*, 196–204. [CrossRef] [PubMed]
52. Malik, H.S.; Henikoff, S. Positive selection of *Iris*, a retroviral envelope-derived host gene in *Drosophila melanogaster*. *PLoS Genet.* **2005**, *1*, e44. [CrossRef] [PubMed]

![viruses logo]

viruses

MDPI

Review

Epigenetic Control of Human Endogenous Retrovirus Expression: Focus on Regulation of Long-Terminal Repeats (LTRs)

Tara P. Hurst [1] and Gkikas Magiorkinis [1,2,*]

[1] Department of Zoology, University of Oxford, Oxford OX1 3PS, UK; tara.hurst@zoo.ox.ac.uk
[2] Department of Hygiene, Epidemiology and Medical Statistics, Medical School, National and Kapodistrian University of Athens, 11527 Athens, Greece
* Correspondence: gkikasmag@gmail.com; Tel.: +306973687010

Academic Editors: David J. Garfinkel and Katarzyna J. Purzycka
Received: 10 March 2017; Accepted: 22 May 2017; Published: 31 May 2017

Abstract: Transposable elements, including endogenous retroviruses (ERVs), comprise almost 45% of the human genome. This could represent a significant pathogenic burden but it is becoming more evident that many of these elements have a positive contribution to make to normal human physiology. In particular, the contributions of human ERVs (HERVs) to gene regulation and the expression of noncoding RNAs has been revealed with the help of new and emerging genomic technologies. HERVs have the common provirus structure of coding open reading frames (ORFs) flanked by two long-terminal repeats (LTRs). However, over the course of evolution and as a consequence of host defence mechanisms, most of the sequences contain INDELs, mutations or have been reduced to single LTRs by recombination. These INDELs and mutations reduce HERV activity. However, there is a trade-off for the host cells in that HERVs can provide beneficial sources of genetic variation but with this benefit comes the risk of pathogenic activity and spread within the genome. For example, the LTRs are of critical importance as they contain promoter sequences and can regulate not only HERV expression but that of human genes. This is true even when the LTRs are located in intergenic regions or are in antisense orientation to the rest of the gene. Uncontrolled, this promoter activity could disrupt normal gene expression or transcript processing (e.g., splicing). Thus, control of HERVs and particularly their LTRs is essential for the cell to manage these elements and this control is achieved at multiple levels, including epigenetic regulations that permit HERV expression in the germline but silence it in most somatic tissues. We will discuss some of the common epigenetic mechanisms and how they affect HERV expression, providing detailed discussions of HERVs in stem cell, placenta and cancer biology.

Keywords: Endogenous retroviruses; HERVs; LTR; epigenetics; Krüppel-associated box zinc finger protein; KRAB-ZFP

1. Introduction

The human genome is littered with endogenous retroelements, including non-long-terminal repeat (LTR) elements such as long interspersed nuclear repeats (LINEs) and short interspersed nuclear repeats (SINEs), as well as the long-terminal repeats (LTR)-containing endogenous retroviruses (ERVs). It is widely accepted that retroelements are subject to repression by both genetic and epigenetic mechanisms. However, much of what is known has been elucidated through studies in other species, such as mice. While informative, such work is limited by differences between the species, particularly differences in ERV activity. For example, ERVs in mice are much more active than in humans and produce infectious particles [1] which have not yet been demonstrated in humans. Indeed, there seems to be a negative correlation between ERV proliferation and body size, suggesting that ERV numbers scale with the number of cells of the host so a mouse is more likely to have active ERVs relative to humans based on body size. Further, tumorigenic insertional mutagenesis is more likely to happen when more cells have ERV proliferation [2]. It seems that a complete knock-out of the ERVs (e.g., through deleterious mutations) would be the safest option for the host but this would lead to a complete extinction of ERVs. Intuitively, the "sweet spot" of ERV activity that allows both ERVs and host survival could be a window of activity near early life stages (e.g., germline and embryonic stem cells) where the number of cells is irrelevant to the final body size of the host, followed by silencing in somatic tissues. This arrangement is likely to be served through epigenetic silencing of ERVs. Here, we will describe some of the recent findings on the regulation of ERVs by epigenetics, emphasising studies on human ERVs (HERVs) in normal tissues and in diseases.

2. Human Endogenous Retroviruses

HERVs were detected in human cells in the 1970s, with early descriptions of retrovirus-like particles in placentae [3] and germ line cancers [4]. The Human Genome Project revealed that HERVs comprise 8% of the genome [5]. HERVs ERVs belong to a number of distinct families that integrated independently during evolution [6]. One of these families, HERV-K HML-2 (HK2), emerged prior to the divergence of hominids from Old World monkeys. However, a number of human-specific integrations have been identified [7]. Further, there is evidence of recent activity of HK2 within the human genome, such as the discovery of polymorphic integrations of HK2 [8,9], including an intact provirus located on the X chromosome [10]. The vast majority of HERVs are rendered inactive by an accumulation of mutations, as well as by epigenetic mechanisms. Despite this, there are numerous reports implicating HERV expression, particularly of HK2, in autoimmune diseases and cancer [11–14]. Thus, the presence, activity and expression of HERVs is of great interest.

Control of HERV expression depends upon regulation at the level of the LTRs. These function as promoters for HERV expression [15], have strong RNA Polymerase II regulatory sequences [16,17], and contain a plethora of transcription factor binding sites [18]. The LTRs can bind nuclear transcription factors [19] and more recently have been shown to be responsive to pro-inflammatory cytokines in a cellular model of amyotrophic lateral sclerosis (ALS) [20]. Importantly, solo LTRs are present in the genome due to recombination that excises the rest of the provirus [21]. Indeed, up to 85% of HERVs have undergone this recombinatorial deletion [22], making most HERV loci solo LTRs. Solo LTRs can serve as promoters in both sense and antisense orientations [23] and can alter the expression of host genes [24,25]. Further, the expression of very long intergenic RNAs (vlincRNAs) which control pluripotency and malignancy was HERV LTR-driven [26], suggesting a role for HERV LTRs in regulating not only protein-coding genes but also the expression of long non-coding RNAs. Thus, the LTRs are an important site for epigenetic modifications to control HERV and human gene expression.

3. Epigenetics

Epigenetic regulation includes the modification of both DNA and the histones around which DNA is wound to create chromatin [27]. The formation and packing of the chromatin is itself an epigenetic mechanism; tightly-packed chromatin is associated with gene silencing and vice versa. Regulation is also achieved by the modification of nucleotides and proteins by the addition of chemical groups, such as methyl or acetyl groups. For example, modification of the histone H3 by trimethylation of lysine 4 (H3K4me3) is associated with gene activity, while that of lysine 9 (H3K9me2/3) or 27 (H3K27me3) defines condensed chromatin packing and gene silencing [27]. There is also a strong association between DNA methylation and the H3K9me3 mark [28].

One way to understand epigenetic system is by considering it to comprise writers, readers and erasers of the modifications; these are enzymes that add, bind to or remove chemical groups e.g., methyl groups from DNA [29]. Strategies to study epigenetics include the use of drugs which inhibit

DNA or histone modifying enzymes (the 'writers' or 'erasers'). For example, the nucleoside analogue 5′-azacytidine (5′-aza) is incorporated into cellular DNA and inhibits the DNA methyltransferase 1 (DNMT1), resulting in passive demethylation and reactivation of silenced genes [30]. Similarly, the use of histone deactylase inhibitors (HDACi) results in the retention of acetyl groups on the histones and therefore of active gene expression [31]. We will first describe current epigenetic mechanisms of control of HERVs and then discuss specific examples in more detail.

4. Epigenetic Regulation of Human Endogenous Retroviruses

The studies that have been done on HERVs suggest that multiple control strategies are used: localisation of proviruses to heterochromatin, chromatin packing to block access to the LTRs, CpG methylation and histone deacetylation contribute to the control of HERVs in the genome. The predominant view is that these epigenetic mechanisms keep HERVs silenced [32,33]. However, it is also possible that a more nuanced view allows epigenetics a role in transcriptional regulation rather than silencing alone. This idea is suggested by transcriptome studies which report that up to one-third of all HERV loci are transcribed [34], a number that would not make sense if the epigenetic repression were not somewhat 'leaky'.

4.1. CpG Methylation

It is usual for CpG nucleotides to be methylated throughout the human genome, including those found in HERVs; exceptions to this, referred to as CpG islands (CGIs), are sites of low methylation that are frequently found near active genes and enhancer elements [35]. The methylation of CpGs is carried out by DNMTs, with DNMT1 being the maintenance methyltransferase which is important for fidelity of methylation during DNA replication [35]. A microarray study analysing HERV families throughout the genome found that HERVs are heavily methylated in normal tissues [36]. Further, the age of the HERVs correlates with their methylation status, with a loss of methylation appearing in older families such as HERV-H [36]. CpG methylation is a critical mechanism of silencing and has been demonstrated for the HK2 5′ LTR in germ cell tumours (GCTs) [37]. In this study, methylation of 5′ LTRs correlated with transcriptional suppression in the Tera-1 cell line [37]. Importantly, the effect is cell-type dependent, implying that other factors are also critical for regulation of HERV expression. These can include transcription factors, as well as other types of epigenetic modifications.

4.2. Histone Acetylation

Acetylation of lysine residues in histones is catalysed by histone acetyltransferases (HATs) and removed by histone deacetylases (HDACs) [38]. Histone acetylation blocks the positive charges on lysine residues which destabilises chromatin and favours transcriptional activation; deacetylation stabilises chromatin and thus leads to transcriptional repression [38]. This has led to the use of HDAC inhibitors (HDACi) to reverse HIV-1 latency as part of a 'kick and kill' approach to curing HIV-1 infection [39]. We were interested in whether the transcriptional activation resulting from HDACi treatment of HIV-1-infected cells would also activate HERVs [40]. To test this, we examined the expression of particular HERV families (HK2, HERV-W, HERV-FRD) following HDACi treatment using quantitative RT-PCR with Molecular Beacons probes. Indeed, we found that HDACi treatment did not significantly up-regulate the HERVs in either latency cell lines or primary T cells infected with HIV-1 [40]. This implies that histone deacetylation alone is not responsible for HERV repression, a finding consistent with the importance of other factors, particularly CpG methylation, in silencing HERVs. For example, the combination of the HDACi trichostatin A (TSA) and 5′-aza increased HERV-Fc1 expression in HEK 293s, whereas TSA alone did not [30]. However, the same study did find that TSA alone or in combination with 5-aza increased in HERV-Fc1 expression in peripheral blood mononuclear cells (PBMCs) [30]. The different results could be due to the distinct cell types used, with cancer cell lines being expected to differ from PBMCs.

4.3. Histone Methylation, Heterochromatin and Krüppel-associated box zinc finger proteins (KRAB-ZFP)

The differential methylation of histones is critical to the activation or repression of genes. In particular, methylation of histone H3 at different lysine residues is an indicator of activity; the predominant marks are H3K9me3 (activity) and H3K4me3 (silencing). In a bioinformatics study, HERV-K was found predominantly in areas of repressed chromatin and there was a strong association with H3K9me3 [41]. In comparison, the localisation of genomic HERVs was in sites of inactive chromatin (older HERV proviruses) or an intermediate position (younger HERVs) [42]. This may be evidence of purifying selection [42], with HERVs that are found in active genetic regions being selected against over time and HERVs found within heterochromatin being retained. The reconstituted HERV (HERV-Kcon) was found to preferentially integrate near active chromatin marks including H3K4me1 and 2 as well as CpG islands [42]. HERV-Kcon is a lab reconstruction of a potential progenitor of HERV-K (HML-2), derived from a consensus sequence. By being integrated near active chromatin marks, it is behaving as a 'young' virus.

One critical system that contributes to histone methylation and heterochromatin formation early in the embryo are Krüppel associated box zinc finger proteins (KRAB-ZFP). The zinc finger domain binds to DNA in a sequence-specific manner, allowing the recruitment of other proteins via the KRAB domain; in particular, the scaffold protein TRIM28/KAP1, forming part of a larger protein complex that modifies the histones [43]. These proteins include DNMT1, and DNMT3a/b, as well as the histone lysine methyltransferase, SETDB1, which is responsible for the H3K9me3 modification [28]. The majority of human KRAB-ZFP binding sites were located within transposons, mainly retrotransposons including HERVs [44]. The KRAB-ZFP bind to HERVs and silence them by burying them in heterochromatin.

Indeed, the LTR-containing retrotransposons have been found to co-evolve with KRAB-ZFP genes [43]. The authors propose a model in which the threat to the genome of each new integration leads to the emergence of new KRAB-ZFP genes [43]. This was supported by a later study showing that the integration of each family of HERVs coincided with a new KRAB-ZFP [45]. This occurs via positive selection of divergent KRAB-ZFP genes, particularly in the region coding for DNA contact residues in the protein [43]. Thus, evolution of KRAB-ZFP genes allows the protein to bind to novel DNA sequences found in newly-integrated retroelements. More recently, it was found that the KRAB-ZFPs that recognise HERVs and LINEs emerged in the same last common ancestor as their target retrotransposons [44], providing further support for the co-evolution hypothesis.

In addition, it has been suggested that genomic imprinting emerged from the use of epigenetics to deal with the retroviral burden and this is supported by the existence of KRAB-ZFP genes involved in imprinting [43]. For example, Zfp57 recruits TRIM28/KAP1 complexes to imprinted differentially methylated regions (DMRs) and maintains methylation during the pre-implantation period in the embryo [28]. Zfp57 is highly expressed in embryonic stem cells but then down-regulated in adult tissues except the ovaries and testes [28]. Zfp57 is also predicted to bind to motifs present in the HERVS71-int family [45]. One hypothesis is that this initial suppression by DNA methylation in embryonic development obviates the need for further involvement of the KRAB-ZFP in HERV silencing in adult tissues. However, not all HERVs may be silenced during imprinting. For example, not all LTRs in mice are suppressed by KRAB-ZFP in oocytes and during embryonic development [46] and this is thought to possibly allow transcripts of ERVs or chimeric ERV/host gene transcripts to persist and assist in development [47]. One possibility is that this escape from repression permits the novel use of the LTRs as alternative promoters of genes [46]. Whether this occurs in human oocytes and embryos is not clear; the ethics of studying human embryos also makes this difficult to determine and hence we have included mouse studies here.

Likewise, the function of another effector protein recruited to KRAB-ZFP complexes, SETDB1, has been studied more intensively in mice, where it was found to be critical for global repression of ERVs [48]. The loss of SETDB1 resulted in up-regulation of ERVs but this depended upon the presence of the particular transcription factors; there was a 'functional match' between transcription factor expression and the ERV LTRs [48]. Further, the de-repressed ERVs could have a causal effect on altered gene expression of proximal genes [48]. This study showed that SETDB1 was responsible for histone methylation and ERV repression in lineage-committed adult cells, in this case B cells [48]. Similarly, KRAB-ZFPs and KAP1 were found to control transposable elements in adult tissues [49]. Thus, there is a role for the KRAB-ZFPs beyond imprinting in the ongoing epigenetic regulation of ERVs. More research into the function of these proteins in human tissues is needed to fully understand the regulation of HERVs.

4.4. Nucleosomal Positioning

There is a growing appreciation of the role of nucleosome position in the regulation of gene expression [50]. The regulation of HERV expression by nucleosomal positioning was postulated almost 20 years ago [51]. The HERV-K LTR was found to lack the TATA box promoter and to not use an initiator sequence in its place; thus, initiation of transcription was by a distinct mechanism involving the cellular transcription factors Sp1 and Sp3 [52]. The HERV-K LTR contains multiple transcription start sites (TSS), with one of these forming the major TSS and the others being dispersed sites [52]. The authors hypothesised that the Sp1 and Sp3 binding to the LTR freed the TSS from nucleosomes, allowing transcription [52]. Recently, the use of alternative transcription start sites (TSS) was described as a positive mechanism to regulate LTR-directed transcription [53]. Critical to this was altered nucleosomal occupancy; in normal cells, this functions to keep retrotransposons silenced by the production of truncated transcripts [53]. In contrast, reduced nucleosomal occupancy in stressed cells altered the usage of TSS to permit full-length transcripts [53]. Thus, while nucleosomal positioning can suppress HERV expression, it is also possible for HERVs to get around this obstacle under conditions of cell stress or by the use of alternative transcription factors.

5. Examples of HERV Regulation

5.1. Embryonic and Induced Pluripotent Stem Cells

The regulation of HERVs is particularly critical during embryonic development and thus the expression of HERVs in stem cells is of great interest. There are two times when the genome undergoes an epigenetic 'reset' by the loss and subsequent re-establishment of DNA and histone methylation; the first is following fertilisation (partial reset, as some imprinted loci are protected from demethylation) and the second is during gametogenesis (full reset) [28]. These periods of global demethylation theoretically favour HERV activity since transcriptional repression is lost. Thus, a role for HERVs is likely during the embryonic stage as this is when there is de-repression of the proviral loci.

HERVs belonging to the HK2 and HERV-H families have been implicated in stem cell identity and embryonic development. Human embryonic carcinoma cells, such as the NCCIT cell line, have been used to model early development. In NCCIT cells, hypomethylation at a particular LTR belonging to the youngest HERV-K elements (LTR5HS), coupled with Oct4 binding, increased HERV-K expression [54]. The HERV-K ORFs were expressed and viral-like particles were produced [54]. In particular, the expression of the HERV accessory protein, Rec, modulates cellular mRNA expression and nuclear translocation [54]. Selective hypomethylation of HERV LTRs is therefore essential to regulate expression of HERVs required in early human development. In addition, HERV-H is associated with the active chromatin mark H3K4me3 in embryonic stem cells [55]. Further, the HERV-H LTR contains binding sites for the stem cell factor NANOG and sites for Oct4 and Sox2 are in close proximity [55]. Finally, HERV-H RNAs act as long non-coding RNAs (lncRNAs), which are important for the pluripotent identity of stem cells [56]. Thus, there is evidence of a role for the epigenetic regulation of HERVs in stem cells and embryonic development.

Given ethical considerations, it is difficult to study human embryonic development other than by using cell lines. It is thus not surprising that much more is known about the epigenetic regulation of ERVs during early developmental stages in mice. The KRAB-ZFPs are critical in this process, restoring methylation lost during fertilisation and thereby silencing ERV expression. ERVs are silenced early in embryonic development by the action of TRIM28/KAP1 [57]. The TRIM28 complex preserves imprinting marks and restores the methylation in the early mouse embryo [47]. In neural progenitor cells, TRIM28-mediated histone modifications repressed ERV expression; deletion of TRIM28 in these cells resulted in increased ERV expression as well as decreased H3K9me3 [58]. The knockdown of TRIM28/KAP1 also resulted in a loss of the repression and increased expression of ERVs in murine embryonic fibroblasts in an OCT4-GFP transgenic mouse model [59]. Similarly, in the production of induced pluripotent stem cells (iPSCs), TRIM28/KAP1 and SETDB1 act as barriers to reprogramming [59]. There is thus a critical role for the KRAB-ZFPs in suppressing ERVs which then leads to the loss of pluripotency.

5.2. Placenta and Pregnancy

It is well-established that HERVs contribute to formation of the placenta. The HERV-W env protein has been co-opted to serve as a fusion protein (called syncytin-1) critical to the formation of the syncytiotrophoblast [60]. Another co-opted *env* gene belonging to HERV-FRD encodes syncytin-2, which contributes to syncytiotrophoblast formation [61] and has a role in immune tolerance of the foetus [62]. A higher risk of pre-eclampsia was associated with reduced expression of both syncytin-1 and -2, with the reduction in syncytin-2 being more important [63]. Moreover, problems during gestational diabetes are linked to aberrant expression of syncytin-2 and its receptor, MFSD2 [64]. The role of syncytins in the placenta is discussed in detail in a recent review [65]; here, we are concerned with the epigenetic regulation of syncytin expression.

Both the HERV-W and HERV-FRD LTRs are controlled by histone H3 acetylation in placental tissues [66]. In addition, control of syncytin-1 expression is mediated by differential methylation. There is a global reduction in methylation levels in the placenta relative to other tissues, consistent with a high proportion of HERV LTRs acting as tissue-specific promoters in the placenta [67]. In particular, a CpG island in the 5′ LTR is hypomethylated in placental cells and hypermethylated in other tissues [68]. Over the course of a pregnancy, this CpG island becomes progressively more methylated [69]. Altered methylation of the HERV-W env locus and decreased expression of syncytin-1 have been observed in placentae from pre-eclampsia [61]. Exposure to oestrogens in the environment causes changes in the methylation of HERVs and this is linked to effects particularly on male children [70]. Aberrant expression of syncytin-1 in hydatiform moles has recently been described to contribute to malignant transformation [71]. Thus, altered epigenetic regulation of HERVs can lead to aberrant pregnancy and development.

There are lesser known roles for HERVs in fertility and pregnancy that merit further study. High levels of syncytin-2 are detected in the testes, which is another tissue that displays global hypomethylation and this is thought to favour HERV expression [65]. Syncytin-1 is also thought to be involved in fertilisation, possibly contributing to the fusion of gametes. Sperm express syncytin-1 on the cell surface whereas oocytes do not; instead, oocytes express the syncytin-1 receptor SLC1A5 [65]. Finally, HERV-K particles have also been detected in human placenta [72] but the functional significance of this, if any, remains unclear.

5.3. Cancer

There are numerous types of cancer and this makes it difficult to generalise about the contribution, if any, of HERVs to tumorigenesis. A number of papers do report a positive correlation between HERVs and cancers, while others find a lack of association [73]. Critically, the accessory proteins of HERV-K, Rec and Np9, have been associated with cancers [74–76] but also found to be expressed in normal tissues [77]. This illustrates some of the uncertainty in determining a causal relation between HERVs and cancer. A bystander effect might be useful in itself, permitting a HERV-based biomarker [78] or the use of HERV proteins as surrogate tumour antigens for therapeutic purposes [79,80]. A detailed analysis of the evidence for HERVs in cancer is beyond the scope of the current review; we will here limit ourselves to the contribution of epigenetic regulation.

One common feature of cancer is a global hypomethylation of the genome; moreover, certain genes may be locally hypomethylated in cancer relative to normal tissues [81]. It is thus feasible that global and/or local hypomethylation leads to loss of repression of HERVs and there is evidence for this in a number of cancers. For example, there is a global hypomethylation of HERV-W and the LINE-1 retroelement in ovarian cancers [82] and hypomethylation of the HERV-K 5′ LTR is observed in melanomas [83]. Global hypomethylation does not necessarily correlate with expression of all HERVs. For example, the treatment of neuroblastoma cell lines with 5′-aza induced expression of multiple HERV-W loci [84], showing that a cancer cell line could still have HERVs that are suppressed by CpG methylation. Further, de-repression of HERV LTRs could lead to activation of otherwise silent oncogenes, a process referred to as 'onco-exaptation' [85]. Examples of such oncogenes induced following onco-exaptation of LTRs include tyrosine kinase receptors (ALK, ERBB4); in these examples, LTR-driven expression results in truncated proteins being produced and these are associated with cancers including lymphoma and melanoma [85].

Global hypomethylation could play a role in the expression of HERVs in GCT cell lines, such as the Tera-1 and NCCIT. These cell lines are known to produce HERV particles, with those of the NCCIT being mature particles that bud from the cells [86], while those from Tera-1 cells appear to lack the env glycoprotein [87]. In our lab, we found that the NCCIT cell line was particularly permissive to HK2 expression (manuscript in preparation), consistent with HERV expression and particle production by these cells. The NCCIT have a methylation pattern reminiscent of the pluripotent state [88] and, as they are an embryonic carcinoma cell line, they have been used to model early embryonic development [54]. While cancer cells and embryonic stem cells are clearly different, there is growing recognition of the common features of pluripotency and malignancy [26]. In particular, the global hypomethylation in tumour cells could be similar to that observed during early developmental stages. It is feasible that a stem cell-like phenotype is found in GCT [88], at least among the subset of cancer stem cells that are hypothesised to exist in most tumours to sustain cancer progression [89]. Since HERVs contribute to the identity of embryonic stem cells, they might also contribute to the formation of cancer stem cells.

In addition, altered histone methylation or acetylation in cancer may contribute to de-repression of HERVs. This has been described in a recent analysis of repetitive elements in cancer cell lines using ENCODE ChIP-Seq data [16]. For example, increased HERV-Fc1 expression from a locus on chromosome 7 was found to be associated with active histone methylation [16]. One HERV LTR seems particularly sensitive to HDACi: the ERV9 LTR is present in thousands of copies in the human genome that is highly expressed in the male testes [90]. The expression of a testes-specific tumour suppressor protein, GTAp63, is under the regulation of the ERV9 LTR. This protein is absent in GCT but its expression can be induced in these cells by the use of the HDACi TSA and vorinostat [91]. Expression of GTAp63 induces apoptosis in these cells and is thought to be protective against GCT formation; the silencing of this protein contributes to tumour formation by preventing GTAp63-induced apoptosis [91]. This clearly indicates a role for histone acetylation in the control of GTAp63 expression and thus of ERV9 LTR promoter activity. In contrast, treatment with 5′-aza did not induce expression of GTAp63 and therefore CpG methylation is not involved in repression of this LTR.

Interestingly, the ERV9 LTR was subsequently found to control the expression of several genes, many of which are involved in immunity or apoptosis [90]. The pro-apoptotic genes included TNFRSF10B, which encodes the death receptor 5 (DR5/Killer) protein [90]. The ERV9 LTR control of tumour suppressor genes in male germ cells reveals a protective effect of HERV LTR activity in preventing cancer. All of the ERV9 LTRs were activated by treatment with HDACi, leading the authors to propose the therapeutic use of HDACi to restore tumour suppression and induce apoptosis in GCT [90]. Of further note is the fact that HDACi did not increase the expression of other HERV subfamilies [90], suggesting that repression by histone deacetylation is not universal or, at least, not the sole mechanism of silencing. However, it is well-documented that HERVs are expressed in cell lines derived from GCTs [86] and we have measured HK2 expression in NCCITs. In contrast to Beyer and colleagues, we did detect a modest increase in HK2 expression with HDACi treatment (vorinostat, panobinostat) (manuscript in preparation).

While HERVs are thought to have a contributory role in tumorigenesis, it is unlikely to be simply a matter of expression being on or off. The complexity of this is revealed by the analysis of genomic and transcriptomic data using new technologies. For example, the ENCODE ChIP-seq data revealed de-repression of certain repetitive elements including HERV-Fc1. However, of the seven loci of HERV-Fc1, only one was identified as having altered expression in cancer cell lines [16]. Moreover, the treatment of cells with inhibitors of such as 5'-aza or HDACi does not necessarily lead to increased HERV expression [30]. A fascinating twist in this story is that the use of DNA methylation inhibitors allows HERVs to be expressed and to trigger an innate immune response [92,93]. For example, the demethylation of HERVs leads to an immune response to dsRNA, producing exogenous interferon that could then prime neighbouring cells for immune checkpoint (anti-CTLA4) therapy [93]. Thus, the use of inhibitors of epigenetic modifications could prove beneficial in treating human cancers by harnessing HERV expression.

6. Conclusions

Recent developments have revealed some of the complexities of HERV regulation by epigenetics. HERVs are not universally silenced; in normal physiology, there is a real need for HERV expression but this seems to be limited to particular tissues and times, with the key examples being placentation and embryonic development. As discussed, the expression of syncytin-1 contributes to the formation of the placenta and normal pregnancy. HERV-W, which encodes syncytin-1, has also been associated with neurological disorders and autoimmune diseases. This could be due to de-repression of the LTRs which could permit syncytin-1 expression in adult cells and this expression could be further enhanced by other stimuli. For example, cytokine-mediated transactivation of HERV expression has been described for HERV-W and HERV-K in ALS [20]. These data suggest a susceptibility of HERV LTRs to pro-inflammatory stimuli, allowing them to act in a positive way to amplify the immune response in the right context but possibly contributing to diseases such as ALS and multiple sclerosis (MS). A further example is the finding that HERV-W loci showed decreased association with H3K9me3 in the context of influenza virus expression, as well as transactivation by the transcription factor glial cells missing 1 (GCM1) [94]. It is thus feasible that exogenous virus infection could precipitate altered epigenetic marks and aberrant HERV expression in tissues where it is normally silenced.

In addition, HERVs that are expressed may have a beneficial role in preventing cancer onset, such as by tumour suppression as in the case of the ERV9 LTR in male germ cells. Alternatively, it is still unclear to what extent HERVs may also contribute to tumour formation. For example, the expression of HERVs in GCT may be merely a consequence of the global hypomethylation but it is also possible that HERV expression somehow contributes to cancer onset or progression, such as through the action of the HERV-K accessory protein Rec. The altered epigenetic regulation in cancer cells may favour HERV expression, which could then have knock-on effects. The loss of epigenetic regulation at the level of the LTRs could allow the binding of transcription factors to consensus sites that are normally occluded. This has been described for the activation of oncogenes following onco-exaptation of HERV LTRs. These examples show that harnessing the potential of HERVs, particularly the HERV LTRs, comes at a potential cost should the epigenetic regulation be disrupted.

Acknowledgments: This work was supported by an MRC Clinician Scientist Fellowship awarded to GM.

Conflicts of Interest: The authors declare that they have no conflicts of interest.

References

1. Stocking, C.; Kozak, C.A. Endogenous retroviruses. *Cell. Mol. Life Sci.* **2008**, *65*, 3383–3398. [CrossRef] [PubMed]
2. Katzourakis, A.; Magiorkinis, G.; Lim, A.G.; Gupta, S.; Belshaw, R.; Gifford, R. Larger Mammalian Body Size Leads to Lower Retroviral Activity. *PLoS Pathog.* **2014**, *10*, e1004214. [CrossRef] [PubMed]
3. Vernon, M.L.; McMahon, J.M.; Hackett, J.J. Brief Communication: Additional Evidence of Type-C Particles in Human Placentas. *JNCI J. Natl. Cancer Inst.* **1974**, *52*, 987–989. [CrossRef] [PubMed]
4. Bronson, D.L.; Fraley, E.E.; Fogh, J.; Kalter, S.S. Induction of retrovirus particles in human testicular tumor (Tera-1) cell cultures: An electron microscopic study. *J. Natl. Cancer Inst.* **1979**, *63*, 337–339. [PubMed]
5. Lander, E.S.; Linton, L.M.; Birren, B.; Nusbaum, C.; Zody, M.C.; Baldwin, J.; Devon, K.; Dewar, K.; Doyle, M.; FitzHugh, W.; et al. Initial sequencing and analysis of the human genome. *Nature* **2001**, *409*, 860–921. [CrossRef] [PubMed]
6. Tristem, M. Identification and characterization of novel human endogenous retrovirus families by phylogenetic screening of the human genome mapping project database. *J. Virol.* **2000**, *74*, 3715–3730. [CrossRef] [PubMed]
7. Shin, W.; Lee, J.; Son, S.-Y.; Ahn, K.; Kim, H.-S.; Han, K. Human-specific HERV-K insertion causes genomic variations in the human genome. *PLoS ONE* **2013**, *8*, e60605. [CrossRef] [PubMed]
8. Marchi, E.; Kanapin, A.; Magiorkinis, G.; Belshaw, R. Unfixed endogenous retroviral insertions in the human population. *J. Virol.* **2014**, *88*, 9529–9537. [CrossRef] [PubMed]
9. Macfarlane, C.M.; Badge, R.M. Genome-wide amplification of proviral sequences reveals new polymorphic HERV-K(HML-2) proviruses in humans and chimpanzees that are absent from genome assemblies. *Retrovirology* **2015**, *12*, 35. [CrossRef] [PubMed]
10. Wildschutte, J.H.; Williams, Z.H.; Montesion, M.; Subramanian, R.P.; Kidd, J.M.; Coffin, J.M. Discovery of unfixed endogenous retrovirus insertions in diverse human populations. *Proc. Natl. Acad. Sci. USA* **2016**, *113*, E2326–E2334. [CrossRef] [PubMed]
11. Volkman, H.E.; Stetson, D.B. The enemy within: Endogenous retroelements and autoimmune disease. *Nat. Immunol.* **2014**, *15*, 415–422. [CrossRef] [PubMed]
12. Tugnet, N.; Rylance, P.; Roden, D.; Trela, M.; Nelson, P. Human Endogenous Retroviruses (HERVs) and Autoimmune Rheumatic Disease: Is There a Link? *Open Rheumatol. J.* **2013**, *7*, 13–21. [CrossRef] [PubMed]
13. Gonzalez-Cao, M.; Iduma, P.; Karachaliou, N.; Santarpia, M.; Blanco, J.; Rosell, R. Human endogenous retroviruses and cancer. *Cancer Biol. Med.* **2016**, *13*, 483–488. [PubMed]
14. Downey, R.F.; Sullivan, F.J.; Wang-Johanning, F.; Ambs, S.; Giles, F.J.; Glynn, S.A. Human endogenous retrovirus K and cancer: Innocent bystander or tumorigenic accomplice? *Int. J. Cancer* **2015**, *137*, 1249–1257. [CrossRef] [PubMed]
15. Kovalskaya, E.; Buzdin, A.; Gogvadze, E.; Vinogradova, T.; Sverdlov, E. Functional human endogenous retroviral LTR transcription start sites are located between the R and U5 regions. *Virology* **2006**, *346*, 373–378. [CrossRef] [PubMed]
16. Criscione, S.W.; Zhang, Y.; Thompson, W.; Sedivy, J.M.; Neretti, N. Transcriptional landscape of repetitive elements in normal and cancer human cells. *BMC Genom.* **2014**, *15*, 583. [CrossRef] [PubMed]
17. Thompson, P.J.; Macfarlan, T.S.; Lorincz, M.C. Long Terminal Repeats: From Parasitic Elements to Building Blocks of the Transcriptional Regulatory Repertoire. *Mol. Cell* **2016**, *62*, 766–776. [CrossRef] [PubMed]
18. Manghera, M.; Douville, R.N. Endogenous retrovirus-K promoter: A landing strip for inflammatory transcription factors? *Retrovirology* **2013**, *10*, 16. [CrossRef] [PubMed]
19. Akopov, S.B.; Nikolaev, L.G.; Khil, P.P.; Lebedev, Y.B.; Sverdlov, E.D. Long terminal repeats of human endogenous retrovirus K family (HERV-K) specifically bind host cell nuclear proteins. *FEBS Lett.* **1998**, *421*, 229–233. [CrossRef]
20. Manghera, M.; Ferguson-Parry, J.; Lin, R.; Douville, R.N. NF-κB and IRF1 Induce Endogenous Retrovirus K Expression via Interferon-Stimulated Response Elements in Its 5′ Long Terminal Repeat. *J. Virol.* **2016**, *90*, 9338–9349. [CrossRef] [PubMed]
21. Hughes, J.F.; Coffin, J.M. Human endogenous retrovirus K solo-LTR formation and insertional polymorphisms: Implications for human and viral evolution. *Proc. Natl. Acad. Sci. USA* **2004**, *101*, 1668–1672. [CrossRef] [PubMed]

22. Belshaw, R.; Watson, J.; Katzourakis, A.; Howe, A.; Woolven-Allen, J.; Burt, A.; Tristem, M. Rate of recombinational deletion among human endogenous retroviruses. *J. Virol.* **2007**, *81*, 9437–9442. [CrossRef] [PubMed]
23. Cohen, C.J.; Lock, W.M.; Mager, D.L. Endogenous retroviral LTRs as promoters for human genes: A critical assessment. *Gene* **2009**, *448*, 105–114. [CrossRef] [PubMed]
24. Dunn, C.A.; Romanish, M.T.; Gutierrez, L.E.; van de Lagemaat, L.N.; Mager, D.L. Transcription of two human genes from a bidirectional endogenous retrovirus promoter. *Gene* **2006**, *366*, 335–342. [CrossRef] [PubMed]
25. Romanish, M.T.; Lock, W.M.; van de Lagemaat, L.N.; Dunn, C.A.; Mager, D.L.; Lander, E.; Linton, L.; Birren, B.; Nusbaum, C.; Zody, M.; et al. Repeated Recruitment of LTR Retrotransposons as Promoters by the Anti-Apoptotic Locus NAIP during Mammalian Evolution. *PLoS Genet.* **2007**, *3*, e10. [CrossRef] [PubMed]
26. Laurent, G.S.; Shtokalo, D.; Dong, B.; Tackett, M.R.; Fan, X.; Lazorthes, S.; Nicolas, E.; Sang, N.; Triche, T.J.; McCaffrey, T.A.; et al. VlincRNAs controlled by retroviral elements are a hallmark of pluripotency and cancer. *Genome Biol.* **2013**, *14*, R73. [CrossRef] [PubMed]
27. Brookes, E.; Shi, Y. Diverse Epigenetic Mechanisms of Human Disease. *Annu. Rev. Genet.* **2014**, *48*, 237–268. [CrossRef] [PubMed]
28. Voon, H.P.J.; Gibbons, R.J. Maintaining memory of silencing at imprinted differentially methylated regions. *Cell. Mol. Life Sci.* **2016**, *73*, 1871–1879. [CrossRef] [PubMed]
29. Allis, C.D.; Jenuwein, T. The molecular hallmarks of epigenetic control. *Nat. Rev. Genet.* **2016**, *17*, 487–500. [CrossRef] [PubMed]
30. Laska, M.J.; Brudek, T.; Nissen, K.K.; Christensen, T.; Møller-Larsen, A.; Petersen, T.; Nexø, B.A. Expression of HERV-Fc1, a human endogenous retrovirus, is increased in patients with active multiple sclerosis. *J. Virol.* **2012**, *86*, 3713–3722. [CrossRef] [PubMed]
31. Hull, E.E.; Montgomery, M.R.; Leyva, K.J. HDAC Inhibitors as Epigenetic Regulators of the Immune System: Impacts on Cancer Therapy and Inflammatory Diseases. *BioMed Res. Int.* **2016**, *2016*, 1–15. [CrossRef] [PubMed]
32. Maksakova, I.A.; Mager, D.L.; Reiss, D. Keeping active endogenous retroviral-like elements in check: The epigenetic perspective. *Cell. Mol. Life Sci.* **2008**, *65*, 3329–3347. [CrossRef] [PubMed]
33. Leung, D.C.; Lorincz, M.C. Silencing of endogenous retroviruses: When and why do histone marks predominate? *Trends Biochem. Sci.* **2012**, *37*, 127–133. [CrossRef] [PubMed]
34. Pérot, P.; Mugnier, N.; Montgiraud, C.; Gimenez, J.; Jaillard, M.; Bonnaud, B.; Mallet, F. Microarray-Based Sketches of the HERV Transcriptome Landscape. *PLoS ONE* **2012**, *7*, e40194. [CrossRef] [PubMed]
35. Kazanets, A.; Shorstova, T.; Hilmi, K.; Marques, M.; Witcher, M. Epigenetic silencing of tumor suppressor genes: Paradigms, puzzles, and potential. *Biochim. Biophys. Acta Rev. Cancer* **2016**, *1865*, 275–288. [CrossRef] [PubMed]
36. Szpakowski, S.; Sun, X.; Lage, J.M.; Dyer, A.; Rubinstein, J.; Kowalski, D.; Sasaki, C.; Costa, J.; Lizardi, P.M. Loss of epigenetic silencing in tumors preferentially affects primate-specific retroelements. *Gene* **2009**, *448*, 151–167. [CrossRef] [PubMed]
37. Lavie, L.; Kitova, M.; Maldener, E.; Meese, E.; Mayer, J. CpG methylation directly regulates transcriptional activity of the human endogenous retrovirus family HERV-K(HML-2). *J. Virol.* **2005**, *79*, 876–883. [CrossRef] [PubMed]
38. Bannister, A.J.; Kouzarides, T. Regulation of chromatin by histone modifications. *Cell Res.* **2011**, *21*, 381–395. [CrossRef] [PubMed]
39. Rasmussen, T.A.; Søgaard, O.S.; Brinkmann, C.; Wightman, F.; Lewin, S.R.; Melchjorsen, J.; Dinarello, C.; Østergaard, L.; Tolstrup, M. Comparison of HDAC inhibitors in clinical development: Effect on HIV production in latently infected cells and T-cell activation. *Hum. Vaccines Immunother.* **2013**, *9*, 993–1001. [CrossRef] [PubMed]
40. Hurst, T.; Pace, M.; Katzourakis, A.; Phillips, R.; Klenerman, P.; Frater, J.; Magiorkinis, G. Human endogenous retrovirus (HERV) expression is not induced by treatment with the histone deacetylase (HDAC) inhibitors in cellular models of HIV-1 latency. *Retrovirology* **2016**, *13*, 10. [CrossRef] [PubMed]
41. Campos-Sánchez, R.; Cremona, M.A.; Pini, A.; Chiaromonte, F.; Makova, K.D. Integration and Fixation Preferences of Human and Mouse Endogenous Retroviruses Uncovered with Functional Data Analysis. *PLoS Comput. Biol.* **2016**, *12*, e1004956. [CrossRef] [PubMed]

42. Brady, T.; Lee, Y.N.; Ronen, K.; Malani, N.; Berry, C.C.; Bieniasz, P.D.; Bushman, F.D. Integration target site selection by a resurrected human endogenous retrovirus. *Genes Dev.* **2009**, *23*, 633–642. [CrossRef] [PubMed]

43. Thomas, J.H.; Schneider, S. Coevolution of retroelements and tandem zinc finger genes. *Genome Res.* **2011**, *21*, 1800–1812. [CrossRef] [PubMed]

44. Imbeault, M.; Helleboid, P.-Y.; Trono, D. KRAB zinc-finger proteins contribute to the evolution of gene regulatory networks. *Nature* **2017**, *543*, 550–554. [CrossRef] [PubMed]

45. Lukic, S.; Nicolas, J.-C.; Levine, A.J. The diversity of zinc-finger genes on human chromosome 19 provides an evolutionary mechanism for defense against inherited endogenous retroviruses. *Cell Death Differ.* **2014**, *21*, 381–387. [CrossRef] [PubMed]

46. Evsikov, A.V.; de Evsikova, C.M. Friend or Foe: Epigenetic Regulation of Retrotransposons in Mammalian Oogenesis and Early Development. *Yale J. Biol. Med.* **2016**, *89*, 487–497. [PubMed]

47. Lim, A.K.; Knowles, B.B. Controlling Endogenous Retroviruses and Their Chimeric Transcripts During Natural Reprogramming in the Oocyte. *J. Infect. Dis.* **2015**, *212* (Suppl. S1), S47–S51. [CrossRef] [PubMed]

48. Collins, P.L.; Kyle, K.E.; Egawa, T.; Shinkai, Y.; Oltz, E.M. The histone methyltransferase SETDB1 represses endogenous and exogenous retroviruses in B lymphocytes. *Proc. Natl. Acad. Sci. USA* **2015**, *112*, 8367–8372. [CrossRef] [PubMed]

49. Ecco, G.; Cassano, M.; Kauzlaric, A.; Duc, J.; Coluccio, A.; Offner, S.; Imbeault, M.; Rowe, H.M.; Turelli, P.; Trono, D. Transposable Elements and Their KRAB-ZFP Controllers Regulate Gene Expression in Adult Tissues. *Dev. Cell* **2016**, *36*, 611–623. [CrossRef] [PubMed]

50. Jiang, C.; Pugh, B.F. Nucleosome positioning and gene regulation: Advances through genomics. *Nat. Rev. Genet.* **2009**, *10*, 161–172. [CrossRef] [PubMed]

51. Sverdlov, E.D. Perpetually mobile footprints of ancient infections in human genome. *FEBS Lett.* **1998**, *428*, 1–6. [CrossRef]

52. Fuchs, N.V.; Kraft, M.; Tondera, C.; Hanschmann, K.-M.; Löwer, J.; Löwer, R. Expression of the human endogenous retrovirus (HERV) group HML-2/HERV-K does not depend on canonical promoter elements but is regulated by transcription factors Sp1 and Sp3. *J. Virol.* **2011**, *85*, 3436–3448. [CrossRef] [PubMed]

53. Persson, J.; Steglich, B.; Smialowska, A.; Boyd, M.; Bornholdt, J.; Andersson, R.; Schurra, C.; Arcangioli, B.; Sandelin, A.; Nielsen, O.; et al. Regulating retrotransposon activity through the use of alternative transcription start sites. *EMBO Rep.* **2016**, *17*, 753–768. [CrossRef] [PubMed]

54. Grow, E.J.; Flynn, R.A.; Chavez, S.L.; Bayless, N.L.; Wossidlo, M.; Wesche, D.J.; Martin, L.; Ware, C.B.; Blish, C.A.; Chang, H.Y.; et al. Intrinsic retroviral reactivation in human preimplantation embryos and pluripotent cells. *Nature* **2015**, *522*, 221–225. [CrossRef] [PubMed]

55. Santoni, F.A.; Guerra, J.; Luban, J. HERV-H RNA is abundant in human embryonic stem cells and a precise marker for pluripotency. *Retrovirology* **2012**, *9*, 111. [CrossRef] [PubMed]

56. Lu, X.; Sachs, F.; Ramsay, L.; Jacques, P.-É.; Göke, J.; Bourque, G.; Ng, H.-H. The retrovirus HERVH is a long noncoding RNA required for human embryonic stem cell identity. *Nat. Struct. Mol. Biol.* **2014**, *21*, 423–425. [CrossRef] [PubMed]

57. Rowe, H.M.; Jakobsson, J.; Mesnard, D.; Rougemont, J.; Reynard, S.; Aktas, T.; Maillard, P.V.; Layard-Liesching, H.; Verp, S.; Marquis, J.; et al. KAP1 controls endogenous retroviruses in embryonic stem cells. *Nature* **2010**, *463*, 237–240. [CrossRef] [PubMed]

58. Fasching, L.; Kapopoulou, A.; Sachdeva, R.; Petri, R.; Jönsson, M.E.; Männe, C.; Turelli, P.; Jern, P.; Cammas, F.; Trono, D.; et al. TRIM28 Represses Transcription of Endogenous Retroviruses in Neural Progenitor Cells. *Cell Rep.* **2015**, *10*, 20–28. [CrossRef] [PubMed]

59. Miles, D.C.; de Vries, N.A.; Gisler, S.; Lieftink, C.; Akhtar, W.; Gogola, E.; Pawlitzky, I.; Hulsman, D.; Tanger, E.; Koppens, M.; et al. TRIM28 is an Epigenetic Barrier to Induced Pluripotent Stem Cell Reprogramming. *Stem Cells* **2016**, *35*, 147–157. [CrossRef] [PubMed]

60. Frendo, J.-L.; Olivier, D.; Cheynet, V.; Blond, J.-L.; Bouton, O.; Vidaud, M.; Rabreau, M.; Evain-Brion, D.; Mallet, F. Direct involvement of HERV-W Env glycoprotein in human trophoblast cell fusion and differentiation. *Mol. Cell. Biol.* **2003**, *23*, 3566–3574. [CrossRef] [PubMed]

61. Denner, J. Expression and function of endogenous retroviruses in the placenta. *APMIS Acta Pathol. Microbiol. Immunol. Scand.* **2016**, *124*, 31–43. [CrossRef] [PubMed]

62. Blaise, S.; Ruggieri, A.; Dewannieux, M.; Cosset, F.-L.; Heidmann, T. Identification of an envelope protein from the FRD family of human endogenous retroviruses (HERV-FRD) conferring infectivity and functional conservation among simians. *J. Virol.* **2004**, *78*, 1050–1054. [CrossRef] [PubMed]

63. Vargas, A.; Toufaily, C.; LeBellego, F.; Rassart, E.; Lafond, J.; Barbeau, B. Reduced Expression of Both Syncytin 1 and Syncytin 2 Correlates With Severity of Preeclampsia. *Reprod. Sci.* **2011**, *18*, 1085–1091. [CrossRef] [PubMed]

64. Soygur, B.; Sati, L.; Demir, R. Altered expression of human endogenous retroviruses syncytin-1, syncytin-2 and their receptors in human normal and gestational diabetic placenta. *Histol. Histopathol.* **2016**, *31*, 1037–1047. [PubMed]

65. Soygur, B.; Sati, L. The role of syncytins in human reproduction and reproductive organ cancers. *Reproduction* **2016**, *152*, R167–R178. [CrossRef] [PubMed]

66. Trejbalová, K.; Blazková, J.; Matouskova, M.; Kucerová, D.; Pecnová, L.; Vernerová, Z.; Herácek, J.; Hirsch, I.; Hejnar, J. Epigenetic regulation of transcription and splicing of syncytins, fusogenic glycoproteins of retroviral origin. *Nucleic Acids Res.* **2011**, *39*, 8728–8739. [CrossRef] [PubMed]

67. Reiss, D.; Zhang, Y.; Mager, D.L. Widely variable endogenous retroviral methylation levels in human placenta. *Nucleic Acids Res.* **2007**, *35*, 4743–4754. [CrossRef] [PubMed]

68. Matouskova, M.; Blazková, J.; Pajer, P.; Pavlícek, A.; Hejnar, J. CpG methylation suppresses transcriptional activity of human syncytin-1 in non-placental tissues. *Exp. Cell Res.* **2006**, *312*, 1011–1020. [CrossRef] [PubMed]

69. Huang, Q.; Chen, H.; Li, J.; Oliver, M.; Ma, X.; Byck, D.; Gao, Y.; Jiang, S.-W. Epigenetic and non-epigenetic regulation of syncytin-1 expression in human placenta and cancer tissues. *Cell Signal.* **2014**, *26*, 648–656. [CrossRef] [PubMed]

70. Vilahur, N.; Bustamante, M.; Byun, H.-M.; Fernandez, M.F.; Marina, L.S.; Basterrechea, M.; Ballester, F.; Murcia, M.; Tardón, A.; Fernández-Somoano, A.; et al. Prenatal exposure to mixtures of xenoestrogens and repetitive element DNA methylation changes in human placenta. *Environ. Int.* **2014**, *71*, 81–87. [CrossRef] [PubMed]

71. Bolze, P.-A.; Patrier, S.; Cheynet, V.; Oriol, G.; Massardier, J.; Hajri, T.; Guillotte, M.; Bossus, M.; Sanlaville, D.; Golfier, F.; et al. Expression patterns of ERVWE1/Syncytin-1 and other placentally expressed human endogenous retroviruses along the malignant transformation process of hydatidiform moles. *Placenta* **2016**, *39*, 116–124. [CrossRef] [PubMed]

72. Kämmerer, U.; Germeyer, A.; Stengel, S.; Kapp, M.; Denner, J. Human endogenous retrovirus K (HERV-K) is expressed in villous and extravillous cytotrophoblast cells of the human placenta. *J. Reprod. Immunol.* **2011**, *91*, 1–8. [CrossRef] [PubMed]

73. Kessler, A.; Wiesner, M.; Denner, J.; Kämmerer, U.; Vince, G.; Linsenmann, T.; Löhr, M.; Ernestus, R.-I.; Hagemann, C. Expression-analysis of the human endogenous retrovirus HERV-K in human astrocytic tumors. *BMC Res. Notes* **2014**, *7*, 159. [CrossRef] [PubMed]

74. Chen, T.; Meng, Z.; Gan, Y.; Wang, X.; Xu, F.; Gu, Y.; Xu, X.; Tang, J.; Zhou, H.; Zhang, X.; et al. The viral oncogene Np9 acts as a critical molecular switch for co-activating β-catenin, ERK, Akt and Notch1 and promoting the growth of human leukemia stem/progenitor cells. *Leukemia* **2013**, *27*, 1469–1478. [CrossRef] [PubMed]

75. Gonzalez-Hernandez, M.J.; Swanson, M.D.; Contreras-Galindo, R.; Cookinham, S.; King, S.R.; Noel, R.J.; Kaplan, M.H.; Markovitz, D.M. Expression of human endogenous retrovirus type K (HML-2) is activated by the Tat protein of HIV-1. *J. Virol.* **2012**, *86*, 7790–7805. [CrossRef] [PubMed]

76. Singh, S.; Kaye, S.; Francis, N.; Peston, D.; Gore, M.; McClure, M.; Bunker, C. Human endogenous retrovirus K (HERV-K) *rec* mRNA is expressed in primary melanoma but not in benign naevi or normal skin. *Pigment Cell Melanoma Res.* **2013**, *26*, 426–428. [CrossRef] [PubMed]

77. Schmitt, K.; Heyne, K.; Roemer, K.; Meese, E.; Mayer, J. HERV-K(HML-2) rec and np9 transcripts not restricted to disease but present in many normal human tissues. *Mob. DNA* **2015**, *6*, 4. [CrossRef] [PubMed]

78. Wang-Johanning, F.; Li, M.; Esteva, F.J.; Hess, K.R.; Yin, B.; Rycaj, K.; Plummer, J.B.; Garza, J.G.; Ambs, S.; Johanning, G.L. Human endogenous retrovirus type K antibodies and mRNA as serum biomarkers of early-stage breast cancer. *Int. J. Cancer* **2014**, *134*, 587–595. [CrossRef] [PubMed]

79. Zhou, F.; Krishnamurthy, J.; Wei, Y.; Li, M.; Hunt, K.; Johanning, G.L.; Cooper, L.J.; Wang-Johanning, F. Chimeric antigen receptor T cells targeting HERV-K inhibit breast cancer and its metastasis through downregulation of Ras. *Oncoimmunology* **2015**, *4*, e1047582. [CrossRef] [PubMed]

80. Rycaj, K.; Plummer, J.B.; Yin, B.; Li, M.; Garza, J.; Radvanyi, L.; Ramondetta, L.M.; Lin, K.; Johanning, G.L.; Tang, D.G.; et al. Cytotoxicity of Human Endogenous Retrovirus K-Specific T Cells toward Autologous Ovarian Cancer Cells. *Clin. Cancer Res.* **2015**, *21*, 471–483. [CrossRef] [PubMed]

81. Feinberg, A.P.; Vogelstein, B. Hypomethylation distinguishes genes of some human cancers from their normal counterparts. *Nature* **1983**, *301*, 89–92. [CrossRef] [PubMed]

82. Menendez, L.; Benigno, B.B.; McDonald, J.F. L1 and HERV-W retrotransposons are hypomethylated in human ovarian carcinomas. *Mol. Cancer* **2004**, *3*, 12. [CrossRef] [PubMed]

83. Stengel, S.; Fiebig, U.; Kurth, R.; Denner, J. Regulation of human endogenous retrovirus-K expression in melanomas by CpG methylation. *Genes. Chromosomes Cancer* **2010**, *49*, 401–411. [CrossRef] [PubMed]

84. Hu, L.; Uzhameckis, D.; Hedborg, F.; Blomberg, J. Dynamic and selective HERV RNA expression in neuroblastoma cells subjected to variation in oxygen tension and demethylation. *APMIS Acta Pathol. Microbiol. Immunol. Scand.* **2016**, *124*, 140–149. [CrossRef] [PubMed]

85. Babaian, A.; Mager, D.L. Endogenous retroviral promoter exaptation in human cancer. *Mob. DNA* **2016**, *7*, 24. [CrossRef] [PubMed]

86. Bieda, K.; Hoffmann, A.; Boller, K. Phenotypic heterogeneity of human endogenous retrovirus particles produced by teratocarcinoma cell lines. *J. Gen. Virol.* **2001**, *82 Pt 3*, 591–596. [CrossRef] [PubMed]

87. Bhardwaj, N.; Montesion, M.; Roy, F.; Coffin, J.M. Differential Expression of HERV-K (HML-2) Proviruses in Cells and Virions of the Teratocarcinoma Cell Line Tera-1. *Viruses* **2015**, *7*, 939–968. [CrossRef] [PubMed]

88. You, J.S.; Kang, J.K.; Seo, D.-W.; Park, J.H.; Park, J.W.; Lee, J.C.; Jeon, Y.J.; Cho, E.J.; Han, J.-W. Depletion of Embryonic Stem Cell Signature by Histone Deacetylase Inhibitor in NCCIT Cells: Involvement of Nanog Suppression. *Cancer Res.* **2009**, *69*, 5716–5725. [CrossRef] [PubMed]

89. Aponte, P.M.; Caicedo, A. Stemness in Cancer: Stem Cells, Cancer Stem Cells, and Their Microenvironment. *Stem Cells Int.* **2017**, *2017*. [CrossRef] [PubMed]

90. Beyer, U.; Krönung, S.K.; Leha, A.; Walter, L.; Dobbelstein, M. Comprehensive identification of genes driven by ERV9-LTRs reveals TNFRSF10B as a re-activatable mediator of testicular cancer cell death. *Cell Death Differ.* **2016**, *23*, 64–75. [CrossRef] [PubMed]

91. Beyer, U.; Moll-Rocek, J.; Moll, U.M.; Dobbelstein, M. Endogenous retrovirus drives hitherto unknown proapoptotic p63 isoforms in the male germ line of humans and great apes. *Proc. Natl. Acad. Sci. USA* **2011**, *108*, 3624–3629. [CrossRef] [PubMed]

92. Saito, Y.; Nakaoka, T.; Sakai, K.; Muramatsu, T.; Toshimitsu, K.; Kimura, M.; Kanai, T.; Sato, T.; Saito, H. Inhibition of DNA Methylation Suppresses Intestinal Tumor Organoids by Inducing an Anti-Viral Response. *Sci. Rep.* **2016**, *6*, 25311. [CrossRef] [PubMed]

93. Chiappinelli, K.B.; Strissel, P.L.; Desrichard, A.; Li, H.; Henke, C.; Akman, B.; Hein, A.; Rote, N.S.; Cope, L.M.; Snyder, A.; et al. Inhibiting DNA Methylation Causes an Interferon Response in Cancer via dsRNA Including Endogenous Retroviruses. *Cell* **2015**, *162*, 974–986. [CrossRef] [PubMed]

94. Li, F.; Nellaker, C.; Sabunciyan, S.; Yolken, R.H.; Jones-Brando, L.; Johansson, A.-S.; Owe-Larsson, B.; Karlsson, H. Transcriptional Derepression of the ERVWE1 Locus following Influenza A Virus Infection. *J. Virol.* **2014**, *88*, 4328–4337. [CrossRef] [PubMed]

viruses

MDPI

Review

Type W Human Endogenous Retrovirus (HERV-W) Integrations and Their Mobilization by L1 Machinery: Contribution to the Human Transcriptome and Impact on the Host Physiopathology

Nicole Grandi [1] and Enzo Tramontano [1,2,*]

[1] Department of Life and Environmental Sciences, University of Cagliari, Cittadella Universitaria di Monserrato SS554, 09042 Monserrato, Cagliari, Italy; nicole.grandi2@gmail.com
[2] Istituto di Ricerca Genetica e Biomedica, Consiglio Nazionale delle Ricerche (CNR), 09042 Monserrato, Cagliari, Italy
* Correspondence: tramon@unica.it; Tel.: +39-070-6754538

Academic Editors: David J. Garfinkel and Katarzyna J. Purzycka
Received: 16 March 2017; Accepted: 20 June 2017; Published: 27 June 2017

Abstract: Human Endogenous Retroviruses (HERVs) are ancient infection relics constituting ~8% of our DNA. While HERVs' genomic characterization is still ongoing, impressive amounts of data have been obtained regarding their general expression across tissues. Among HERVs, one of the most studied is the W group, which is the sole HERV group specifically mobilized by the long interspersed element-1 (LINE-1) machinery, providing a source of novel insertions by retrotransposition of HERV-W processed pseudogenes, and comprising a member encoding a functional envelope protein coopted for human placentation. The HERV-W group has been intensively investigated for its putative role in several diseases, such as cancer, inflammation, and autoimmunity. Despite major interest in the link between HERV-W expression and human pathogenesis, no conclusive correlation has been demonstrated so far. In general, (i) the absence of a proper identification of the specific HERV-W sequences expressed in a given condition; and (ii) the lack of studies attempting to connect the various observations in the same experimental conditions are the major problems preventing the definitive assessment of the HERV-W impact on human physiopathology. In this review, we summarize the current knowledge on the HERV-W group presence within the human genome and its expression in physiological tissues as well as in the main pathological contexts.

Keywords: HERV-W; endogenous retroviruses; Syncytin; autoimmunity; cancer

1. Introduction

In the last 15 years, great efforts have been made to provide a complete assembled sequence of the human genome, progressively revealing an unexpected, highly repetitive composition. Transposable elements (TEs) account, in fact, for >50% of our genetic material, while protein-coding regions constitute only the ~2% [1]. TEs can be broadly divided in two general classes, based on whether DNA or RNA serves as the intermediate in the process of transposition. Human Endogenous Retroviruses (HERVs) belong to class-I TEs, also called retrotransposons, which are characterized by a RNA intermediate that is reverse-transcribed into a double stranded DNA (dsDNA). This dsDNA, commonly called a provirus, is competent for the subsequent integration into the host cell genome [2]. In addition to HERVs, which have 5′ and 3′ long terminal repeats (LTRs), retrotransposons also comprise elements devoid of LTRs and characterized by 3′ poly(A) repeats that are critical for their retroposition, namely long and short interspersed nuclear elements (LINEs and SINEs, respectively) [3].

HERVs are remnants of ancient retroviral infections acquired by the host genome in several waves that occurred mostly between 100 and 40 million years ago [4]. HERVs were once exogenous retroviruses and, in contrast to the retroviruses currently threatening humans, their infection not only affected somatic cells, but also involved, in particular, the germline. Hence, at the time, the proviral integration into the germline cells' DNA made HERV sequences stable components of our genome (Figure 1). Such a process of endogenization and the further fixation in the human population have allowed HERVs to be vertically transmitted to offspring in a Mendelian fashion, constituting up to ~8% of the human genome [1]. In general, HERV sequences have been formed by a traditional process of reverse transcription and integration, and thus show a classical proviral structure. The latter is characterized by an internal portion, including the main retroviral genes (*gag*, *pro*, *pol*, and *env*), flanked by the two LTRs. Owing to their long-time persistence in the host genome, however, individual HERV sequences have independently accumulated nucleotide substitutions, deletions, and insertions, often leading to the loss of coding capacity. In several cases, the homologous recombination between the two LTRs of a same provirus led to the elimination of the whole internal portion [5], a phenomenon reflected by the several thousands of solitary LTRs widespread in the human genome.

Despite the abundant presence of HERVs in the human DNA, their general classification at the genomic level has been incomplete and sometimes controversial, due to the increasing amount of bioinformatics data and the concomitant absence of precise taxonomic rules [6]. Based on sequence similarity with respect to the exogenous retroviruses, HERVs were originally divided into three main classes: class I (*Gammaretrovirus-* and *Epsilonretrovirus*-like), class II (*Betaretrovirus*-like), and class III (*Spumaretrovirus*-like). Each class encloses a variable number of HERV groups, which have been named with discordant criteria in over the years, e.g., based on the human tRNA putatively recognized by the primer binding site (PBS) (e.g., HERV-K for Lysine, HERV-W for Tryptophan, and so on) or according to the name of a nearby gene (e.g., HERV-ADP) or a particular amino acid motif (e.g., HERV-FRD). Only very recently, the human genome assembly GRCh37/hg19 has been analyzed with RetroTector program (ReTe) [7], leading to the recognition and global characterization of >3000 HERV insertions [8]. A multi-step classification approach, based on similarity image analysis, *pol* gene phylogeny and taxonomic feature identification, allowed characterization of 39 "canonical" well defined groups of HERVs, and 31 additional "non-canonical" clades [8]. Interestingly, HERV sequences included in the latter showed several degrees of mosaicism that mainly occurred as the consequence of recombination and secondary integration events [8]. Moreover, this comprehensive classification provided a reliable background for the exhaustive characterization of individual HERV groups at the genomic level, which still remains a major genetic and bioinformatics goal [9].

In contrast to the genomic characterization, which is still ongoing for most of the HERV groups, there are many studies—mainly based on microarrays, hybridization assays or reverse transcriptase polymerase chain reaction (RT-PCR) approaches—that assessed HERV expression in healthy human tissues and cell lines [10–17]. These reports suggested that HERVs are stable components of the human transcriptome, and display differential expression among the diverse human tissues. In particular, variability of HERV transcription between healthy and pathological samples acted as a driving force to determine HERV's role in several human disorders, such as cancer, inflammation, autoimmunity, and infectious diseases. Overall, even if the relevance of HERVs expression to the human physiopathological transcriptome is undeniable, its association with the diverse pathological conditions has lacked, until now, sufficient support. In fact, due to the absence of an unequivocal cause-effect relationship between HERV expression and any human disease [18–20], a number of studies unfortunately ended in the field of "rumor-virology" [18]. As mentioned above, the failure to establish cause–effect relationships primarily depends on the lack of proper characterization of the HERV single groups at the genomic level. The latter is essential to understand which precise HERV sequence is expressed in a given circumstance [21], and if its expression is beneficial, detrimental, or just functionally linked to a specific condition. It is also important to consider that many of the diseases tentatively linked to HERV expression are chronic conditions with a poorly understood etiology, in which several other factors (either genetic or environmental) could potentially produce a causal association [18]. All these aspects have to be considered in relation to the wide panorama of disparate HERVs expression studies, which includes many disparate data and very few studies attempting to assess the various observations in the proper standardized experimental conditions [18]. Thus, once it is established that HERV transcripts are stable signatures of many pathological conditions, the reliable assessment of their specificity and causality to various diseases will be required to explore HERVs as both etiological contributors and innovative therapeutic targets.

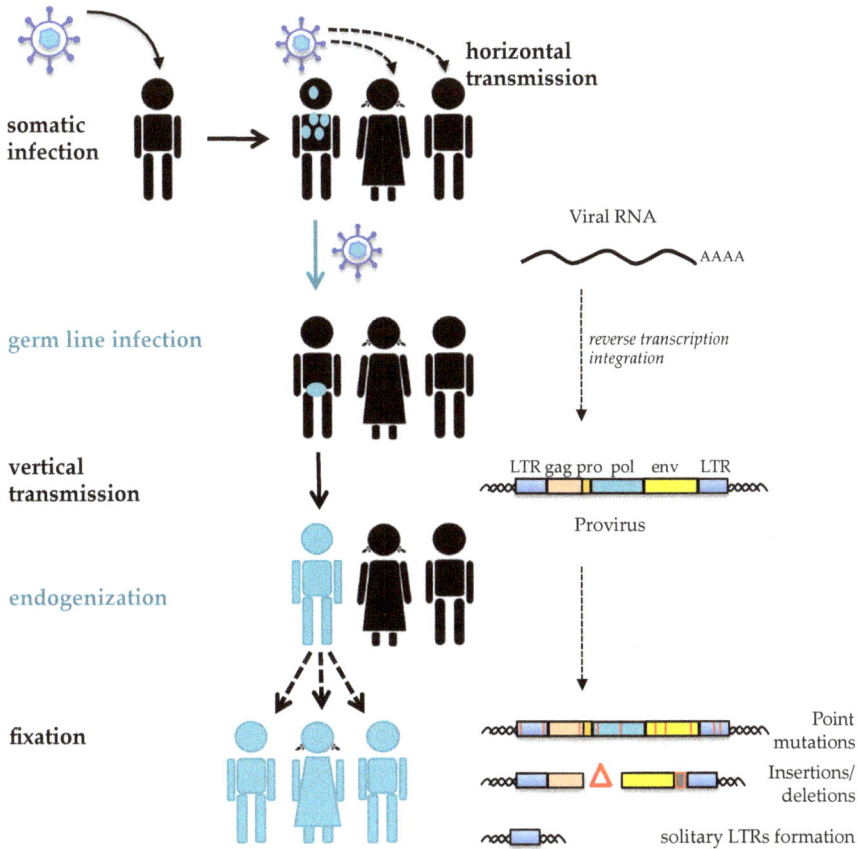

Figure 1. Retrovirus endogenization and human endogenous retroviruses (HERV) formation. During replication, retroviral RNA is reverse-transcribed into a double stranded DNA (dsDNA) provirus and integrated into the cellular genome. All current human retroviruses target somatic cells, showing a horizontal transmission from an infected individual to new hosts. The exogenous retroviruses that gave rise to HERVs were also able to infect germ line cells. In this way, the integrated HERV sequences have been inherited in a Mendelian fashion, being vertically transmitted through the offspring and fixed into the human genome. During evolution, the majority of HERVs accumulated mutations that generally compromised their coding capacity. In several cases, the homologous long terminal repeat (LTR)-LTR recombination has led to the elimination of the whole internal portion, leaving only a solitary LTR as a relic.

Among HERVs, the HERV-W group is one of the most intensively investigated for its possible physiopathological effects on the host. After its initial identification as putative causative agent for multiple sclerosis (MS) [22], strong expression of the HERV-W group was found in placental tissues [23]. This observation led to the identification of a single HERV-W member (ERVWE1, locus 7q21.2) still able to encode a functional Envelope (Env) protein, which, during evolution, has been coopted for an important function in placentation [24,25]. On the one hand, this individual HERV-W element and its physiological role have been described in great detail [26–31]; on the other hand, the general expression of the HERV-W group has been broadly investigated in a variety of tissues, mainly to find correlations with human diseases. In this way, the HERV-W group hyperexpression has been reported in a great number of pathological contexts, making it one of the most promising endogenous elements to be exploited for novel therapeutic and diagnostic strategies. However, in the great majority of cases, the observed expression profiles were yet not linked to any specific HERV-W sequence, preventing so far a definitive association with human pathology.

It is noteworthy that in contrast to all other known HERV groups, HERV-W transcripts have the unique capacity to be mobilized by LINE-1 (L1) human retrotransposons [32,33]. HERV-W colonization of primate genomes was, in fact, mainly sustained by the L1-mediated formation of processed pseudogenes. This occurred through the reverse transcription of RNA transcripts originating from preexisting HERV-W proviral insertions, and their subsequent integration in new chromosomal positions [32,33] (see Figure 2 and paragraph 2). Considering that the human genome contains about 80–100 L1 elements still competent for retrotransposition [34–36], the expression of integrated HERV-W sequences could represent an indirect source of ongoing insertions. This would be more likely to happen in those pathological contexts characterized by an altered epigenetic environment, which could strongly liberate HERV expression, such as cancer and autoimmunity. In this way the general abundance of HERV-W transcripts reported in many tissues could provide a great number of RNA sequences suitable for L1 mobilization, possibly contributing to the intra- and inter-individual genetic variability and being responsible, occasionally, for sporadic insertional mutagenesis and genetic disorders [34–36].

The present review focuses on the HERV-W group as an example of the multifaceted effects that retrotransposon movement can exert on the host. In fact, although TEs have been considered as mere genomic parasites for a long time, the presence of such a wide proportion of mobile elements in eukaryotic DNA suggests that they cannot be only detrimental to the host [37]. Hence, we provide a comprehensive overview of the HERV-W group potential impact on human biology, summarizing its contribution to the human genome and the current knowledge of its expression in physiological conditions and, above all, in pathological contexts. We also briefly discuss the current needs for the definitive assessment of the HERV-W expression biological significance, and the future perspectives for its specific exploitation as innovative biomarkers and/or therapeutic targets for a wide range of human diseases.

2. HERV-W Group Contribution to the Human Genome

As for the other HERV groups, HERV-W integration in human germ line cells resulted from traditional retroviral infection (Figure 2). In general, it is still not clear whether the exogenous retrovirus progenitor of HERVs had germ line cells as their specific target or infected such population by chance [38]. In any case, after the entry into germ line cells, the viral RNA was reverse transcribed into proviral dsDNA, flanked by identical LTRs and competent for the insertion into the host cell genome. Repeated integration events determined the initial spread of HERV-W within human chromosomes, with new provirus formation possibly occurring even in the absence of an infectious phase [35] by intracellular reverse transcription and integration of the proviral RNA transcripts [38] (Figure 2). In addition, differently from all other known HERV groups, the HERV-W acquisition by primate genomes has been for the most part sustained by the L1-mediated formation of processed pseudogenes [32,33] (Figure 2). L1 elements encode for a protein with both reverse transcriptase and endonuclease activities [39]. Through these proteins, L1 sequences can copy and paste into new genomic positions not only their RNA, but also the transcripts generated by non-autonomous retrotransposons (*Alu* and SINE–VNTR–*Alu*, or SVA) and by HERV-W proviruses. According to this model, RNA transcripts originating from preexisting HERV-W proviral insertions were reverse transcribed and integrated into additional chromosomal loci by the L1 machinery. These HERV-W processed pseudogenes are characterized by specific signatures that structurally resemble viral mRNA: (i) truncated 5′ and 3′ LTRs, showing a R-U5 and U3-R structure instead of the traditional U3-R-U5 structure, respectively; (ii) a poly(A) tail of variable length, and (iii) a TT/AAAA insertion motif and a 5–15 nucleotides target site duplication [21,32]. As reported in the few studies aimed at characterizing the group at the genomic level [21,32,33], L1-mediated processed pseudogenes acquisition has not been a minor event in the HERV-W diffusion, having formed >2/3 of the group members. The molecular model of L1-mediated HERV-W transcripts retrotransposition, as well as the specific determinants that limited the process to this HERV group, still remain to be clarified.

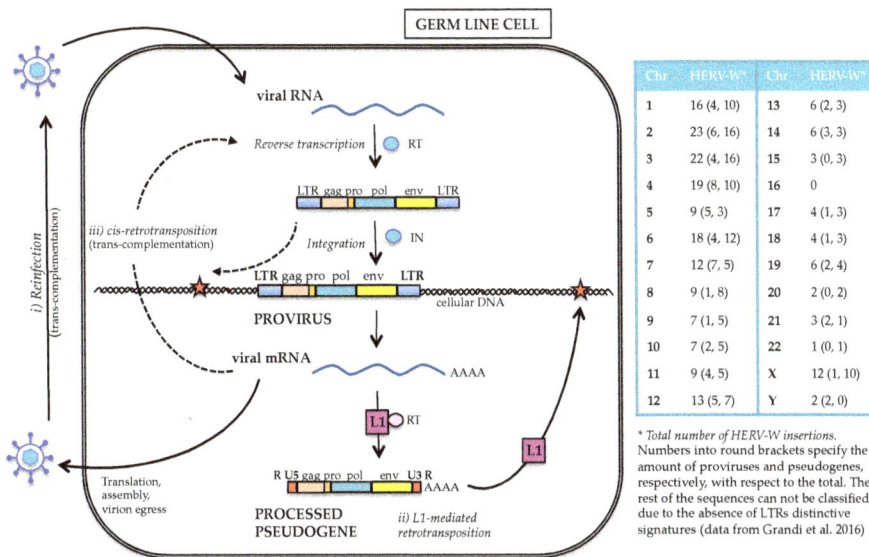

Chr	HERV-W*	Chr	HERV-W*
1	16 (4, 10)	13	6 (2, 3)
2	23 (6, 16)	14	6 (3, 3)
3	22 (4, 16)	15	3 (0, 3)
4	19 (8, 10)	16	0
5	9 (5, 3)	17	4 (1, 3)
6	18 (4, 12)	18	4 (1, 3)
7	12 (7, 5)	19	6 (2, 4)
8	9 (1, 8)	20	2 (0, 2)
9	7 (1, 5)	21	3 (2, 1)
10	7 (2, 5)	22	1 (0, 1)
11	9 (4, 5)	X	12 (1, 10)
12	13 (5, 7)	Y	2 (2, 0)

* Total number of HERV-W insertions. Numbers into round brackets specify the amount of proviruses and pseudogenes, respectively, with respect to the total. The rest of the sequences can not be classified due to the absence of LTRs distinctive signatures (data from Grandi et al. 2016)

Figure 2. HERV-W sequence amplification in germline cells. The initial acquisition of HERV-W sequences has been due to a traditional retroviral infection process. The viral RNA was reverse transcribed and the proviral dsDNA was integrated into the host cell genome by reverse transcriptase (RT) and integrase (IN) viral enzymes, respectively. Integrated provirus expression provided viral mRNAs, which generated new HERV-W insertions (red stars) through (i) L1-mediated retrotransposition: copy and paste mechanism in which viral mRNAs were reverse-transcribed by L1 RT and inserted into a new genomic position, generating HERV-W processed pseudogenes; (ii) reinfection: proviral mRNAs were translated and the deriving proteins assembled into a mature viral particle, that after its egress could have re-infected the same cell; (iii) *cis*-retrotransposition: HERV-W mRNAs could have been used as templates for further reverse transcription–integration events, leading to the acquisition of new insertions in the absence of an extracellular phase. Owing to the accumulation of mutations over time, the last two mechanisms could have required proteins provided in *trans* by a helper virus. As shown in the table that reports the number of HERV-W insertions in each chromosome, the L1-mediated processed pseudogenes formation was responsible for the acquisition of about the 2/3 of the HERV-W sequences.

Three independent studies performed a number of years ago on either isolated human chromosomes [40] or incomplete draft versions of the human genome characterized the HERV-W group at the genomic level [32,33]. Although these studies certainly represent milestones in the analysis of the HERV-W group, the use of different methodologies led to discordant results, currently difficult to retrieve and, especially, to correlate with modern expression data. Recently, a new bioinformatics analysis of the human genome assembly GRCh37/hg19 defined, for the first time, the precise localization and detailed description of 213 HERV-W insertions. The latter were classified, according to their general structure, into proviruses (65), L1-mobilized processed pseudogenes (135), and undefined sequences lacking both LTRs (13) [21]. In this study, the exhaustive characterization of each single HERV-W member in terms of estimated time of integration, genomic context of insertion, and nucleotide sequence provided a dataset that distinguishes the uniqueness of each HERV-W sequence. This dataset is particularly valuable for determining the link between the observed HERV-W expressed products (RNA and proteins) and their specific locus of origin [21]. In addition, some insights regarding the HERV-W group taxonomy were reported for the first time, such as the presence of a second Gag nucleocapsid Zinc finger and the classification of the HERV-W members in two phylogenetic subgroups (named 1 and 2) based on both LTR phylogeny and key mutations [21]. Interestingly, even if the transcripts originated from both HERV-W subgroups were mobilized by the L1 machinery to generate processed pseudogenes, the mechanism was more frequent for subgroup 1 than for subgroup 2 proviruses (1:2.5 and 1:1 ratio, respectively) [21]. The reason for is unclear, but it is possible that the presence of specific sequences made the retrotransposition of subgroup 1 transcripts more efficient or, alternatively, that subgroup 2 elements were expressed at lower levels. Hence, the analysis of the single HERV-W insertions could provide information about L1-mediated HERV-W processed pseudogene formation, which is important for assessing novel retrotransposition events in disease contexts. The molecular elucidation of such events could be important to finally establish the possible pathogenic role of HERV-W processed pseudogenes that, due to their defective structure, have been often disregarded in expression studies, and are thus still poorly investigated in the human physiopathological environment.

In the following sections, we summarize the current knowledge about the HERV-W group expression in both physiological and pathological tissues, based on the many studies performed in the past twenty years. With very few exceptions, these studies investigated the HERV-W group general expression, i.e., without any information on the transcripts genomic origin (specific HERV-W locus) and the structure (provirus or processed pseudogene). If, on the one hand, the available data suggest that HERV-W sequences are differentially expressed in almost all analyzed tissues, being often hyperexpressed in the presence of diseases, on the other hand, these observations deserve more specific investigation aimed at finally identifying which HERV-W loci are selectively deregulated in various conditions.

3. HERV-W Placental Expression and Syncytin-1 Env Protein Cooption for Human Physiology

The reported presence of retroviral particles with reverse transcriptase (RT) activity in MS patients samples [41,42] led to the first description of the so called "MS Retrovirus" (MSRV). Subsequent Southern blot analysis using MSRV-derived probes allowed detection of a previously undescribed HERV multicopy family [22], formally named as HERV-W group [23]. Interestingly, the molecular characterization of the group coding capacity revealed a strong expression restricted to placenta [24] (apart from minor expression in testis [25]), showing the presence of a complete open reading frame (ORF) encoding for two major *env* transcripts (4 and 8 kb) [24,25]. Such ORF was shown to produce a 538 amino acids functional Env protein [23] that was mapped to a HERV-W locus on chromosome 7q21.2 (ERVWE1) [25]. ERVWE1 harbored a 5'LTR functional promoter, exhibiting several binding sites for transcriptional regulators involved in the control of proliferation and differentiation [26,43]. ERVWE1 Env protein was expressed in a panel of different species cell lines, interacting with the type D mammalian retrovirus receptor (hASCT2, human sodium-dependent neutral amino acid transporter type 2) strongly inducing syncytia formation [24,25]. It was therefore named Syncytin-1 [25]. The evidence that syncytia formation could be specifically impaired by Syncytin-1 inhibition (through both specific antibodies [24,25] and anti-sense transcripts [44]) confirmed a central role of this Env protein in the homo- and heterotypic fusogenicity [24,25]. Even if the cell-cell fusion mediated by Syncytin-1 primary depends on its interaction with hASCT2 receptor [24], this Env can efficiently also bind to hASCT1 [45] and even the highly divergent mouse orthologous transporters, mASCT1 and mASCT2, after the elimination of their N-linked glycosylation sites [45]. This flexibility indicates a lower restriction of receptor usage as compared to the other retroviral Env proteins, probably due to the strong selective pressure acting on Syncytin-1 throughout evolution [45]. Syncytin-1 placental expression has been specifically confirmed in the villous [25] and extravillous trophoblasts [46], and its strong fusogenic activity has been associated with the formation of the villous syncytiotrophoblast, the main site for trophic exchanges and other placental functions essential for fetal growth and development [24,25]. Beside its central fusogenic role, Syncytin-1 is also directly involved in cytotrophoblast differentiation and proliferation, which is essential for syncytiotrophoblast homeostasis [44,47]. In fact, Syncytin-1 siRNA knockdown in BeWo cultures reduced cell growth and proliferation, probably through cell cycle arrest in G1 phase [47]. In contrast, ectopic overexpression of Syncytin-1 stimulated, as expected, trophoblast proliferation, confirming its critical role in promoting the G1/S transition during syncytiotrophoblast formation, and emphasizing a subtle balance of fusogenic and non-fusogenic functions in the co-regulation of the cytotrophoblast pool [47]. Moreover, cyclic AMP (cAMP), known to regulate cAMP-dependent protein kinases acting in trophoblast fusion and differentiation [48], is also able to control Syncytin-1 promoter [44]. The latter is a bipartite element formed by (i) the ERVWE1 5′LTR, which contains cAMP-responsive elements for placental basal expression; adjacent to (ii) a placenta-restricted cellular enhancer, located within a MaLR (Mammalian apparent LTR retrotrasposon) solitary LTR and acting as URE (Upstream Regulatory Element), to confer high tissue-specific expression [29]. In addition to trophoblast cell-cycle regulation, Syncytin-1 seems to play a role also in the control of trophoblast survival, since the knockdown of its expression in BeWo cells triggered the death pathway mediated by apoptosis-inducing factor (AIF) [49].

In addition to the above mentioned functions in placental morphogenesis and homeostasis, Syncytin-1 was also hypothesized to have a role in maternal immunotolerance to the fetus [24,25,46] through its immunosuppressive domain [25], as previously shown for a murine [50] and a simian [51] retrovirus. Subsequent studies reported the absence of such activity in a mouse model, suggesting a genetic disjunction between fusogenic and immunosuppressive functions (at least in mice) [52]. In human blood, however, Syncytin-1 was effectively able to inhibit the production of Th1 cytokines known to be important modulators of several immunological functions. This suggests a possible role of Syncytin-1 in mediating the shift from cytokine Th1 to Th2 observed during pregnancy that may also contribute to immunomodulation of the maternal system [53].

4. General HERV-W Expression in Healthy Tissues Other than the Placenta

The Syncytin-1 locus is exceptional, since it retains a residual protein-coding capacity, while the great majority of HERV sequences have accumulated mutations affecting protein production. For this reason, and due to their multi-copy nature, HERVs have often been disregarded in large-scale expression studies and, consequently, have not been exhaustively characterized in terms of functional significance [54]. A number of studies, however, investigated their expression across human tissues and cells. Overall, the results show that the various HERV groups display differential global expression profiles, which could be tissue/cell type-specific and vary depending on tissue state changes (e.g., differentiation, pathogenesis) as well as on environmental and individual conditions.

As stated above, the HERV-W group is strongly expressed in normal placenta [23–25] and shows significant transcriptional activity in testis [25]. In addition, HERV-W transcription in healthy tissues has been monitored using RT-PCR protocols amplifying *gag*, *pol* or *env* genes with primers that, in general, were specifically designed for the placental Syncytin-1 ERVWE1 locus (Table 1). A few other studies investigated expressed sequence tags (ESTs) databases using Syncytin-1 proviral sequence as a query, or analyzed group expression using *pol* probes. In this way, general HERV-W expression has been detected in various human cell lines and healthy tissues—often lacking, however, any information about the transcript's origin from specific loci (Table 1).

Table 1. General type W human endogenous retroviruses (HERV-W) group expression in non-placental healthy tissues.

Tissue	Method	Ref.	Possible Biases [a]
Blood	Search of Syncytin query in EST data	[11]	Low total HERV EST counts, could not detect HERV-Ws divergent from Syncytin, no information on LTR activity, number of cDNA/EST libraries great variability across tissues, under-representation of poorly expressed genes in small libraries (1)
Brain	Search of Syncytin query in EST data	[11]	(1)
	RT-PCR (*gag+*, *pol+*, *env+*)	[55]	Primers specific for single expressed sequences (placental Syncytin (*gag:* AF072500, *env:* AF072506), MSRV clones (*pol:* AF009668)) could not detect divergent HERV-Ws, no information on full-length HERVs expression and LTR activity, samples amount is poorly representative (2)
Brain (cortex and pons)	*env* real time qRT-PCR	[56]	Primers specific for placental Syncytin (NM_014590.3) can could not detect *env* defective or highly divergent HERV-Ws, no information on full-length HERVs expression and LTR activity, samples amount is poorly representative (3)
Breast	Search of Syncytin query in EST data	[11]	(1)
	env real time qRT-PCR	[56]	(3)
Colon	*env* real time qRT-PCR	[56]	(3)
Heart	RT-PCR (*gag−*, *pol−*, *env+*)	[55]	(2)
Endometrium	GammaHERV and HERV-W *pol*-based probe and probe-less real time qPCRs	[57] [14]	Could not detect transcripts defective or highly divergent for *pol* gene, no information about full-length sequences expression and LTR activity, samples amount is poorly representative (4)
Kidney	*pol*-expression arrays hybridization	[10]	Cross-amplification/hybridization of related HERV groups; could not detect transcripts defective for *pol* gene, no information about full-length sequences expression and LTR activity, no quantitative information, samples amount is poorly representative (5)
	RT-PCR (*gag−*, *pol+*, *env+*)	[55]	(2)
Liver	*pol*-expression arrays hybridization	[10]	(5)
	RT-PCR (*gag−*, *pol+*, *env+*)	[55]	(2)
	env real time qRT-PCR	[56]	(3)
Liver-spleen (fetal)	Search of Syncytin query in EST data	[11]	(1)
Lung	RT-PCR (*gag−*, *pol+*, *env+*)	[55]	(2)
Ovary	Search of Syncytin query in EST data	[11]	(1)
	GammaHERV and HERV-W *pol*-based probe and probe-less real time qPCRs	[57] [14]	(4)
PBMC	*pol* RT-PCR and *env* real time PCR	[17]	Low sensitivity and cross-amplification of related HERV groups by RT-PCR degenerate primers, real time PCR primers specific for placental Syncytin (NM_014590.3) could not detect divergent HERV-Ws and transcripts defective for *pol/env* genes, no information on full-length sequences expression and LTR activity, incomplete characterization of individuals health status

Table 1. *Cont.*

Tissue	Method	Ref.	Possible Biases [a]
Prostate	RT-PCR (*gag*−, *pol*+, *env*+)	[55]	(2)
Skel. Muscle	RT-PCR (*gag*−, *pol*+, *env*+)	[55]	(2)
Spleen	RT-PCR (*gag*+, *pol*+, *env*+)	[55]	(2)
Stomach	*env* real time qRT-PCR	[56]	(3)
Testis	RT-PCR (*gag*+, *pol*+, *env*+)	[55]	(2)
Thymus	RT-PCR (*gag*−, *pol*+, *env*+)	[55]	(2)
Uterus	RT-PCR (*gag*−, *pol*−, *env*+)	[55]	(2)
	env real time qRT-PCR	[56]	(3)

General HERV-W expression was reported by Stauffer et al. in the blood, brain, breast, liver/spleen, ovary and placenta, and subsequent analysis confirmed such results for placental and breast tissues only [11]. The physiological HERV-W *env* transcription in healthy brain and breast was detected also by Kim et al. [56]. Yi et al. investigated the HERV-W *gag*, *pol* and *env* genes expression within 12 tissues (brain, prostate, testis, heart, kidney, liver, lung, placenta, skeletal muscle, spleen, thymus, and uterus), detecting *env* transcripts in all the analyzed samples and reporting also some tissue-specific expression for *gag* and *pol* [55]. HERV-W RNA expression was also reported in the normal endometrium and ovary [14,57] and in the colon, liver, stomach, and uterus [56]. The HERV-W group was found to be transcriptionally active in peripheral blood mononuclear cells (PBMC) since early childhood [17]. High resolution melting temperature analysis [58] assessed the occurrence of systematic variations in the HERV-W *gag* sequences expression in primary fibroblasts, depending on both tissues and individuals considered [59]. [a] Methodological biases potentially affecting the effective and specific detection and characterization of the expressed HERV-W sequences. After the first citation, biases with multiple citations are reported as a number into round brackets. EST: expressed sequence tags; LTR: long terminal repeats; MSRV: multiple sclerosis retrovirus; qRT-PCR: quantitative reverse transcriptase PCR; PBMC: peripheral blood mononuclear cells.

In summary, global HERV-W transcriptional activity in healthy conditions was reported by at least one study in the brain, breast, skeletal muscle, spleen, lungs, digestive trait (stomach, liver, colon), genitourinary apparatus (ovary, endometrium, uterus, prostate, testis, and kidneys), and cardiovascular system (heart, whole blood, peripheral blood mononuclear cells (PBMC)). It is noteworthy that all these reports assessed the HERV-W group generic expression, i.e., without connecting the observed transcripts to a specific locus. Moreover, significant biases could derive from the use of Syncytin-1 provirus and/or MSRV cDNA clone sequences as a query and for the design of primers and probes. This could lead, in fact, to the lack of detection of HERV-W expressed loci with divergent nucleotide sequence as compared to the query, or defective for the single genes analyzed. Another possible bias is due to the potential contamination with genomic DNA, possibly representing a further complication in the analysis of multicopy repetitive elements, if not prevented through a correct treatment of RNA samples, e.g., with DNase. Finally, in the majority of cases, no information on the full-length HERV-W sequences expression and the LTR residual regulatory activity are available, and the samples are often limited in number and sometimes incompletely characterized for the individual's health status.

An attempt to connect HERV transcriptome to specific loci of origin was performed by Pérot et al. through a dedicated microarray designed on a collection of >5500 HERVs (including both proviruses and solitary LTRs) that could be reasonably allocated to unique genomic loci [15] (Table 2). Based on their results, the HERV-W group showed, as expected, predominant expression in placenta and testis, attributable to Syncytin-1 locus activity. In addition, five other specific HERV-W loci (one provirus, one processed pseudogene, and three solitary LTRs) were also transcribed in the same tissues, showing in two cases a concomitant LTR promoter activity [15]. Despite the fact that the tissues considered by Pérot et al. were limited (colon, lung, breast, ovary, prostate, testis, uterus and placenta) and that all expressed HERVs are co-localized within human genes that could influence their transcription, the analysis is a remarkable effort to match HERV transcriptome to its specific genomic contributors, taking into account relevant aspects such as promoter activity and tissue specificity.

Table 2. Specific HERV-W loci for which an expression in healthy tissues has been reported.

Locus	Chr:start-end (Strand) [a]	Type	Genomic Context [b]	Tissue	Method	Ref.
2q22.1	2:139030735-139031481 (−)	Solo LTR	*LTR8* (+)	Testis	Microarray	[15]
2q24.3	2:165514421-165516121 (−)	Pseudogene	COBLL1 (−) *TCONS_00004484* (−)	Placenta	Microarray	[15]
5q12.1 *	5:59954322-59962280 (+)	Provirus	DEPDC1B (−)	Placenta	Microarray	[15]
7q21.2 *	7:92097313:92107506 (−)	Provirus	-	Placenta Testis	Northern blot	[23] [25]
15q21.2	15:51552784-51553570 (+)	Solo LTR	CYP19A1 (−)	Placenta	Microarray	[15]
Xq21.33	X:93824238-93824702 (−)	Solo LTR	MER4A (−)	Placenta	Microarray	[15]

[a] Chromosomal positions are referred to genome assembly GRCh37/hg19. The Syncytin locus is highlighted in bold. [b] Localization of HERV-W element within a human gene (italic names correspond to non-coding elements). For sequences marked with an * LTR promoter activity has been also reported.

5. Syncytin-1 Expression in Placental Pathologies

Consistent with its proven role in human placentation, Syncytin-1 abnormal expression has been observed in various pathological conditions affecting placental and maternal-fetal physiology, i.e., Pre-eclampsia (PE); hemolysis elevated liver enzymes and low platelet count (HELLP) syndrome; Trisomy 21; intrauterine growth restriction (IUGR) and endometriosis. The main findings in these pathological contexts are summarized below and in the Supplementary Materials (Table S1).

In general, it is worth noting that in many of the diseases affecting placental tissues, hypoxia is a common pathological trait able to influence Syncytin-1 expression (Table S1). In light of this, Syncytin-1 downregulation, commonly observed in diseased placentas, and the consequent reduction in trophoblasts fusion and differentiation, is likely to result from the pathological hypoxic environment.

PE is a multisystem condition affecting ~5% of pregnant women [60]. It is clinically characterized by hypertension, proteinuria and hypoxia, and is associated with adverse perinatal outcome and preterm birth. A significant fusion reduction in trophoblast cells isolated from PE placentas was reported [61] and, in line with this, the placentas of women affected by PE showed a marked decrease in Syncytin-1 expression [61–65]. Such reduction seems to be correlated with PE severity [61] and depends on Syncytin-1 promoter hypermethylation [65], leading to a consequent decrease in cytotrophoblast differentiation [43].

Similar Syncytin-1 expression reduction was found in HELLP syndrome [62,63]. Considering that experimental hypoxia reduces Syncytin-1 expression by 80% in BeWo cells in vitro and in isolated placental cotyledons ex vivo [66], it has been suggested that such reduction in Syncytin-1 expression might arise due to the HELLP failure in trophoblasts arterial transformation and the consequent poor placental perfusion [67].

Reduced Syncytin-1 expression was also observed in trophoblast cells from placentas bearing a trisomy 21 fetus. Trophoblast cells were still able to aggregate, but fused poorly or late in culture [68–70], and showed increased levels of superoxide dismutase encoded on chromosome 21 [71]. When this antioxidant enzyme was overexpressed in normal cytotrophoblasts, impairment in syncytiotrophoblast formation as well as abnormal cell fusion and Syncytin-1 downregulation were observed, further suggesting that oxidative states are able to influence trophoblasts Syncytin-1 production [68,71]. Since it is known that hypoxia can activate the caspase apoptotic pathway, the hypoxic environments common to many placental diseases could possibly lead to trophoblast cell death via both this mechanism and the above mentioned AIF pathway [47], specifically triggered by Syncytin-1 decreased expression [47,60].

IUGR is another important cause of perinatal morbidity and mortality for both mother and fetus, and it is also related to hypoxia and abnormal trophoblast development. In line with this, IUGR placentas showed significantly lower Syncytin-1 RNA and protein amounts with respect to control placentas [64,72], although still sufficient to mediate trophoblast cells fusion [72].

Finally, two studies reported a high HERV-W expression in endometriotic tissues, even though no great differences were found with respect to control tissues [14,57]. This Syncytin-1 upregulation, dependent on the hypomethylation of its promoter, has been proposed to be involved in endometriotic lesion development [73].

Overall, these findings have confirmed a pivotal role of Syncytin-1 expression in placental physiology, and showed how its deregulation could contribute to maternal systemic disorders [60].

6. HERV-W Expression in Tumorigenesis and Cancer Progression

Tumorigenesis is a complex multistep process involving both inherited and environmental factors and possible association with HERV expression. Of course, this link has been greatly sustained by the well-described transforming nature of exogenous animal retroviruses, which were originally designated as "RNA tumor viruses". However, the high copy number and repetitive nature of HERVs may also trigger additional tumorigenesis mechanisms that do not require the production of infectious viral particles, as summarized in Figure 3. In particular, HERV mobilization and integration could be responsible for insertional mutagenesis events (panel a), which could disrupt or deregulate host genes (e.g., oncosuppressors, transcriptional regulators). The presence of repetitive elements could also trigger chromosomal rearrangements by non-allelic homologous recombination (panel b). HERV transcriptional de-repression, possibly prompted by the altered epigenetic environment commonly associated with cancerous tissues, can lead to uncontrolled activation of downstream cellular genes (e.g., oncogenes, transcription factors) (panel c). Even in the absence of protein production, HERVs transcription could stimulate the accumulation of incomplete replication intermediates, which can activate innate immunity pathways and deregulate non-coding RNA networks (panel d). Finally, if a HERV protein is produced, its activities (e.g., fusogenic and/or immunosuppressive functions) and/or abilities (e.g., interaction with cellular proteins) may contribute to tumor development (panel e).

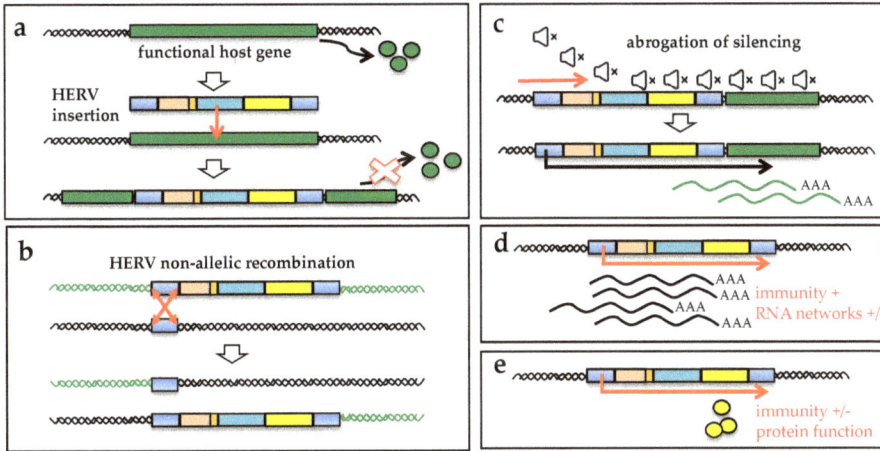

Figure 3. Potential mechanisms of HERV-mediated transformation in tumorigenesis. (**a**) Insertional mutagenesis could disrupt/deregulate host genes; (**b**) non-allelic homologous recombination could induce chromosomal rearrangements; (**c**) transcriptional silencing abrogation could trigger LTR promoter activity; (**d**) accumulation of replication intermediates could evoke immunity and/or deregulate RNA networking; (**e**) protein production could evoke immunity and/or provide oncogenic functions.

Remarkably, despite several studies that reported the general increase—or even the de novo appearance—of HERV-W transcripts in tumors as compared to healthy tissues (Table 3), it is yet to be understood whether such HERV overexpression is the cause or just a consequence of transformation. In fact, HERV expression is generally silenced by epigenetic mechanisms in normal cells, yet abnormal hypomethylation of CpG dinucleotides is commonly observed in tumor cells. This dysregulation could possibly lead to increased levels of HERV expression as an indirect product of the altered epigenetic environment, instead of a main determinant of the disease onset. Unfortunately, despite the central role of epigenetic changes in influencing HERV transcription, very few studies to date have analyzed the HERV-W sequences methylation status in tumor tissues. One study investigating the epigenetic state of L1 and HERV-W sequences in human ovarian carcinomas reported a consistent reduction in promoter methylation, corresponding to an increase in expression [74]. Upregulation of L1 and HERV-W expression could contribute to tumor progression by de novo mobilization of the abundant HERV-W transcripts. Interestingly, despite an overall increase in hypomethylation, some L1 and HERV-W sequences remained hypermethlyated in malignant samples [74]. This result raises the possibility of more specific regulation of HERV expression leading to a beneficial or detrimental effect on disease progression. HERV-W transcriptional increase in ovarian carcinomas has been reported by Hu et al., but a similar expression level was similarly observed in healthy ovaries, and the number of samples was too low to be statistically significant [57]. HERV-W expression has also been investigated in endometrial carcinomas, due to the presence of giant syncytial cells possibly mediated by Env fusogenic activity. Results showed that Syncytin-1 was upregulated in both benign and malignant tissues; however, the highest expression was detected in endometrial carcinomas [75].

In contrast to the above mentioned studies that reported an increase of HERV-W expression in tumor tissues, other studies reported no significant upregulation of HERV-W transcriptional activity in human cancers. Stauffer et al. investigated HERV-W expression in placenta, breast, colon, and kidney cancers, observing that the HERV-W transcription levels in healthy breast and placenta were higher than in corresponding tumor samples [11]. Similarly, Kim et al. reported no significant differences in HERV-W expression between paired tumor and normal adjacent tissues from breast, colon, liver, stomach, and uterus [56].

In addition to the analyses performed on tumor samples from patients, a number of studies investigated the HERV-W group transcriptional activity in cancer cell lines. These studies, however, could not reliably measure the HERV-W expression in cancers, showing a lack of correlation between the expression levels observed in normal tissues and the corresponding cancer cell lines [55]. Moreover, the observed upregulation of HERV-W expression could be, at least in part, a consequence of the tumor cell line environment instead of a specific signature of cancer. For instance, HERV-W RNA levels were increased in three neuroblastoma cell lines (SH-SY5Y, SK-N-DZ, and SK-N-AS), with a selective upregulation during hypoxia recovery and after the treatment with demethylating agents. Both treatments are known to influence HERV transcription with no specificity for tumor environment [78]. Similarly, Díaz-Caballo et al. reported a HERV-W hyperexpression in HCT8 colon carcinoma cells, and proposed a correlation with the induction of a chemotherapy-refractory state [77]. Such increased transcription, however, is possibly the consequence of the experimental induction of a cytostatic stress.

Studies performed in additional cancer cell lines reported more specific effects associated with HERV-W hyperexpression. SH-SY5Y and another neuroblastoma cell line transfected with HERV-W *env* resulted in increased expression of SK3 (small conductance Ca^{2+}-activated K^+ channel protein 3), an ion channel relevant for neuronal excitotoxicity and linked to various diseases of the nervous system [79]. Such upregulation was proposed to depend on the activation of the SK3 promoter cAMP responsive elements (CRE) as direct consequence of the HERV-W Env-mediated increased phosphorylation of the activating transcription factor CREB (CRE-binding protein) [79]. Similarly, Bjerrgarden et al. hypothesized a direct role of Syncytin-1 fusogenic activity in breast cancer based on the fact that MCF-7 and MDA-MB-231 cells express Syncytin-1 on the cell surface and hence are able to fuse with endothelial cells presenting hASCT-2 receptor [76].

As previously described for HERV-W physiological expression, many reports assessed the altered HERV-W transcription in different tumor tissues, however, very few studies attempted to connect transcription to specific HERV-W loci (Table 4). These studies, even if not conclusive for the definitive association of HERV-W expression to tumor development, provide a more reliable picture of the single HERV-W elements upregulated in different human cancers, and suggest further investigations are warranted to determine HERV-W's epigenetic status and specific roles in pathogenesis. Moreover, the identification of specific HERV-W loci expressed in cancer tissues also allows evaluation of their structural characteristics. It is interesting to note that, besides 9 HERV-W proviruses, a number of L1-generated processed pseudogenes (6) and solitary LTRs (8) are specifically upregulated in cancer tissues (Table 4). This suggests that highly defective HERV-W elements, especially in the presence of an altered epigenetic control, can be actively transcribed and differentially expressed in cancerous tissues, possibly contributing to the disease progression.

Table 3. General HERV-W group expression in tumoral tissues.

Tumoral Tissue	Ref.	Method [a]	Physiol. Expression [b]	Possible Biases of HERV-W Members Underrepresentation [c]
B cells	[55] *	RT-PCR ($gag-$, $pol-$, $env+$)	[17] °	Primers specific for single expressed sequences (placental Syncytin-1 gag AF072500 and env AF072506; MSRV clones pol AF009668) could not detect divergent HERV-Ws, no information on full-length HERVs expression and LTR activity, samples amount is poorly representative (2)
Bladder	[55] *	RT-PCR ($gag-$, $pol+$, $env+$)	-	(2)
Breast	[11]	Search of Syncytin-1 in EST data	[11,56]	Low total HERV EST counts, could not detect HERV-Ws divergent from Syncytin-1, no information on LTR activity, number of cDNA/EST libraries great variability across tissues, under-representation of poorly expressed genes in small libraries (1)
	[76] *	RT-PCR, real time qRT-PCR,		Specific detection of a Syncytin-1 env portion only, could not detect transcripts divergent/defective for env, no information on full-length sequences expression and LTR activity
	[56]	env real time qRT-PCR		Primers specific for placental Syncytin-1 (NM_014590.3) could not detect env defective or highly divergent HERV-Ws, no information on full-length HERVs expression and LTR activity, samples amount is poorly representative (3)
	[55] *	RT-PCR ($gag-$, $pol+$, $env+$)		(2)
Brain	[55] *	RT-PCR ($gag-$, $pol+$, $env+$)	[11,55]	(2)
Colon	[11]	Search of Syncytin-1 in EST data	[56]	(1)
	[56]	env real time qRT-PCR		(3)
	[55] *	RT-PCR ($gag-$, $pol+$, $env+$)		(2)
	[77] *	qPCR		Specific detection of a Syncytin-1 env portion only, could not detect transcripts divergent/defective for env, no information on full-length sequences expression and LTR activity
Endometrium	[75]	qPCR, RT-PCR, NB, WB	[14,57,75]	Specific detection of a small portion of Syncytin-1 env only, samples amount is poorly representative, expression values are highly heterogeneous
Esophagus	[55] *	RT-PCR ($gag-$, $pol+$, $env+$)	-	(2)
Histiocyte	[55] *	RT-PCR ($gag-$, $pol+$, $env+$)	-	(2)
Kidney	[11]	Search of Syncytin-1 in EST data	[10,55]	(1)
	[55] *	RT-PCR ($gag-$, $pol+$, $env+$)		(2)
Neuroblasts	[78] *	pol real time qPCRs	-	Could not detect transcripts defective or highly divergent for pol gene, no information about full-length sequences expression and LTR activity, samples amount is poorly representative (4)
Ovary	[74]	Real time qRT-PCR	[57,74]	Primers designed on Syncytin-1 locus (AC000064) could not detect divergent HERV-Ws, samples amount is poorly representative
	[57]	pol real time qPCRs		(4)
	[55] *	RT-PCR ($gag-$, $pol+$, $env+$)		(2)

123

Table 3. *Cont.*

Tumoral Tissue	Ref.	Method [a]	Physiol. Expression [b]	Possible Biases of HERV-W Members Underrepresentation [c]
Pancreas	[55] *	RT-PCR (*gag*−, *pol*+, *env*+)	-	(2)
Placenta	[11]	Search of Syncytin-1 in EST data	[23–25]	(1)
Prostate	[55] *	RT-PCR (*gag*−, *pol*−, *env*+)	[55]	(2)
Skin	[55] *	RT-PCR (*gag*−, *pol*−, *env*+)	-	(2)
Stomach	[56] [55] *	*env* real time qRT-PCR RT-PCR (*gag*−, *pol*+, *env*+)	[56]	(3) (2)
T-cells	[55] *	RT-PCR (*gag*−, *pol*+, *env*+)	[17] °	(2)
Uterus	[56] [55] *	*env* real time qRT-PCR RT-PCR (*gag*−, *pol*+, *env*+)	[55,56]	(3) (2)

[a] NB = Northern Blot, WB = Western Blot; [b] Studies that reported the general group expression in healthy tissues; [c] Methodological biases that potentially affected the effective and specific detection and characterization of the expressed HERV-W sequences. After the first mention, biases with multiple citations are reported as a number; ° data obtained in total PBMC; * data obtained in cancer cell lines.

Table 4. Specific HERV-W loci reported as hyperexpressed in tumoral tissues.

Locus	Chr:start-end (Strand) [a]	Type [b]	Genomic Context [c]	Tissue [d]	LTR [e]	Method [f]	Ref.
1q31.2	1:192855545-192856320 (−)	LTR	MER21C (−)	Testis	-	MA	[15]
2p24.2	2:17520208-17527981 (+)	PV	L3 (−)	Testis	Pro °	MA, qRT-PCR	[15,80]
2p12	2:76098816-76106624 (+)	PV	-	Testis	Pro	MA	[15]
3p12.3	3:74921984-74927237 (−)	PG	-	Prostate	-	MA	[15]
3q28	3:191376573-191383381 (+)	PG	-	Testis	-	MA	[15]
4p13	4:42287455-42294913 (−)	PV	TCONS_00007753 (−)	Testis	Pro °	MA, qRT-PCR	[15,80]
4q26	4:114965536-114972972 (+)	PG	-	Testis	-	MA	[15]
5p13.3	5:31109366-31109859 (−)	LTR	-	Ovary	-	MA	[15]
6q21	6:106676012-106683689 (+)	PG	*ATG5* (−)	Skin T cells Testis*	- Pro °	MA, qRT-PCR MA, qRT-PCR	[81] [80]
7q21.2	7:92097313:92107506 (−)	PV	-	Bladder Skin T cells	Pro -	qRT-PCR MA, qRT-PCR	[82] [81]
7q21.3	7:95987661-95988433 (−)	LTR	Alu Sx (−)	Testis	-	MA	[15]
7q31.1b	7:114019143-114026368 (−)	PG	*FOXP2* (+)	Testis	-	MA	[15]
7q36.3	7:155177752-155178503 (−)	LTR	BC150495 (+)	Testis	PA	MA	[15]
8q24.13	8:125912007-125919468 (−)	PV	-	Prostate	Pro	MA	[15]
13q21.1	13:55627766-55635877 (+)	PV	-	Testis	-	MA	[15]
13q21.33	13:69795752-69799468 (+)	PV	LINC00383 (+) (Ex)	Testis	Pro °	MA, qRT-PCR	[80]
16p12.3	16:18124951-18125494 (−)	LTR	-	Testis	-	MA	[15]
17q22	17:53088886-53095859 (−)	PG	*STXBP4* (+)	Testis	-	MA	[15]
21q21.1	21:20125060-20132866 (−)	PV	MIR548XHG (−) (Ex)	Testis	-	MA	[15]
21q21.3	21:28226756-28234297 (+)	PV	-	Testis	Pro °	MA, qRT-PCR	[15,80]
Xq21.1	X:82517449-82517774 (−)	LTR	L1 PA11 (+), L1 MA2 (+)	Testis	-	MA	[15]
Xq23	X:113140352-113141135 (−)	LTR	L1 (−), XACT (−)	Testis	Pro °	MA, qRT-PCR	[80]
Xq24	X:120490096-120490859 (+)	LTR	-	Testis	PA	MA	[15]

[a] Chromosomal positions are referred to genome assembly GRCh37/hg19. Syncytin locus is highlighted in bold; [b] PV: provirus; PG: processed pseudogene, LTR: solitary LTR; [c] Elements co-localized with HERV-W loci: italics indicates coding genes, (Ex) indicates HERVs within an exon; [d] Tissues for which the HERV-W sequence expression was reported also in physiological conditions are marked with *; [e] Reported activity of the sequences LTRs: Pro: promoter; PA: PolyA signal. The mark ° indicates a hypomethylated status with respect to normal samples; [f] MA: microarray.

In particular, the majority of studies reported HERV-W sequence specific expression both in tumoral testis, along with the previously reported Syncytin-1 expression [25], and in a number of other cancers mostly affecting the genitourinary trait. Pérot and coworkers have analyzed paired normal and tumoral tissues through a dedicated microarray (see also paragraph 4 and Table 2), and reported a number of HERV-W loci that were differentially expressed in testis (16), prostate (2) and ovary (1) cancer samples [15]. Similarly, Gimenez and coworkers identified six HERV-W loci, including Syncytin-1, whose expression was upregulated in testicular cancer [80]. The precise localization of these expressed HERV-W sequences allowed comparison of their epigenetic status in normal and tumoral tissues, revealing, in the latter a U3 promoters hypomethylation in at least five out of the six loci [80]. As is the case for ovarian cancer [74], some sequences remained unmethylated in the tumor environments but not in the normal counterparts [80], suggesting the presence of different levels of HERV transcriptional control. When considering bladder urothelial cell carcinomas, Syncytin-1 was significantly hyperexpressed in >75% of the analyzed tumor tissues ($n = 82$) as compared to the 6% of the matched adjacent tissues, increasing proliferation and viability of human immortalized uroepithelial cells [82]. In this case, the identification of specific HERV-W sequences significantly upregulated in tumor tissues also allowed detection of single nucleotide substitutions. The latter were found in positions 142 and 277 of the Syncytin-1 3'LTR, in ~88% tumoral tissues while they were observed only in a small proportion (~5%) of healthy controls. Interestingly, the T142C mutation apparently resulted in selective binding of the c-Myb transcription factor to ERVWE1 LTR, and was possibly associated with the selective enhancement of Syncytin-1 promoter activity in bladder urothelial cell carcinoma [82]. In addition, the expression of specific HERV-W loci has been assessed in mycosis fungoides, the most common Cutaneous T-Cell Lymphoma (CTCL) [81]. In fact, two HERV-W loci in chromosomes 6 (6q21) and 7 (7q21.2—Sincytin-1), which frequently harbor abnormalities and rearrangements in CTCL, were predominantly and significantly upregulated in mycosis fungoides lesions as compared to the same patient intact skin [81].

Despite the number of studies investigating HERV-W expression (either general or associated with specific loci), no human cancer has been unequivocally related to this or any other HERV group. This greatly depends on the lack of definitive evidence that specific HERV sequences are effectively able to induce tumors through the so far proposed mechanisms. Although HERV expression in tumors may contribute to the disease's clinical outcome, the currently available results suggest only that the HERV-W group has variable expression profiles in both normal and cancerous tissues [11]. As in previous cases, the use of different experimental approaches often affected by potential methodological biases, together with the lack of connection between the observed transcripts and the specific originating locus, currently impedes effective assessment of the biological significance of the HERV-W group expression in tumors. Moreover, the current lack of exhaustive information on HERV-W loci basal expression in healthy tissues clearly limits complete evaluation of their effective dysregulation in the corresponding tumors, which are further complicated by an altered epigenetic regulation. In light of this, even with clear evidence of differential HERV-W expression between tumoral and healthy tissues, further studies are needed to establish which HERV-W loci are actively involved in tumorigenesis and which ones constitute an "epiphenomenon" due to the altered tumoral environment [83].

7. HERV-W Expression in MS and Other Autoimmune Diseases

Autoimmune diseases comprise a heterogeneous group of complex multifactorial disorders, all sharing the incorrect recognition of healthy cells and/or the loss of immune tolerance to self-Antigens (Ags) by the immune system. Clinically, such loss of tolerance leads to Antibody (Abs) production and/or cytotoxic T cells responses against body components, resulting in chronic inflammation and tissue destruction. A role for HERV in autoimmune disorders was primarily suggested by (i) the presence of retroviral Ags and/or specific Abs at the site of disease and in patients' sera, respectively [84,85]; and (ii) an increased HERV expression in patients with autoimmune disorders as compared to healthy individuals [86]. Theoretically, given that HERVs are stable components of the human genome, the immune tolerance to them should have been established during development. Despite this, HERVs still show the ability to induce, or at least to influence, both innate and adaptive immunity [84–89] (Figure 4). Currently, the most accepted theory is that HERV expression can evoke autoimmunity by molecular mimicry between common auto-Ags and exogenous retroviral proteins [86,90–93]. HERV RNAs and proteins may, in fact, be recognized as PAMPs (Pathogen Associated Molecular Patterns) by innate immunity pathogen recognition receptors (PRRs) (recently reviewed in [89]), that determine inflammation and auto-Ab production. Moreover, HERV proteins may act as super-Ags, triggering the non-specific polyclonal activation of auto-reactive T lymphocytes and inducing massive cytokine release. Besides the direct immunogenic effects of retroviral products, HERV proteins may affect the host immune response in additional ways, such as by *trans*-activating/suppressing genes involved in immune modulation. Even in the absence of any expressed product, the mere presence of HERVs can contribute to autoimmunity through insertional mutagenesis events and/or *cis*-regulation of adjacent immune regulatory gene expression at the transcriptional and post-transcriptional level.

Importantly, as described for cancer, autoimmunity is also influenced by abnormal hypomethylation, which can eventually release HERVs expression [94]. Such an occurrence, even as the mere consequence of epigenetic alterations, could contribute to immunopathogenesis by providing nucleic acids or proteins acting as PAMPs. Furthermore, the loss of epigenetic control can provide HERV-W transcripts suitable for de novo mobilization by L1. Therefore, the proper identification and characterization of the expressed HERV loci is essential to assess their effective involvement in the disease onset and progression.

Focusing now on the HERV-W group, the major field of investigation in autoimmunity is certainly MS, although few studies have been reported for other autoimmune or immune-related disorders.

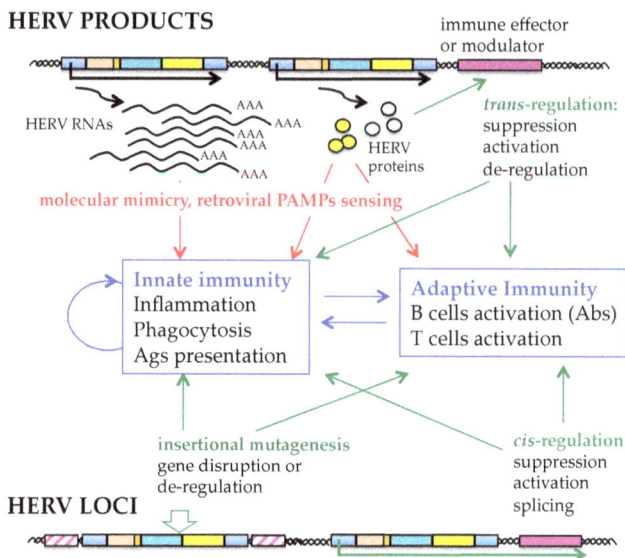

Figure 4. Potential mechanisms of HERV contribution to autoimmunity. HERVs can trigger autoimmunity through the direct sensing of their expression products by pathogen recognition receptors (PRRs) (red) as well as by mediating the deregulation of the host immune effectors and modulators (green). In both cases, the eventual hypomethylated status associated with autoimmunity can upregulate HERVs that are normally silenced in healthy tissues. HERV expressed RNAs and proteins (upper part) can act as pathogen associated molecular patterns (PAMPs) prompting the innate immunity effectors, and, consequently, evoking an adaptive response. HERV proteins can either act as super antigens Ags activating a polyclonal expansion of autoreactive T cells, or deregulate immunity genes. These mechanisms can be also based on the molecular mimicry of HERVs products, due to their identity with the exogenous elements. HERV integrated sequences, or even their sole LTR, (lower part) can affect the host immunity even in the absence of any expressed product, if their insertion disrupts or deregulates genes involved in immune response and its control.

7.1. Multiple Sclerosis

MS is an autoimmune disorder with poorly understood etiology, and characterized by progressive demyelination of the central nervous system (CNS). Both innate and adaptive immunity dysregulation contributes to MS immunopathogenesis, although adaptive immunity may predominate in the disease onset with selective T and B cell activation accompanying clinical relapses [95]. The precise causes of axon demyelination and damage remain unclear, even if inflammatory molecules such as cytokines, chemokines, prostaglandins, reactive oxygen species and matrix metalloproteinases contribute to MS [95]. In addition, different infectious agents have been investigated for a possible association with MS [95–101].

As previously mentioned, the HERV-W group was initially related to MS due to its nucleotide identity with MSRV [102,103], a putative retroviral element detected in some MS patients [22,41,42, 104,105], and proposed as an exogenous competent member of the HERV-W group [22,106–110]. The origin of MSRV is, however, still highly debated [18,111,112], and recent findings suggest that the previously identified MSRV sequences could have arisen from the expression of a single HERV-W locus, or the in vitro recombination of many HERV-W transcripts [21,113]. In the last twenty years, many studies investigated the HERV-W/MSRV involvement in MS, mainly by (i) the detection of HERV-W/MSRV nucleic acids in MS samples; (ii) the presence of HERV-W/MSRV Ags in MS lesions; (iii) the onset of an immune response against these elements; and (iv) the use of some animal models of MS. Even if all these types of investigation are taken into account, the evidence strongly suggests that the presence of HERV-W/MSRV sequences (i) and Ags (ii) could contribute to a higher prevalence in MS. The clear immunopathogenic potential of these HERV products on cellular-mediated immunity, as shown in both humans (iii) and animal models (iv), could indeed take part, together with other individual factors, to cause MS disease.

7.1.1. Detection of HERV-W/MSRV Nucleic Acids in MS Samples

Regarding the presence of HERV-W/MSRV nucleic acids, most studies focused on the detection of expressed HERV-W/MSRV RNA transcripts, while a few of them investigated the differential amounts of integrated DNA sequences copy-number in MS samples.

HERV-W/MSRV *pol* RNA sequences have been detected by RT-PCR approaches in MS patient brain [114], leptomeningeal, choroid plexus and B cells [22], peripheral blood lymphocytes [115], cerebrospinal fluid (CSF) [22,115,116], serum [116,117] and plasma [115]. Overall, HERV-W/MSRV *pol* amplicons were found in a variable proportion of the MS samples (~50 to 100%) as well as in some healthy controls (0–50%) and non-MS pathological samples (0–65%). These results suggest that HERV-W expression may be associated with the pathological environment and have a role in a particular subset of susceptible individuals. Unfortunately, the expressed RNA sequences were not attributable to specific HERV-W loci [97]. Also MRSV/HERV-W *env* RNA expression was reported to be upregulated in MS patient brains [118,119] and PBMC [120]. Finally, a significantly higher accumulation of both HERV-W/MSRV *pol* and *env* RNAs was reported in MS brains [107] and CSF [121] with respect to healthy and pathological controls, even if all samples tested contained the HERV-W/MSRV transcripts regardless of health/disease status.

HERV-W/MSRV DNA copy-number was reported to increase in PBMCs of MS patients as compared to controls, and copy number also correlated with disease severity [110,120]. Considering that active HERV-W proliferation ended several millions of years ago, before the evolutionarily speciation of humans [21], it is unlikely that such variation could depend on the presence of unfixed proviral integrations in the modern population, as shown for younger HERV groups. It is indeed more probable that, as described above, the additional HERV-W copies found in MS patients could be due to processed pseudogenes derived from novel L1-mediated retrotransposition events triggered by the autoimmune hypomethylated environment. A positive relationship between HERV-W/MSRV DNA copy number and female gender has been also hypothesized, which is consistent with the higher incidence of MS in women. In particular, the one proviral copy and 10 L1-generated processed pseudogenes of HERV-W on the X chromosome could possibly play a role in MS sex-based variants, similarly to other X chromosome abnormalities [110]. Finally, MSRV *pol* sequences have been detected by fluorescence in situ hybridization (FISH) in the peripheral blood cell DNA from all patients with active MS and healthy controls tested, which supports an endogenous origin of MSRV [116,122].

7.1.2. Presence of HERV-W/MSRV Ags in MS Lesions

As is the case for HERV-W/MSRV nucleic acids, the presence of HERV-W/MSRV proteins has been reported in both normal and MS brain tissues, thus questioning their direct role in MS pathogenesis. However, the presence of HERV-W/MSRV proteins in diseased sites suggests that they may contribute to MS immunopathogenicity and clinical manifestations. In fact, Syncytin-1 protein was present in MS patient brains and in specific cell types involved in lesions, neuroinflammation, and were expressed at a low level [118] or absent [107,120] in controls. In addition, Syncytin-1 in vitro expression mediated the production of proinflammatory molecules, potentially involved in astrocytes and oligodendrocyte damage [95], and an accumulation of HERV-W Gag Ags was shown in MS demyelinated brain lesions [123]. Also, HERV-W Env epitopes were detected in higher quantities on the surface of B cells and monocytes from patients with active MS with respect to stable MS patients and healthy controls [124]. Finally, HERV-W/MSRV Env protein abundance in MS brain lesions was recently associated with areas of active demyelination, being predominantly expressed by macrophages and microglia, while moderate expression was observed in reactive astrocytes within active lesions [125].

7.1.3. Onset of an Immune Response against These Elements

Growing evidence suggests that HERV-W/MSRV Env may act as super-Ag that triggers an abnormal innate immune response independently of a specific recognition pathway. This immune activation could lead to the overproduction of cytokines, which are known to play a major role in MS inflammatory demyelinating process. MSRV Env induced in both healthy donors and MS patients the in vitro polyclonal activation of Vβ16 T-lymphocytes [126] and the subsequent increase of multiple pro-inflammatory cytokines [126,127]. These HERV-W/MSRV Env pro-inflammatory properties have been attributed to the protein's ability to trigger the Toll-Like Receptor 4 (TLR4) [126, 128,129], leading to the overexpression of the same proinflammatory cytokines involved in MS, such as interleukins 1 and 6 (IL-1, IL-6) and Tumor Necrosis Factor α (TNF-α), and inducing lymphocyte Th-1 polarization [127,128,130]. Moreover, HERV-W Env interaction with TLR4 and the consequent upregulation of proinflammatory factors, in particular inducible nitric oxide synthase, led to the formation of nitrotyrosine groups, which directly affected myelin expression and remyelination by blocking oligodendrocyte precursor differentiation [131]. In support of these findings, HERV-W Env neutralization by monoclonal Ab GNbAC1 reduced such stress reactions and rescuing myelin expression [132], and MSRV Env was recently confirmed to be a potent agonist of human TLR4 in vitro and in vivo [133].

However, some reports assessed the development of a specific humoral response against HERV-W/MSRV in MS patients, and found weaker support for its role in MS pathogenesis. Ruprecht et al. reported the presence of Syncytin-1 Abs in only 1 of 50 MS patients and in none in 59 controls, whereas MSRV Gag or Env Abs were not detected [112]. In a follow up study in MS patients that monitored Abs titers against HERV Env proteins, a decrease in anti-HERV-W Env reactivity as a consequence of interferon (IFN)-β therapy was reported but the decrease was not statistically significant [134], as previously observed for circulating Env RNA [135]. Finally, a study assessing the humoral response against selected HERV-W Env peptides showed that two peptides were strongly recognized by MS patients IgG as compared to controls, and a decrease in recognition after six months of IFN-β therapy was also reported [136].

7.1.4. Use of Some Animal Models of MS

The potential link between HERV-W/MSRV immunopathogenic properties and MS has been investigated through mice models, which generally supported the active involvement of these elements in disease development. Intraperitoneal injection of MSRV virions in immunodeficient mice transplanted with human lymphocytes led to the onset of acute neurological symptoms, causing death by massive brain hemorrhage [137]. Further analysis confirmed the presence of circulating MSRV RNA and splenic overexpression of proinflammatory cytokines [137]. In another study, Syncytin-1 induced neuroinflammation, neurobehavioral abnormalities and oligodendrocyte and myelin injury principally evoked by redox reactant–mediated cellular brain damage [118,138]. Always in mice, MSRV-Env was able to activate the TLR4- and CD14-mediated release of proinflammatory cytokines and, when associated to the myelin oligodendrocyte glycoprotein (MOG) Ag, induced a specific T cells IFN-C production. Such combined innate and acquired responses promoted the development of experimental allergic encephalomyelitis, and was proposed as a suitable MS animal model [139].

In addition to the high number of studies that all assessed the whole group general expression in MS, a limited number of studies were dedicated to the investigation of individual transcribed HERV-W loci. The HERV-W processed pseudogene in locus Xq22.3 (ERVWE2) was among the most highly investigated due to the presence of an almost complete *env* ORF, interrupted only by a premature stop at codon 39. Noteworthy is that this L1-retrotransposed HERV-W element is transcribed in human PBMCs [13,140,141], producing ex vivo an N-terminally truncated Env protein (N-Trenv) [142]. In addition, the evidence that a monoclonal Ab previously used to detect HERV-W Env in MS lesions (13H5A5) [123] was able to bind N-Trenv, but not Syncytin-1, allowed speculation that this and other expressed defective proteins may exert some effects in vivo [142]. Also, the ERVWE2 locus in chromosome X has been proposed as the hypothetical genomic origin of MSRV Env proteins [108,110] and investigated for its potential role in MS and its higher incidence in women. However, analysis of ERVWE2 DNA sequences in MS patients and healthy individuals PBMC revealed that all harbored the stop codon at site 39, and assessed whether genetic polymorphisms could possibly allow the production of a full-length protein in vivo [143]. The authors also identified 5 ERVWE2 DNA regions similar to the MOG Ig-like domain that, together with other co-factors, could trigger the immune cross-reaction against myelin in MS [143]. García-Montojo et al. genotyped the ERVWE2 insertion in a wide group of individuals, and reported a significant association with female MS susceptibility and polymorphisms rs6622139 and rs1290413, which are more frequent in controls than MS affected women [144]. A similar analysis was performed for an HERV-W insertion in chromosome 20, but the two identified polymorphisms were not significantly linked to MS susceptibility based on case–control studies [145]. Other work addressed HERV-W loci expression in MS. Laufer at al. tried to clarify the origin of the HERV-W/MSRV *env* sequences detected in MS samples by evaluating expression from single HERV-W loci. Interestingly, expressed HERV sequences was shown to be often complicated by in vitro recombination between HERV transcripts, probably caused by RT template switches and/or PCR-mediated recombination [13,141]. In particular, the authors proposed that some previously published MSRV *env* sequences, as well as a high number of HERV-W *env* cDNA clones, had actually arisen from the recombination of different HERV-W *env* transcripts, detecting up to four recombination events involving up to five HERV-W loci for the same sequence. It was also shown that the primers commonly used for HERV-W expression studies were similar enough to anneal with multiple HERV-W loci, underlining the importance of precisely assessing the transcripts genomic origin when studying HERV RNA expression [83,141]. Of note, similar individual HERV-W *env* loci expression levels were found in PBMCs from MS patients and healthy controls [141], further supporting the low specificity of RNA transcripts for MS disease. Another comprehensive analysis of HERV-W loci brain transcription was performed by high-throughput sequencing of *env*-specific RT-PCR products, identifying >100 HERV-W loci transcribed at very similar levels in MS patients and healthy individuals [113]. Interestingly, while the deregulated expression of HERV-W *env* in MS brain lesions was refuted, the authors reported an inter-individual variability in HERV-W transcript levels, and a residual promoter activity for many HERV-W LTRs, even if incomplete [113]. A third study analyzing age- and disease-dependent HERV-W *env* RNA diversity showed that HERV-W *env* transcripts originated from multiple loci in primary human neurons, while astrocytes and microglia showed lower diversity in HERV-W transcript chromosomal origin [146]. Similarly, while multiple loci encoding HERV-W *env* RNA sequences were detected in both fetal and adult healthy brains, transcripts cloned from neurologic patients mostly mapped to Syncytin-1 locus (7q21.2), and their abundance was highly correlated with pro-inflammatory gene expression in diseased brains [146]. This could indicate a wide and complex scenario, poorly clarified by the mere upregulation of general HERV-W group expression.

Taken together, HERV-W/MSRV expression analyses in MS patients do not definitively confirm a specific association of these retroviral elements to MS etiology, but strongly suggest a possible role of group expression, especially at the protein level, in disease immunopathogenesis. Variable HERV-W/MSRV expression found in both MS patients and healthy individuals could probably constitute a normal physiological phenomenon, possibly with higher prevalence in MS due to an altered epigenetic and immunological environment [94,147]. This hypothesis opens the possibility of a co-contributing role or predisposing factors that will require additional studies on the HERV-W/MSRV brain proteomic profile in different ethnic populations [147]. In fact, considering the non-specific super-Ag activity of HERV-W/MSRV Env showed neuropathogenic effects coincident with the major hallmarks of MS inflammation [125], the HERV-W group could participate in a complex inflammatory interplay with other not fully understood factors, including genetic predisposition and exogenous infections [97,148]. It is noteworthy that a therapeutic treatment targeting HERV-W/MSRV has been proposed as a possible innovative approach for MS using the GNbAC1 monoclonal Ab developed to selectively recognize MSRV Env. It showed neutralizing effects in vitro and in MS mouse models and is currently in phase II clinical development [149,150].

7.2. Other Autoimmune Diseases

Besides MS, a few studies investigated HERV-W group expression in other disorders with poorly understood etiology, in which autoimmunity mechanisms play a major pathogenic role, such as rheumatoid arthritis (RA), osteoarthritis (OA), chronic inflammatory demyelinating polyradiculoneuropathy (CIDP), psoriasis and lichen planus (LP). In all these disorders, no evidence of an etiological link between HERV-W expression and pathogenesis has been reached yet, mostly due to the poor sample representation and failure to assign the observed transcripts to individual HERV-W loci.

RA is characterized by the progressive destruction of articular components and leads to severe disability. A common sign of the RA autoimmune response is the presence of a cellular infiltrate of neutrophils, lymphocytes, and macrophages in the synovial tissue, accompanied by the increased production of metalloproteinases contributing to extracellular matrix erosion [151]. Based on preliminary results reporting HERV-W/MSRV RNA in the 50% of RA patient plasma samples, Gaudin et al. determined if particle-associated HERV-W/MSRV RNA were present in patient samples [151]. The results showed that neither the patients nor the controls had HERV-W/MSRV RNA in plasma, while such RNA was detected in the synovial fluid samples of two out of nine RA patients and one control, suggesting its lack of specificity with respect to RA etiology [151].

OA is another common form of arthritis characterized by the progressive destruction of articular cartilage, in which many factors, including viral infections, seem to play a role [152]. Bendiksen et al. analyzed cartilage and chondrocytes from advanced OA as compared to early/non-OA individuals. While all samples were negative for a number of exogenous infections, a HERV-W *env* gene was commonly expressed in advanced OA patient cartilage (88% of patients) while expression was detected in a lower proportion of controls (0–38%) [152]. The authors also reported the abundant expression of Env proteins in OA-derived chondrocytes, and the occurrence of viral budding and virus-like particles. However, the particles were neither isolated nor characterized [152].

Another pathology tentatively linked to HERV-W/MSRV is CIDP, a rare immune disease of the peripheral nervous system characterized by inflammatory and demyelinating lesions in nerve roots [153]. Driven by the presence of MSRV-Env in a small number of CIDP patients (5 out of 8) [120], Faucard et al. confirmed an upregulation of MSRV *env* and/or *pol* mRNAs in ~50–65% of CIDP patients PBMC [153]. The authors also reported the presence of MSRV Env protein in 5 out of 7 CIDP patients nerve lesions and dominant expression in Schwan cells [153]. Moreover, Schwan cell cultures exposed to MSRV-Env displayed a potent induction of IL-6 and CXCL10 chemokines, which could be significantly inhibited by GNbAC1 MSRV-Env mAb [153].

Finally, HERV-W/MSRV expression may contribute to some skin diseases with unclear etiology. Psoriasis is a chronic disease characterized by epidermal proliferation and abnormal keratinocytes differentiation and shows similarities to systemic immunological disorders closely related to autoimmunity [154]. Considering that HERV expression has been reported in human skin, being either activated or repressed by UV irradiation [155,156], Molès et al. assessed HERV expression in psoriatic lesions. Their work detected various *pol* amplicons comprising HERV-W sequences in both psoriatic and control skin samples [154]. Another pathology taken into account was LP, a chronic skin inflammatory disease characterized by lichenoid papules and possibly involving also microbial agents in its unclear etiology [157]. de Sousa Nogueira et al. observed a downregulation of some HERV groups, including HERV-W *env*, in skin biopsies of LP patients, with a concomitant activation of antiviral restriction genes APOBEC3G, MxA, and IFN-inducible genes that may be involved in immune control of HERV transcription [157].

8. HERV-W Expression in Neurological and Neuropsychiatric Disorders

HERV-W neuropathogenic effects have also been investigated in a number of neurological and neuropsychiatric diseases with poorly understood etiology, such as Motor Neuron Disease (MND), sporadic Creutzfeldt–Jakob disease (sCJD), autistic spectrum disorder (ASD), attention deficit hyperactivity disorder (ADHD), and schizophrenia. In general, the available information does not yet support a direct role of HERV-W group in any neurological or neuropsychiatric diseases. In fact, a proportion of HERV-W-negative patients is reported in the majority of the studies, while a significant upregulation of HERV-W expression was shown in a subset of cases, strongly suggesting the presence of other major factors contributing to a complex and poorly understood etiology. It is also worth noting that many of these pathologies could be concomitant with behavioral variables, such as drug and alcohol abuse [158], which could be confounding if they are able to influence HERV brain expression [159].

MND is a heterogeneous group of neurologic disorders characterized by the progressive degeneration of motor neurons. Elevated levels of HERV-W *env* transcripts were observed in biopsies from MND patients limbs as compared to control tissues from the same individual and from healthy donors [160]. The authors also detected a parallel upregulation of the SOD1 (oxidative stress-responsive) gene, a marker for oxidative stress, suggesting that its activation could be due to the primary loss of motor neurons instead of being a direct consequence of HERV-W Env neurotoxic effects [160].

sCJD is a rare form of prion disease, causing fatal neurodegeneration and having as key event the conformational change of cellular prion protein to an abnormal protease-resistant isoform. Joang et al. examined the expression of 10 HERV groups in sCJD patients CSFs, detecting transcripts of all analyzed groups and reporting the highest incidence for HERV-W *pol* (82.5% positivity), with a significant increase with respect to controls [161]. Based on subsequent subcloning analysis, all observed transcripts showed non-identical nucleotide sequence, and none had specificity for sCJD [161].

ASD and ADHD are two neurodevelopmental diseases caused by complex interactions with not fully clarified genetic and environmental factors. ASD patients PBMC showed higher positivity for HERV-H and HERV-W mRNAs as compared to controls [162]. Subsequent quantification showed that HERV-H and HERV-W were, respectively, more and less abundantly expressed in ADS patients [162]. Similarly, the HERV-H transcript level in ADHD patients PBMC was significantly higher, while no differences were found in HERV-W expression [163].

Among neuropsychiatric disorders, the field of greatest interest for HERV-W involvement is schizophrenia. The first findings were provided by Deb-Rinker et al. in monozygotic twins discordant for schizophrenia, presenting one sequence (schizophrenia associated retrovirus, SZRV-1, AF135487) similar to both a MSRV (AF009668) and a HERV-9 (S77575) sequences expressed in placenta [164]. Karlsson et al. then detected HERV-W/MSRV *pol* sequences in the cell-free CSF from ~29% acute onset schizophrenia patients and 5% individuals in later stages of the disease, but not in patients with non-inflammatory neurological diseases and healthy controls [165]. Similarly, HERV-W/MSRV expression was up-regulated in the brain frontal cortex regions of schizophrenia patients when compared with control tissues from healthy individuals [165]. In subsequent studies, the same authors reported the presence of HERV-W RNA in the plasma of a subgroup (9 out of 54) of recent-onset schizophrenia patients, 5 of which harbored HERV-W/MSRV sequences in CSF [166]. They detected an elevated level of HERV-W *gag* (but not *env*) transcripts in PBMC of patients with schizophrenia-related psychosis, and reported an upregulation of HERV-W sequences from locus 11q13.5 [167]. HERV-W *env* plasmatic mRNA was found in 36% of recent-onset schizophrenia patients and in none of the 106 controls, and also RT activity was significantly increased in patient sera [168]. At the protein level, HERV-W Env hyperexpression in U251 human glioma cells triggered the production of the dopamine receptor D3 and brain-derived neurotrophic factor (BDNF), both associated with schizophrenia, and increased the phosphorylation of CREB protein, which is necessary for BDNF expression [168] and confirmed in human neuroblastoma cells [79]. Moreover, recent findings also suggested that phosphorylation of Glycogen Synthase Kinase 3β might be involved in HERV-W Env-mediated BDNF induction [169]. A study detecting HERV-W Ags in patients reported positive serum antigenemia for Gag and Env in ~50% of schizophrenic patients and in 3–4% of blood donors [170]. Of note, a full-length HERV-W LTR was found in the regulatory region of GABBR1 (GABA receptor B1) gene, which is downregulated in schizophrenic patients [171]. However, the roles of this LTR and GABBR1 in schizophrenia remain to be clarified.

In contrast to the studies reported above, which showed an increased HERV-W expression in Schizophrenia [166–168], a number of investigations reported no specific correlation between the HERV-W transcription and development of neurological diseases [158,172,173]. The comprehensive microarray-based analysis of 20 HERV group's transcriptional activity in 215 brain samples from schizophrenia or bipolar disorders (BD) patients and matched controls failed to show relevant links between HERV brain expression and schizophrenia, suggesting that it could be more likely that HERV transcriptional activity is influenced by the individual genetic background and the presence of immune cell infiltrates and/or medical treatments [172]. Interestingly, the different brain areas of each individual showed a common pattern of HERV expression, where the HERV-W *env* gene was transcriptionally active but did not show significant differences between healthy controls and schizophrenic patients [172]. Weis et al. observed that HERV-W Gag proteins were present in human brain anterior cingulate cortex and hippocampus, mostly associated with neurons and astrocytes, and showed significantly reduced expression in schizophrenia, major depression, and BD patients as compared to controls [158]. HERV-W *env* transcription was increased in schizophrenia and BD patient PBMCs, but the corresponding DNA copy number was paradoxically lower in patients than in healthy controls. Moreover, differences in HERV-W *env* amplicon nucleotide sequences and their relative frequencies were observed in comparisons of patients to controls and in comparisons among Schizophrenia and BD patients to MS patients [173]. The authors hypothesized that when HERV-W genes are hypomethylated during development, environmental stimuli (such as exogenous infections) could prompt lineage-specific HERV-W genomic modifications and determine variable patterns responding differently to subsequent environmental triggers, leading to diverse clinical manifestations [173].

9. HERV-W Expression in the Presence of Exogenous Infections

HERVs have also been proposed to influence exogenous viral infections, and such a role could be either beneficial or harmful. HERV antisense transcripts have been hypothesized as a plausible defense against exogenous infections, in which complementarity between homologous retroviral RNA sequences could form dsRNA and detected as a PAMP by the innate immunity PRRs [89,174]. Another possible HERV-mediated antiviral effect could be the partial resistance to infection, evoked by receptor interference and blocking by HERV proteins [175,176]. However, exogenous viruses and expressed HERVs may also generate cooperative effects, stimulating each other's transcription or leading to the complementation of defective elements. Clearly, some of these interactions require a certain degree of sequence and structural homology, and are most likely to happen between HERVs and exogenous retroviruses.

9.1. Retroviral Infections

Humans are currently threatened by two exogenous retroviruses: Human Immunodeficiency Virus (HIV, *Lentiviridae*) and Human T-cell Lymphotropic Virus (HTLV, *Deltaretroviridae*). HERV-W Env glycoprotein was shown to functionally complement an *env*-defective HIV-1 strain, generating HERV-W-pseudotyped particles infectious for CD4-negative cells, and therefore, possibly expanding HIV-1 tropism [177]. HERV-W elements were upregulated in three persistently HIV-1 infected cell lines, but not in infected cells [178]. Of note, reversal of HIV-1 latency by treatment with histone deacetylase inhibitors caused no substantial increases of HERV-W *env* gene transcription [179]. Significant HERV-W RNA hyperexpression was detected in the brains of AIDS patients suffering from dementia [114]. However, this variation in expression is probably a consequence of increased immune activity linked to monocyte differentiation and macrophage activation, and had no active role in AIDS neuropathy. HIV Tat transactivator protein increased MSRV *env* mRNAs and HERV-W Env protein expression in astrocytes and differentiated macrophages but reduced expression in monocytes [180]. Similarly, HTLV-I Tax homolog of Tat also activates transcription from HERV-W LTRs by interacting with CREB along with other transcription factors [181]. T cell cross-reactivity between HERV and HIV epitopes was tested in vitro, giving negative results [182].

9.2. Herpesviral Infections

The possible interplay between Herpesviruses infection and HERV-W expression has been widely analyzed, especially in the context of MS and autoimmunity, and the ability of various Herpesviruses to influence HERV-W transcription [85–87]. HERV-W/MSRV expression was enhanced by Herpes Simplex Virus 1 (HSV-1) superinfection in MS patients cells [41,96,183]. More specifically, Lafon et al. showed that HERV-W Env protein expression in neuroblastoma cell lines can be reactivated by HSV-1, probably through its infected cell polypeptide 0 and 4 (ICP0 and ICP4, respectively) early proteins [96]. HERV-W Gag and Env proteins were also induced by HSV-1 in neuronal and brain endothelial cells in vitro, and expression was also compatible with an ICP0-mediated activation [184]. Additional evidence has been reported in HeLa cells, in which HSV-1 IE1 protein stimulated LTR-driven transcription of HERV-W elements, probably through the modulation of the Oct-1 transcription factor [185]. The authors proposed that IE1 activation could also involve HERV-W solitary LTRs, potentially promoting possible nearby genes [185].

Besides HSV-1, other Herpesviruses have been analyzed for their ability to activate HERV-W in MS. A hypothetical ERVWE1 Env peptide (29 aa) harbored an epitope predicted to be presented by different HLA class I molecules and possibly acted as a target for effector T-cells in MS. Interestingly, this epitope was partially homologous with all the pathogens against with elevated Abs titers in MS patients, including HSV-1, HHV-6, VZV (Varicella Zoster Virus), EBV (Epstein Barr Virus), and measles virus [186]. Hence, it was claimed that the effector T cell recognizing this putative epitope would most readily cooperate with regulatory T cells to support an immune response, leading either to a prompt resolution of the infection or to tissue damage by autoimmune processes [99,186]. Regarding EBV, the exposure to the virus or to its major Env glycoprotein (gp350) triggered HERV-W/MSRV expression in PBMCs from MS patients and MSRV positive healthy controls, as well as in cultured U87-MG astrocytes, with an activation pathway possibly involving NF-kB [187]. The infection of a number of cancer and non-cancer cell lines with CMV induced RT activity in all cells, and upregulated various HERVs, including HERV-W, in CMV-infected neural tumor stem cells after UV irradiation [188]. Other evidence of a helper role in HERV-W activation came from kidney transplant recipients with high CMV load, who displayed significantly higher HERV-W *pol* expression than patients with moderate/undetectable CMV load or healthy subjects [189].

9.3. Other Exogenous Infections

Although retroviral and herpesviral infections have been most intensively studied for their effects on HERV-W expression, influenza virus, spleen necrosis virus (SNV) and porcine endogenous retrovirus (PERV) infections have also been implicated in modulating HERV-W.

Nellåker et al. described specific expression patterns of HERV-W *gag* and *env* genes (even if encoded by sequences with truncated/no LTRs) in different cell-lines, and observing subsets of elements being transactivated by influenza virus active replication [140]. Similar variations were also observed as a consequence of serum deprivation, suggesting that the cellular stress itself could contribute to HERV-W modulation [140]. Subsequent analysis showed that influenza virus infection induced spliced ERVWE1 transcripts able to encode Syncytin-1 in extra-placental cells by GCM1 overexpression [190], and downregulated the level of repressive histone mark H3K9me3 [191].

The HERV-W Env glycoprotein induced cellular resistance to SNV, whose infectivity was reduced by 1000–10,000-fold in D-17 cells expressing HERV-W Env [192].

Finally, in the field of xenotransplantation, the expression level of HERV-W genes differed in PERV-infected HEK-293 cells in comparison to uninfected cultures [193].

10. The HERV-W Transcriptional Landscape in the Context of Human Physiopathology: Current Needs and Future Perspectives

As described in the previous sections, no definitive link between HERV-W expression and disease onset and progression has been found. The great majority of the studies, in fact, have been based on the detection and the eventual quantification of expression from the entire HERV group. The analysis of bulk HERV expression cannot reveal the contribution of specific HERV-W loci to physiological or pathological disease states. Furthermore, most of the studies have been based on not fully standardized methodologies performed on differently representative samples. Thus, although a great amount of data has been generated, they are frequently discordant and difficult to compare with each other. The main biases that affected, in our opinion, a great number of studies aimed at analyzing the HERV-W group expression are summarized in Table S2, along with their inconclusive nature and suggestions for improvement.

The currently available information strongly suggests that the HERV-W group is commonly transcribed in human cells, and that this happens in both healthy and pathological conditions. Collective expression greatly varies between tissues and reflects individual genetic backgrounds. In particular, the presence of HERV-W RNA and Ags in human tissues and cells seems to be a physiological phenomenon. HERV-W expression is detected in (a subset of) healthy individuals and possibly increases under pathological conditions, without necessarily representing an etiological factor. The direct pathogenic effects of HERV-W RNA and proteins have not been confirmed and still lack a molecular mode of action in vivo. In contrast, the effects that these HERV-W products could trigger in their interplay with the host—including immune stimulation, insertional mutagenesis (also through de novo retrotransposition events) and deregulatory functions—are more likely to contribute to human pathogenesis, and also influence some animal models of human disease. Even in this case, however, the identification and characterization of the specific HERV-W sequences and the exact molecular mechanisms involved remains a major goal and necessary for the exploitation of HERV-W candidates for therapeutic purposes.

Defining the expression patterns of single HERV-W loci through a dedicated microarray or RNA-seq analyses will be essential to provide additional insights into the quantitative changes originating from specific HERV-W sequences and to identify single members possibly linked to human pathogenesis. Indeed, this promising field deserves a deeper investigation aimed, first of all, at characterizing all the possible mechanisms involving HERV-W presence and expression in both physiological and pathological conditions. The latter also includes the possibility that an altered epigenetic environment could prompt de novo mobilization of HERV-W transcripts by active L1 elements. This ongoing process would generate additional HERV-W processed pseudogenes when compared with the ones recently mapped in the human genome reference sequence [21].

The current lack of knowledge of the individual HERV-W loci transcriptional status is in large part due to the previous incomplete characterization of the HERV-W genomic landscape. Much of the current experimental designs have focused on a few HERV-W sequences, above all Syncytin-1 and MSRV clones (Table S2). This has prevented until now (i) the univocal assignment of the HERV-W expressed sequences to the locus of origin; (ii) the characterization of the single HERV-W sequences differential expression in the diverse physiological tissues, essential to assess their effective dysregulation in diseased environments; (iii) the evaluation of the full-length HERV-W sequence transcription, coding capacity and regulatory elements; (iv) the characterization of the single HERV-W sequences epigenetic status in both physiological and pathological contexts; and (v) the study of the HERV-W sequences genomic context of insertion, identification of nearby host genes that can potentially influence (or be influenced by) HERV-W elements even in the absence of a detectable expressed product (Table S2).

We also outline future research investigating HERV-W's contribution to human physiopathology, including dedicated genome-wide high-throughput studies using stringent primers and probes that are able to distinguish the uniqueness of single HERV-W elements conforming to standard conditions. Developing these reagents will help properly define the specific contribution of the different retroelements to the human transcriptome [95]. Importantly, the physiological consequences of individual HERV-W loci expression must be evaluated a priori, in order to have a reliable "basal" level to compare with the same, diseased, tissue [86]. Such specific quantitative analyses must then be performed on a statistically significant population, possibly including paired samples of both healthy tissues and pathological lesions from the same individual. Moreover, these investigations also take into account the influence of the HERV-W loci genomic context of integration, and to analyze the molecular diversity of single insertions within the human population. HERV-W polymorphisms are necessary to understand due to the possibility that different HERV-W allelic variants may exert specific effects on the pathogenesis phenotype, progression and therapeutic response, depending on the host genetic background [95]. Considering that ~80–100 copies of L1 are active in the human genome [36,39,194], the eventual L1-mediated retrotransposition of HERV-W sequences should also be considered, especially in those pathological environments associated with an altered epigenetic control, where the hypomethylation of both L1 and HERV-W sequences have been reported [74]. Finally, since structurally-incomplete LTRs could be still able to drive the transcription of HERV-W proviruses, processed pseudogenes or nearby host genes, the methylation levels of truncated and solitary LTRs should also be evaluated. Indeed, besides detecting HERV-W group overall expression, the identification of the specific encoding locus appears to be mandatory to establish any definitive associations between human diseases and specific retroelements, and also to properly understand the molecular nature of emergent forms that have arisen through recombination events involving different HERV-W loci [95], especially in those contexts where the epigenetics alteration could liberate HERVs expression. Similarly, the molecular determinants responsible for specific HERV-W loci upregulation as well as their role either as a cause or a consequence of disease must also be clarified in detail [97]. Together, these approaches will finally provide well-characterized mechanisms of HERV-mediated pathogenesis.

In conclusion, it is certainly possible and perhaps likely that specific HERV-W sequences may play a role in human pathogenesis, without necessarily being the only etiological determinant of disease. Indeed, in the field of autoimmunity, one or more HERV-W insertions (or even specific allelic variants) and/or their expressed products may be involved in a complex inflammatory and immune interplay with other unknown or not fully understood co-factors. The latter may include individual predispositions, depending on the host genetic background, as well as extrinsic factors such as stress, environmental stimuli or exogenous infections. All these complex relationships must be considered, especially in the field of multifactorial disorders with poorly clarified etiology. In light of this, the identification and characterization of the precise HERV-W loci showing a differential transcription pattern and/or L1-mediated HERV-W de novo mobilization in a specific pathological context appear to be mandatory to definitively demonstrate a cause–effect connection to any disease etiology, and to subsequently identify single HERV-W sequences exploitable as novel therapeutic targets. The latter could be suitable for various, innovative approaches, from the employment of retroviral inhibitors to the administration of passive as well as active immunotherapy directed against specific HERV products, possibly in association with the treatment with DNA-demethylating agents. However, even in the absence of an etiological contribution, the identification of specific HERV-W sequences selectively expressed in a given pathological context could provide novel and reliable sequence-based biomarkers of disease. Also, disease-associated Ags suitable for directing immunotherapeutic approaches to the precise site of pathogenesis could be developed as protein biomarkers. All these HERV-based therapeutic applications could certainly constitute an innovative treatment for many human diseases.

Supplementary Materials: The following are available online at www.mdpi.com/1999-4915/9/5/162/s1. Table S1: Syncytin-1 expression in pathological placentas, Table S2: Major biases and current needs for HERV-W transcriptome studies.

Acknowledgments: Authors would like to thank Geoffrey Michael Gray for valuable critical revision of the manuscript.

Conflicts of Interest: The authors declare no conflict of interest.

References

1. International Human Genome Sequencing Consortium International Human Genome Sequencing Consortium. Finishing the euchromatic sequence of the human genome. *Nature* **2004**, *431*, 931–945.
2. Cordaux, R.; Batzer, M.A. The impact of retrotransposons on human genome evolution. *Nat. Rev. Genet.* **2009**, *10*, 691–703. [CrossRef] [PubMed]
3. Ohshima, K. RNA-Mediated Gene Duplication and Retroposons: Retrogenes, LINEs, SINEs, and Sequence Specificity. *Int. J. Evol. Biol.* **2013**, *2013*, 424726. [CrossRef] [PubMed]
4. Bannert, N.; Kurth, R. The evolutionary dynamics of human endogenous retroviral families. *Annu. Rev. Genom. Hum. Genet.* **2006**, *7*, 149–173. [CrossRef] [PubMed]
5. Mager, D.L.; Goodchild, N.L. Homologous recombination between the LTRs of a human retrovirus-like element causes a 5-kb deletion in two siblings. *Am. J. Hum. Genet.* **1989**, *45*, 848–854. [PubMed]
6. Blomberg, J.; Benachenhou, F.; Blikstad, V.; Sperber, G.; Mayer, J. Classification and nomenclature of endogenous retroviral sequences (ERVs): Problems and recommendations. *Gene* **2009**, *448*, 115–123. [CrossRef] [PubMed]
7. Sperber, G.; Airola, T.; Jern, P.; Blomberg, J. Automated recognition of retroviral sequences in genomic data—RetroTector©. *Nucleic Acids Res.* **2007**, *35*, 4964–4976. [CrossRef] [PubMed]
8. Vargiu, L.; Rodriguez-Tomé, P.; Sperber, G.O.; Cadeddu, M.; Grandi, N.; Blikstad, V.; Tramontano, E.; Blomberg, J. Classification and characterization of human endogenous retroviruses; mosaic forms are common. *Retrovirology* **2016**, *13*, 7. [CrossRef] [PubMed]
9. Magiorkinis, G.; Belshaw, R.; Katzourakis, A. "There and back again": Revisiting the pathophysiological roles of human endogenous retroviruses in the post-genomic era. *Philos. Trans. R. Soc. Lond. B Biol. Sci.* **2013**, *368*, 20120504. [CrossRef] [PubMed]
10. Schön, U.; Seifarth, W.; Baust, C.; Hohenadl, C.; Erfle, V.; Leib-Mösch, C. Cell Type-Specific Expression and Promoter Activity of Human Endogenous Retroviral Long Terminal Repeats. *Virology* **2001**, *279*, 280–291. [CrossRef] [PubMed]
11. Stauffer, Y.; Theiler, G.; Sperisen, P.; Lebedev, Y.; Jongeneel, C.V. Digital expression profiles of human endogenous retroviral families in normal and cancerous tissues. *Cancer Immun. Arch.* **2004**, *4*, 2.
12. Seifarth, W.; Frank, O.; Zeilfelder, U. Comprehensive analysis of human endogenous retrovirus transcriptional activity in human tissues with a retrovirus-specific microarray. *J. Virol.* **2005**, *79*, 341–352. [CrossRef] [PubMed]

13. Flockerzi, A.; Maydt, J.; Frank, O.; Ruggieri, A.; Maldener, E.; Seifarth, W.; Medstrand, P.; Lengauer, T.; Meyerhans, A.; Leib-Mösch, C.; et al. Expression pattern analysis of transcribed HERV sequences is complicated by ex vivo recombination. *Retrovirology* **2007**, *4*, 39. [CrossRef] [PubMed]
14. Hu, L. *Endogenous Retroviral RNA Expression in Humans*; Uppsala University: Uppsala, Sweden, 2007.
15. Pérot, P.; Mugnier, N.; Montgiraud, C.; Gimenez, J.; Jaillard, M.; Bonnaud, B.; Mallet, F. Microarray-based sketches of the HERV transcriptome landscape. *PLoS ONE* **2012**, *7*, e40194.
16. Haase, K.; Mösch, A.; Frishman, D. Differential expression analysis of human endogenous retroviruses based on ENCODE RNA-seq data. *BMC Med. Genom.* **2015**, *8*, 71. [CrossRef] [PubMed]
17. Balestrieri, E.; Pica, F.; Matteucci, C.; Zenobi, R.; Sorrentino, R.; Argaw-Denboba, A.; Cipriani, C.; Bucci, I.; Sinibaldi-Vallebona, P. Transcriptional activity of human endogenous retroviruses in human peripheral blood mononuclear cells. *Biomed Res. Int.* **2015**, *2015*. [CrossRef] [PubMed]
18. Voisset, C.; Weiss, R.A.; Griffiths, D.J. Human RNA "rumor" viruses: The search for novel human retroviruses in chronic disease. *Microbiol. Mol. Biol. Rev.* **2008**, *72*, 157–196, table of contents. [CrossRef] [PubMed]
19. Jern, P.; Coffin, J.M. Effects of Retroviruses on Host Genome Function. *Annu. Rev. Genet.* **2008**, *42*, 709–732. [CrossRef] [PubMed]
20. Christensen, T. Human endogenous retroviruses in neurologic disease. *Apmis* **2016**, *124*, 116–126. [CrossRef] [PubMed]
21. Grandi, N.; Cadeddu, M.; Blomberg, J.; Tramontano, E. Contribution of type W human endogenous retrovirus to the human genome: Characterization of HERV-W proviral insertions and processed pseudogenes. *Retrovirology* **2016**, *13*, 1–25. [CrossRef] [PubMed]
22. Perron, H.; Garson, J.A.; Bedin, F.; Beseme, F.; Paranhos-Baccala, G.; Komurian-Pradel, F.; Mallet, F.; Tuke, P.W.; Voisset, C.; Blond, J.L.; et al. Molecular identification of a novel retrovirus repeatedly isolated from patients with multiple sclerosis. The Collaborative Research Group on Multiple Sclerosis. *Proc. Natl. Acad. Sci. USA* **1997**, *94*, 7583–7588. [CrossRef] [PubMed]
23. Blond, J.L.; Besème, F.; Duret, L.; Bouton, O.; Bedin, F.; Perron, H.; Mandrand, B.; Mallet, F. Molecular characterization and placental expression of HERV-W, a new human endogenous retrovirus family. *J. Virol.* **1999**, *73*, 1175–1185. [PubMed]
24. Blond, J.L.; Lavillette, D.; Cheynet, V.; Bouton, O.; Oriol, G.; Chapel-Fernandes, S.; Mandrand, B.; Mallet, F.; Cosset, F.L. An envelope glycoprotein of the human endogenous retrovirus HERV-W is expressed in the human placenta and fuses cells expressing the type D mammalian retrovirus receptor. *J. Virol.* **2000**, *74*, 3321–3329. [CrossRef] [PubMed]
25. Mi, S.; Lee, X.; Li, X.; Veldman, G.M.; Finnerty, H.; Racie, L.; Lavallie, E.; Tang, X.; Edouard, P.; Howes, S.; et al. Syncytin is a captive retroviral envelope protein involved. *Nature* **2000**, *403*, 785–789. [PubMed]
26. Cheng, Y.-H. Isolation and Characterization of the Human Syncytin Gene Promoter. *Biol. Reprod.* **2003**, *70*, 694–701. [CrossRef] [PubMed]
27. Mallet, F.; Bouton, O.; Prudhomme, S.; Cheynet, V.; Oriol, G.; Bonnaud, B.; Lucotte, G.; Duret, L.; Mandrand, B. The endogenous retroviral locus ERVWE1 is a bona fide gene involved in hominoid placental physiology. *Proc. Natl. Acad. Sci. USA* **2004**, *101*, 1731–1736. [CrossRef] [PubMed]
28. Bonnaud, B.; Bouton, O.; Oriol, G.; Cheynet, V.; Duret, L.; Mallet, F. Evidence of selection on the domesticated ERVWE1 env retroviral element involved in placentation. *Mol. Biol. Evol.* **2004**, *21*, 1895–1901. [CrossRef] [PubMed]
29. Prudhomme, S.; Oriol, G.; Mallet, F. A retroviral promoter and a cellular enhancer define a bipartite element which controls env ERVWE1 placental expression. *J. Virol.* **2004**, *78*, 12157–12168. [CrossRef] [PubMed]
30. Cheynet, V.; Ruggieri, A.; Oriol, G.; Blond, J.-L.; Boson, B.; Vachot, L.; Verrier, B.; Cosset, F.-L.; Mallet, F. Synthesis, assembly, and processing of the Env ERVWE1/syncytin human endogenous retroviral envelope. *J. Virol.* **2005**, *79*, 5585–5593. [CrossRef] [PubMed]
31. Gimenez, J.; Mallet, F. ERVWE1 (endogenous retroviral family W, Env(C7), member 1). *Atlas Genet. Cytogenet. Oncol. Haematol.* **2008**, *12*, 134–148. [CrossRef]
32. Pavlícek, A.; Paces, J.; Elleder, D. Processed Pseudogenes of Human Endogenous Retroviruses Generated by LINEs: Their Integration, Stability, and Distribution. *Genome Res.* **2002**, *12*, 391–399. [CrossRef] [PubMed]
33. Costas, J. Characterization of the intragenomic spread of the human endogenous retrovirus family HERV-W. *Mol. Biol. Evol.* **2002**, *19*, 526–533. [CrossRef] [PubMed]

34. Beck, C.R.; Garcia-Perez, J.L.; Badge, R.M.; Moran, J.V. LINE-1 Elements in Structural Variation and Disease. *Annu. Rev. Genom. Hum. Genet.* **2011**, *12*, 187–215. [CrossRef] [PubMed]

35. Hancks, D.C.; Kazazian, H.H. Active human retrotransposons: Variation and disease. *Curr. Opin. Genet. Dev.* **2012**, *22*, 191–203. [CrossRef] [PubMed]

36. Richardson, S.R.; Narvaiza, I.; Planegger, R.A.; Weitzman, M.D.; Moran, J.V. APOBEC3A deaminates transiently exposed single-strand DNA during LINE-1 retrotransposition. *Elife* **2014**, *3*, e02008. [CrossRef] [PubMed]

37. Lavialle, C.; Cornelis, G.; Dupressoir, A.; Esnault, C.; Heidmann, O.; Vernochet, C.; Heidmann, T. Paleovirology of "syncytins", retroviral env genes exapted for a role in placentation. *Philos. Trans. R. Soc. Lond. B Biol. Sci.* **2013**, *368*, 20120507. [CrossRef] [PubMed]

38. Mayer, J.; Meese, E. Human endogenous retroviruses in the primate lineage and their influence on host genomes. *Cytogenet. Genome Res.* **2005**, *110*, 448–456. [CrossRef] [PubMed]

39. Lavie, L.; Maldener, E.; Brouha, B.; Meese, E.U.; Mayer, J. The human L1 promoter: Variable transcription initiation sites and a major impact of upstream flanking sequence on promoter activity. *Genome Res.* **2004**, *14*, 2253–2260. [CrossRef] [PubMed]

40. Voisset, C.; Bouton, O.; Bedin, F.; Duret, L.; Mandrand, B.; Mallet, F.; Paranhos-Baccala, G. Chromosomal distribution and coding capacity of the human endogenous retrovirus HERV-W family. *AIDS Res. Hum. Retroviruses* **2000**, *16*, 731–740. [CrossRef] [PubMed]

41. Perron, C.; Geny, A.; Laurent, C.; Mouriquand, J.; Pellat, J.; Perret, J.; Seigneurin, J. Leptomeningeal cell line from multiple sclerosis with reverse transcriptase activity and viral particles. *Res. Virol.* **1989**, *140*, 551–561. [CrossRef]

42. Perron, H.; Lalande, B.; Gratacap, B.; Laurent, A.; Genoulaz, O.; Geny, C.; Mallaret, M.; Schuller, E.; Stoebner, P.; Seigneurin, J.J.M. Isolation of retrovirus from patients with multiple sclerosis. *Lancet* **1991**, *337*, 862–863. [CrossRef]

43. Knerr, I.; Huppertz, B.; Weigel, C.; Dötsch, J.; Wich, C.; Schild, R.L.; Beckmann, M.W.; Rascher, W. Endogenous retroviral syncytin: Compilation of experimental research on syncytin and its possible role in normal and disturbed human placentogenesis. *Mol. Hum. Reprod.* **2004**, *10*, 581–588. [CrossRef] [PubMed]

44. Frendo, J.-L.; Olivier, D.; Cheynet, V.; Blond, J.-L.; Bouton, O.; Vidaud, M.; Rabreau, M.; Evain-Brion, D.; Mallet, F. Direct involvement of HERV-W Env glycoprotein in human trophoblast cell fusion and differentiation. *Mol. Cell. Biol.* **2003**, *23*, 3566–3574. [CrossRef] [PubMed]

45. Lavillette, D.; Marin, M.; Ruggieri, A.; Mallet, F.; Cosset, F.L.; Kabat, D. The envelope glycoprotein of human endogenous retrovirus type W uses a divergent family of amino acid transporters/cell surface receptors. *J. Virol.* **2002**, *76*, 6442–6452. [CrossRef] [PubMed]

46. Malassiné, A.; Handschuh, K.; Tsatsaris, V.; Gerbaud, P.; Cheynet, V.; Oriol, G.; Mallet, F.; Evain-Brion, D. Expression of HERV-W Env glycoprotein (syncytin) in the extravillous trophoblast of first trimester human placenta. *Placenta* **2005**, *26*, 556–562. [CrossRef] [PubMed]

47. Huang, Q.; Li, J.; Wang, F.; Oliver, M.T.; Tipton, T.; Gao, Y.; Jiang, S.-W. Syncytin-1 modulates placental trophoblast cell proliferation by promoting G1/S transition. *Cell. Signal.* **2013**, *25*, 1027–1035. [CrossRef] [PubMed]

48. Keryer, G.; Alsat, E.; Tasken, K.; Evain-Brion, D. Cyclic AMP-dependent protein kinases and human trophoblast cell differentiation in vitro. *J. Cell Sci.* **1998**, *111*, 995–1004. [PubMed]

49. Huang, Q.; Chen, H.; Wang, F.; Brost, B.C.; Li, J.; Li, Z.; Gao, Y.; Gao, Y.; Jiang, S.W. Reduced syncytin-1 expression in choriocarcinoma BeWo cells activates the calpain1-AIF-mediated apoptosis, implication for preeclampsia. *Cell. Mol. Life Sci.* **2014**, *71*, 3151–3164. [CrossRef] [PubMed]

50. Mangeney, M.; Heidmann, T. Tumor cells expressing a retroviral envelope escape immune rejection in vivo. *Proc. Natl. Acad. Sci. USA* **1998**, *95*, 14920–14925. [CrossRef] [PubMed]

51. Blaise, S.; Mangeney, M.; Heidmann, T. The envelope of Mason-Pfizer monkey virus has immunosuppressive properties. *J. Gen. Virol.* **2001**, *82*, 1597–1600. [CrossRef] [PubMed]

52. Mangeney, M.; Renard, M.; Schlecht-Louf, G.; Bouallaga, I.; Heidmann, O.; Letzelter, C.; Richaud, A.; Ducos, B.; Heidmann, T. Placental syncytins: Genetic disjunction between the fusogenic and immunosuppressive activity of retroviral envelope proteins. *Proc. Natl. Acad. Sci. USA* **2007**, *104*, 20534–20539. [CrossRef] [PubMed]

53. Tolosa, J.M.; Schjenken, J.E.; Clifton, V.L.; Vargas, A.; Barbeau, B.; Lowry, P.; Maiti, K.; Smith, R. The endogenous retroviral envelope protein syncytin-1 inhibits LPS/PHA-stimulated cytokine responses in human blood and is sorted into placental exosomes. *Placenta* **2012**, *33*, 933–941. [CrossRef] [PubMed]

54. Li, F.; Karlsson, H. Expression and regulation of human endogenous retrovirus W elements. *Apmis* **2016**, *124*, 52–66. [CrossRef] [PubMed]

55. Yi, J.; Kim, H.; Kim, H. Expression of the human endogenous retrovirus HERV-W family in various human tissues and cancer cells. *J. Gen. Virol.* **2004**, *85*, 1203–1210. [CrossRef] [PubMed]

56. Kim, H.S.; Ahn, K.; Kim, D.S. Quantitative expression of the HERV-W env gene in human tissues. *Arch. Virol.* **2008**, *153*, 1587–1591. [CrossRef] [PubMed]

57. Hu, L.; Hornung, D.; Kurek, R.; Ostman, H.; Helen, O.; Blomberg, J.; Bergqvist, A. Expression of human endogenous gammaretroviral sequences in endometriosis and ovarian cancer. *AIDS Res. Hum. Retroviruses* **2006**, *22*, 551–557. [CrossRef] [PubMed]

58. Nellåker, C.; Wållgren, U.; Karlsson, H. Molecular beacon-based temperature control and automated analyses for improved resolution of melting temperature analysis using SYBR I Green chemistry. *Clin. Chem.* **2007**, *53*, 98–103. [CrossRef] [PubMed]

59. Nellåker, C.; Li, F.; Uhrzander, F.; Tyrcha, J.; Karlsson, H. Expression profiling of repetitive elements by melting temperature analysis: Variation in HERV-W gag expression across human individuals and tissues. *BMC Genom.* **2009**, *10*, 532.

60. Roland, C.S.; Hu, J.; Ren, C.E.; Chen, H.; Li, J.; Varvoutis, M.S.; Leaphart, L.W.; Byck, D.B.; Zhu, X.; Jiang, S.W. Morphological changes of placental syncytium and their implications for the pathogenesis of preeclampsia. *Cell. Mol. Life Sci.* **2016**, *73*, 365–376. [CrossRef] [PubMed]

61. Vargas, A.; Toufaily, C.; LeBellego, F.; Rassart, É.; Lafond, J.; Barbeau, B. Reduced expression of both syncytin 1 and syncytin 2 correlates with severity of preeclampsia. *Reprod. Sci.* **2011**, *18*, 1085–1091. [CrossRef] [PubMed]

62. Lee, X.; Keith, J.C.; Stumm, N.; Moutsatsos, I.; McCoy, J.M.; Crum, C.P.; Genest, D.; Chin, D.; Ehrenfels, C.; Pijnenborg, R.; et al. Downregulation of placental syncytin expression and abnormal protein localization in pre-eclampsia. *Placenta* **2001**, *22*, 808–812. [CrossRef] [PubMed]

63. Knerr, I.; Beinder, E.; Rascher, W. Syncytin, a novel human endogenous retroviral gene in human placenta: Evidence for its dysregulation in preeclampsia and HELLP syndrome. *Am. J. Obstet. Gynecol.* **2002**, *186*, 210–213. [CrossRef] [PubMed]

64. Holder, B.S.; Tower, C.L.; Abrahams, V.M.; Aplin, J.D. Syncytin 1 in the human placenta. *Placenta* **2012**, *33*, 460–466. [CrossRef] [PubMed]

65. Zhuang, X.-W.; Li, J.; Brost, B.C.; Xia, X.-Y.; Chen, H.B.; Wang, C.-X.; Jiang, S.-W. Decreased expression and altered methylation of syncytin-1 gene in human placentas associated with preeclampsia. *Curr. Pharm. Des.* **2014**, *20*, 1796–1802. [CrossRef] [PubMed]

66. Knerr, I.; Weigel, C.; Linnemann, K.; Dotsch, J.; Meissner, U.; Fusch, C.; Rascher, W. Transcriptional effects of hypoxia on fusiogenic syncytin and its receptor ASCT2 in human cytotrophoblast BeWo cells and in ex vivo perfused placental cotyledons. *Am. J. Obstet. Gynecol.* **2003**, *189*, 583–588. [CrossRef]

67. Muir, A.; Lever, A.; Moffett, A. Expression and functions of human endogenous retroviruses in the placenta: An update. *Placenta* **2004**, *25* (Suppl. A), S16–S25. [CrossRef] [PubMed]

68. Frendo, J.L.; Vidaud, M.; Guibourdenche, J.; Luton, D.; Muller, F.; Bellet, D.; Giovagrandi, Y.; Tarrade, A.; Porquet, D.; Blot, P.; et al. Defect of villous cytotrophoblast differentiation into syncytiotrophoblast in Down's syndrome. *J. Clin. Endocrinol. Metab.* **2000**, *85*, 3700–3707. [CrossRef] [PubMed]

69. Massin, N.; Frendo, J.L.; Guibourdenche, J.; Luton, D.; Giovangrandi, Y.; Muller, F.; Vidaud, M.; Evain-Brion, D. Defect of syncytiotrophoblast formation and human chorionic gonadotropin expression in Down's syndrome. *Placenta* **2001**, *22*, S93–S97. [CrossRef] [PubMed]

70. Malassiné, A.; Frendo, J.L.; Evain-Brion, D. Trisomy 21—Affected placentas highlight prerequisite factors for human trophoblast fusion and differentiation. *Int. J. Dev. Biol.* **2010**, *54*, 475–482. [CrossRef] [PubMed]

71. Frendo, J.L.; Thérond, P.; Bird, T.; Massin, N.; Muller, F.; Guibourdenche, J.; Luton, D.; Vidaud, M.; Anderson, W.B.; Evain-Brion, D. Overexpression of copper zinc superoxide dismutase impairs human trophoblast cell fusion and differentiation. *Endocrinology* **2001**, *142*, 3638–3648. [CrossRef] [PubMed]

72. Ruebner, M.; Strissel, P.L.; Langbein, M.; Fahlbusch, F.; Wachter, D.L.; Faschingbauer, F.; Beckmann, M.W.; Strick, R. Impaired cell fusion and differentiation in placentae from patients with intrauterine growth restriction correlate with reduced levels of HERV envelope genes. *J. Mol. Med.* **2010**, *88*, 1143–1156. [CrossRef] [PubMed]

73. Zhou, H.; Li, J.; Podratz, K.C.; Tipton, T.; Marzolf, S.; Chen, H.B.; Jiang, S.-W. Hypomethylation and activation of syncytin-1 gene in endometriotic tissue. *Curr. Pharm. Des.* **2014**, *20*, 1786–1795. [CrossRef] [PubMed]

74. Menendez, L.; Benigno, B.B.; McDonald, J.F. L1 and HERV-W retrotransposons are hypomethylated in human ovarian carcinomas. *Mol. Cancer* **2004**, *3*, 12. [CrossRef] [PubMed]

75. Strick, R.; Ackermann, S.; Langbein, M.; Swiatek, J.; Schubert, S.W.; Hashemolhosseini, S.; Koscheck, T.; Fasching, P.A.; Schild, R.L.; Beckmann, M.W.; et al. Proliferation and cell-cell fusion of endometrial carcinoma are induced by the human endogenous retroviral Syncytin-1 and regulated by TGF-β. *J. Mol. Med.* **2007**, *85*, 23–38. [CrossRef] [PubMed]

76. Bjerregaard, B.; Holck, S.; Christensen, I.J.; Larsson, L.I. Syncytin is involved in breast cancer-endothelial cell fusions. *Cell. Mol. Life Sci.* **2006**, *63*, 1906–1911. [CrossRef] [PubMed]

77. Díaz-Carballo, D.; Acikelli, A.H.; Klein, J.; Jastrow, H.; Dammann, P.; Wyganowski, T.; Guemues, C.; Gustmann, S.; Bardenheuer, W.; Malak, S.; et al. Therapeutic potential of antiviral drugs targeting chemorefractory colorectal adenocarcinoma cells overexpressing endogenous retroviral elements. *J. Exp. Clin. Cancer Res.* **2015**, *34*, 81. [CrossRef] [PubMed]

78. Hu, L.; Uzhameckis, D.; Hedborg, F.; Blomberg, J. Dynamic and selective HERV RNA expression in neuroblastoma cells subjected to variation in oxygen tension and demethylation. *Apmis* **2016**, *124*, 140–149. [CrossRef] [PubMed]

79. Li, S.; Liu, Z.C.; Yin, S.J.; Chen, Y.T.; Yu, H.L.; Zeng, J.; Zhang, Q.; Zhu, F. Human endogenous retrovirus W family envelope gene activates the small conductance Ca^{2+}-activated K^+ channel in human neuroblastoma cells through CREB. *Neuroscience* **2013**, *247*, 164–174. [CrossRef] [PubMed]

80. Gimenez, J.; Montgiraud, C.; Pichon, J.-P.; Bonnaud, B.; Arsac, M.; Ruel, K.; Bouton, O.; Mallet, F. Custom human endogenous retroviruses dedicated microarray identifies self-induced HERV-W family elements reactivated in testicular cancer upon methylation control. *Nucleic Acids Res.* **2010**, *38*, 2229–2246. [CrossRef] [PubMed]

81. Maliniemi, P.; Vincendeau, M.; Mayer, J.; Frank, O.; Hahtola, S.; Karenko, L.; Carlsson, E.; Mallet, F.; Seifarth, W.; Leib-Mösch, C.; et al. Expression of human endogenous retrovirus-w including syncytin-1 in cutaneous T-cell lymphoma. *PLoS ONE* **2013**, *8*, e76281. [CrossRef] [PubMed]

82. Yu, H.; Liu, T.; Zhao, Z.; Chen, Y.; Zeng, J.; Liu, S.; Zhu, F. Mutations in 3'-long terminal repeat of HERV-W family in chromosome 7 upregulate syncytin-1 expression in urothelial cell carcinoma of the bladder through interacting with c-Myb. *Oncogene* **2014**, *33*, 3947–3958. [CrossRef] [PubMed]

83. Ruprecht, K.; Mayer, J.; Sauter, M.; Roemer, K.; Mueller-Lantzsch, N. Endogenous retroviruses and cancer. *Cell. Mol. Life Sci.* **2008**, *65*, 3366–3382. [CrossRef] [PubMed]

84. Balada, E.; Ordi-Ros, J.; Vilardell-Tarrés, M. Molecular mechanisms mediated by Human Endogenous Retroviruses (HERVs) in autoimmunity. *Rev. Med. Virol.* **2009**, *19*, 273–286. [CrossRef] [PubMed]

85. Balada, E.; Vilardell-Tarrés, M.; Ordi-Ros, J. Implication of human endogenous retroviruses in the development of autoimmune diseases. *Int. Rev. Immunol.* **2010**, *29*, 351–370. [CrossRef] [PubMed]

86. Trela, M.; Nelson, P.N.; Rylance, P.B. The role of molecular mimicry and other factors in the association of Human Endogenous Retroviruses and autoimmunity. *APMIS* **2016**, *124*, 88–104. [CrossRef] [PubMed]

87. Brodziak, A.; Ziółko, E.; Muc-Wierzgoń, M.; Nowakowska-zajdel, E.; Kokot, T.; Klakla, K. The role of human endogenous retroviruses in the pathogenesis of autoimmune diseases. *Med. Sci. Monit.* **2012**, *18*, RA80–RA88. [PubMed]

88. Volkman, H.E.; Stetson, D.B. The enemy within: Endogenous retroelements and autoimmune disease. *Nat. Immunol.* **2014**, *15*, 415–422. [CrossRef] [PubMed]

89. Hurst, T.P.; Magiorkinis, G. Activation of the innate immune response by endogenous retroviruses. *J. Gen. Virol.* **2015**, *96*, 1207–1218. [CrossRef] [PubMed]

90. Query, C.C.; Keene, J.D. A human autoimmune protein associated with U1 RNA contains a region of homology that is cross-reactive with retroviral p30gag antigen. *Cell* **1987**, *51*, 211–220. [CrossRef]

91. Talal, N.; Flescher, E.; Dang, H. Are endogenous retroviruses involved in human autoimmune disease? *J. Autoimmun.* **1992**, *5* (Suppl. A), 61–66. [CrossRef]

92. Nelson, P.N.; Lever, A.M.; Bruckner, F.E.; Isenberg, D.A.; Kessaris, N.; Hay, F.C. Polymerase chain reaction fails to incriminate exogenous retroviruses HTLV-I and HIV-1 in rheumatological diseases although a minority of sera cross react with retroviral antigens. *Ann. Rheum. Dis.* **1994**, *53*, 749–754. [CrossRef] [PubMed]

93. Mason, A.L.; Xu, L.; Guo, L.; Garry, R.F. Retroviruses in autoimmune liver disease: Genetic or environmental agents? *Arch. Immunol. Ther. Exp.* **1999**, *47*, 289–297.

94. Sun, B.; Hu, L.; Luo, Z.Y.; Chen, X.P.; Zhou, H.H.; Zhang, W. DNA methylation perspectives in the pathogenesis of autoimmune diseases. *Clin. Immunol.* **2016**, *164*, 21–27. [CrossRef] [PubMed]

95. Antony, J.M.; Deslauriers, A.M.; Bhat, R.K.; Ellestad, K.K.; Power, C. Human endogenous retroviruses and multiple sclerosis: Innocent bystanders or disease determinants? *Biochim. Biophys. Acta* **2011**, *1812*, 162–176. [CrossRef] [PubMed]

96. Lafon, M.; Jouvin-Marche, E.; Marche, P.N.; Perron, H.; Woodland, D.L. Human viral superantigens: To be or not to be transactivated? *Trends Immunol.* **2002**, *23*, 238–239. [CrossRef]

97. Christensen, T. Association of human endogenous retroviruses with multiple sclerosis and possible interactions with herpes viruses. *Rev. Med. Virol.* **2005**, *15*, 179–211. [CrossRef] [PubMed]

98. Perron, H.; Bernard, C.; Bertrand, J.-B.; Lang, A.; Popa, I.; Sanhadji, K.; Portoukalian, J. Endogenous retroviral genes, Herpesviruses and gender in Multiple Sclerosis. *J. Neurol. Sci.* **2009**, *286*, 65–72. [CrossRef] [PubMed]

99. Krone, B.; Grange, J.M. Multiple Sclerosis: Are Protective Immune Mechanisms Compromised by a Complex Infectious Background? *Autoimmune Dis.* **2011**, *2011*, 1–8. [CrossRef] [PubMed]

100. Libbey, J.E.; Cusick, M.F.; Fujinami, R.S. Role of Pathogens in Multiple Sclerosis. *Int. Rev. Immunol.* **2013**, *33*, 1–18. [CrossRef] [PubMed]

101. Morandi, E.; Tarlinton, R.E.; Gran, B. Multiple sclerosis between genetics and infections: Human endogenous retroviruses in monocytes and macrophages. *Front. Immunol.* **2015**, *6*, 1–6. [CrossRef] [PubMed]

102. Alliel, P.M.; Périn, J.-P.; Belliveau, J.; Pierig, R.; Nussbaum, J.-L.; Rieger, F. Multiple sclerosis: Clues on the retroviral hypothesis in the human genome (part 1). *Comptes Rendus l'Académie des Sci. Ser. III Sci. la Vie* **1998**, *321*, 495–499. [CrossRef]

103. Alliel, P.M.; Perin, J.P.; Pierig, R.; Rieger, F. An endogenous retrovirus with nucleic acid sequences similar to those of the multiple sclerosis associated retrovirus at the human T-cell receptor alpha, delta gene locus. *Cell. Mol. Biol. (Noisy-le-grand)* **1998**, *44*, 927–931.

104. Haahr, S.; Sommerlund, M.; Møller-Larsen, A.; Nielsen, R.; Hansen, H. Just another dubious virus in cells from a patient with multiple sclerosis? *Lancet* **1991**, *337*, 863–864. [CrossRef]

105. Perron, H.; Firouzi, R.; Tuke, P.; Garson, J.A.; Michel, M.; Beseme, F.; Bedin, F.; Mallet, F.; Marcel, E.; Seigneurin, J.M.; et al. Cell cultures and associated retroviruses in multiple sclerosis. *Acta Neurol. Scand. Suppl.* **1997**, *169*, 22–31. [CrossRef] [PubMed]

106. Komurian-Pradel, F.; Paranhos-Baccala, G.; Bedin, F.; Ounanian-Paraz, A.; Sodoyer, M.; Ott, C.; Rajoharison, A.; Garcia, E.; Mallet, F.; Mandrand, B.; et al. Molecular cloning and characterization of MSRV-related sequences associated with retrovirus-like particles. *Virology* **1999**, *260*, 1–9. [CrossRef] [PubMed]

107. Mameli, G.; Astone, V.; Arru, G.; Marconi, S.; Lovato, L.; Serra, C.; Sotgiu, S.; Bonetti, B.; Dolei, A. Brains and peripheral blood mononuclear cells of multiple sclerosis (MS) patients hyperexpress MS-associated retrovirus/HERV-W endogenous retrovirus, but not Human herpesvirus 6. *J. Gen. Virol.* **2007**, *88*, 264–274. [CrossRef] [PubMed]

108. Mameli, G.; Poddighe, L.; Astone, V.; Delogu, G.; Arru, G.; Sotgiu, S.; Serra, C.; Dolei, A. Novel reliable real-time PCR for differential detection of MSRVenv and syncytin-1 in RNA and DNA from patients with multiple sclerosis. *J. Virol. Methods* **2009**, *161*, 98–106. [CrossRef] [PubMed]

109. Dolei, A.; Perron, H. The multiple sclerosis-associated retrovirus and its HERV-W endogenous family: A biological interface between virology, genetics, and immunology in human physiology and disease. *J. Neurovirol.* **2009**, *15*, 4–13. [CrossRef] [PubMed]

110. Garcia-Montojo, M.; Dominguez-Mozo, M.; Arias-Leal, A.; Garcia-Martinez, Á.; de las Heras, V.; Casanova, I.; Faucard, R.; Gehin, N.; Madeira, A.; Arroyo, R.; et al. The DNA Copy Number of Human Endogenous Retrovirus-W (MSRV-Type) Is Increased in Multiple Sclerosis Patients and Is Influenced by Gender and Disease Severity. *PLoS ONE* **2013**, *8*, e53623. [CrossRef] [PubMed]

111. Blomberg, J.; Ushameckis, D.; Jern, P. Evolutionary Aspects of Human Endogenous Retroviral Sequences (HERVs) and Disease. In *Madame Curie Bioscience Database*; Landes Bioscience: Austin, TX, USA, 2000.

112. Ruprecht, K.; Gronen, F.; Sauter, M.; Best, B.; Rieckmann, P.; Mueller-Lantzsch, N. Lack of immune responses against multiple sclerosis-associated retrovirus/human endogenous retrovirus W in patients with multiple sclerosis. *J. Neurovirol.* **2008**, *14*, 143–151. [CrossRef] [PubMed]

113. Schmitt, K.; Richter, C.; Backes, C.; Meese, E.; Ruprecht, K.; Mayer, J. Comprehensive analysis of human endogenous retrovirus group HERV-W locus transcription in multiple sclerosis brain lesions by high-throughput amplicon sequencing. *J. Virol.* **2013**, *87*, 13837–13852. [CrossRef] [PubMed]

114. Johnston, J.B.; Silva, C.; Holden, J.; Warren, K.G.; Clark, A.W.; Power, C. Monocyte activation and differentiation augment human endogenous retrovirus expression: Implications for inflammatory brain diseases. *Ann. Neurol.* **2001**, *50*, 434–442. [CrossRef] [PubMed]

115. Dolei, A.; Serra, C.; Mameli, G.; Pugliatti, M.; Sechi, G.; Cirotto, M.C.; Rosati, G.; Sotgiu, S. Multiple sclerosis-associated retrovirus (MSRV) in Sardinian MS patients. *Neurology* **2002**, *58*, 471–473. [CrossRef] [PubMed]

116. Nowak, J.; Januszkiewicz, D.; Pernak, M.; Liwen, I.I.; Zawada, M.; Rembowska, J.; Nowicka, K.; Lewandowski, K.; Hertmanowska, H.; Wender, M. Multiple sclerosis-associated virus-related pol sequences found both in multiple sclerosis and healthy donors are more frequently expressed in multiple sclerosis patients. *J. Neurovirol.* **2003**, *9*, 112–117. [CrossRef] [PubMed]

117. Garson, J.; Tuke, P.; Giraud, P.; Paranhos-Baccala, G.; Perron, H. Detection of virion-associated MSRV-RNA in serum of patients with multiple sclerosis. *Lancet* **1998**, *351*, 33. [CrossRef]

118. Antony, J.M.; van Marle, G.; Opii, W.; Butterfield, D.A.; Mallet, F.; Yong, V.W.; Wallace, J.L.; Deacon, R.M.; Warren, K.; Power, C. Human endogenous retrovirus glycoprotein-mediated induction of redox reactants causes oligodendrocyte death and demyelination. *Nat. Neurosci.* **2004**, *7*, 1088–1095. [CrossRef] [PubMed]

119. Antony, J.M.; Zhu, Y.; Izad, M.; Warren, K.G.; Vodjgani, M.; Mallet, F.; Power, C. Comparative Expression of Human Endogenous Retrovirus-W Genes in Multiple Sclerosis. *AIDS Res. Hum. Retroviruses* **2007**, *23*, 1251–1256. [CrossRef] [PubMed]

120. Perron, H.; Germi, R.; Bernard, C.; Garcia-Montojo, M.; Deluen, C.; Farinelli, L.; Faucard, R.; Veas, F.; Stefas, I.; Fabriek, B.O.; et al. Human endogenous retrovirus type W envelope expression in blood and brain cells provides new insights into multiple sclerosis disease. *Mult. Scler.* **2012**, *18*, 1721–1736. [CrossRef] [PubMed]

121. Arru, G.; Mameli, G.; Astone, V.; Serra, C.; Huang, Y.-M.; Link, H.; Fainardi, E.; Castellazzi, M.; Granieri, E.; Fernandez, M.; et al. Multiple Sclerosis and HERV-W/MSRV: A Multicentric Study. *Int. J. Biomed. Sci.* **2007**, *3*, 292–297. [PubMed]

122. Zawada, M.; Liwén, I.; Pernak, M.; Januszkiewicz-Lewandowska, D.; Nowicka-Kujawska, K.; Rembowska, J.; Lewandowski, K.; Hertmanowska, H.; Wender, M.; Nowak, J. MSRV pol sequence copy number as a potential marker of multiple sclerosis. *Pol. J. Pharmacol.* **2003**, *55*, 869–875. [PubMed]

123. Perron, H.; Lazarini, F.; Ruprecht, K.; Péchoux-Longin, C.; Seilhean, D.; Sazdovitch, V.; Créange, A.; Battail-Poirot, N.; Sibaï, G.; Santoro, L.; et al. Human endogenous retrovirus (HERV)-W ENV and GAG proteins: Physiological expression in human brain and pathophysiological modulation in multiple sclerosis lesions. *J. Neurovirol.* **2005**, *11*, 23–33. [CrossRef] [PubMed]

124. Brudek, T.; Christensen, T.; Aagaard, L.; Petersen, T.; Hansen, H.J.; Møller-Larsen, A. B cells and monocytes from patients with active multiple sclerosis exhibit increased surface expression of both HERV-H Env and HERV-W Env, accompanied by increased seroreactivity. *Retrovirology* **2009**, *6*, 104. [CrossRef] [PubMed]

125. Van Horssen, J.; Van Der Pol, S.; Nijland, P.; Amor, S.; Perron, H. Human endogenous retrovirus W in brain lesions: Rationale for targeted therapy in multiple sclerosis. *Mult. Scler. Relat. Disord.* **2016**, *8*, 11–18. [CrossRef] [PubMed]

126. Perron, H.; Jouvin-Marche, E.; Michel, M.; Ounanian-Paraz, A.; Camelo, S.; Dumon, A.; Jolivet-Reynaud, C.; Marcel, F.; Souillet, Y.; Borel, E.; et al. Multiple Sclerosis Retrovirus Particles and Recombinant Envelope Trigger an Abnormal Immune Response in Vitro, by Inducing Polyclonal Vβ16 T-Lymphocyte Activation. *Virology* **2001**, *287*, 321–332. [CrossRef] [PubMed]

127. Rolland, A.; Jouvin-Marche, E.; Saresella, M.; Ferrante, P.; Cavaretta, R.; Créange, A.; Marche, P.; Perron, H. Correlation between disease severity and in vitro cytokine production mediated by MSRV (Multiple Sclerosis associated RetroViral element) envelope protein in patients with multiple sclerosis. *J. Neuroimmunol.* **2005**, *160*, 195–203. [CrossRef] [PubMed]

128. Rolland, A.; Jouvin-Marche, E.; Viret, C.; Faure, M.; Perron, H.; Marche, P.N. The Envelope Protein of a Human Endogenous Retrovirus-W Family Activates Innate Immunity through CD14/TLR4 and Promotes Th1-Like Responses. *J. Immunol.* **2006**, *176*, 7636–7644. [CrossRef] [PubMed]

129. Saresella, M.; Rolland, A.; Marventano, I.; Cavarretta, R.; Caputo, D.; Marche, P.; Perron, H.; Clerici, M. Multiple sclerosis-associated retroviral agent (MSRV)-stimulated cytokine production in patients with relapsing-remitting multiple sclerosis. *Mult. Scler.* **2009**, *15*, 443–447. [CrossRef] [PubMed]

130. Mameli, G.; Astone, V.; Khalili, K.; Serra, C.; Sawaya, B.E.; Dolei, A. Regulation of the syncytin-1 promoter in human astrocytes by multiple sclerosis-related cytokines. *Virology* **2007**, *362*, 120–130. [CrossRef] [PubMed]

131. Kremer, D.; Schichel, T.; Förster, M.; Tzekova, N.; Bernard, C.; Van Der Valk, P.; Van Horssen, J.; Hartung, H.P.; Perron, H.; Küry, P. Human endogenous retrovirus type W envelope protein inhibits oligodendroglial precursor cell differentiation. *Ann. Neurol.* **2013**, *74*, 721–732. [CrossRef] [PubMed]

132. Kremer, D.; Förster, M.; Schichel, T.; Göttle, P.; Hartung, H.-P.; Perron, H.; Küry, P. The neutralizing antibody GNbAC1 abrogates HERV-W envelope protein-mediated oligodendroglial maturation blockade. *Mult. Scler.* **2014**, *21*, 1–4. [CrossRef] [PubMed]

133. Madeira, A.; Burgelin, I.; Perron, H.; Curtin, F.; Lang, A.B.; Faucard, R. MSRV envelope protein is a potent, endogenous and pathogenic agonist of human toll-like receptor 4: Relevance of GNbAC1 in multiple sclerosis treatment. *J. Neuroimmunol.* **2016**, *291*, 29–38. [CrossRef] [PubMed]

134. Petersen, T.; Møller-Larsen, A.; Thiel, S.; Brudek, T.; Hansen, T.K.; Christensen, T. Effects of interferon-beta therapy on innate and adaptive immune responses to the human endogenous retroviruses HERV-H and HERV-W, cytokine production, and the lectin complement activation pathway in multiple sclerosis. *J. Neuroimmunol.* **2009**, *215*, 108–116. [CrossRef] [PubMed]

135. Mameli, G.; Serra, C.; Astone, V.; Castellazzi, M.; Poddighe, L.; Fainardi, E.; Neri, W.; Granieri, E.; Dolei, A. Inhibition of multiple-sclerosis-associated retrovirus as biomarker of interferon therapy. *J. Neurovirol.* **2008**, *14*, 73–77. [CrossRef] [PubMed]

136. Mameli, G.; Cossu, D.; Cocco, E.; Frau, J.; Marrosu, M.G.; Niegowska, M.; Sechi, L.A. Epitopes of HERV-Wenv induce antigen-specific humoral immunity in multiple sclerosis patients. *J. Neuroimmunol.* **2015**, *280*, 66–68. [CrossRef] [PubMed]

137. Firouzi, R.; Rolland, A.; Michel, M.; Jouvin-Marche, E.; Hauw, J.; Malcus-Vocanson, C.; Lazarini, F.; Gebuhrer, L.; Seigneurin, J.; Touraine, J.; et al. Multiple sclerosis–associated retrovirus particles cause T lymphocyte–dependent death with brain hemorrhage in humanized SCID mice model. *J. Neurovirol.* **2003**, *9*, 79–93. [CrossRef] [PubMed]

138. Antony, J.M.; Ellestad, K.K.; Hammond, R.; Imaizumi, K.; Mallet, F.; Warren, K.G.; Power, C. The human endogenous retrovirus envelope glycoprotein, syncytin-1, regulates neuroinflammation and its receptor expression in multiple sclerosis: A role for endoplasmic reticulum chaperones in astrocytes. *J. Immunol.* **2007**, *179*, 1210–1224. [CrossRef] [PubMed]

139. Perron, H.; Dougier-Reynaud, H.-L.; Lomparski, C.; Popa, I.; Firouzi, R.; Bertrand, J.-B.; Marusic, S.; Portoukalian, J.; Jouvin-Marche, E.; Villiers, C.L.; et al. Human Endogenous Retrovirus Protein Activates Innate Immunity and Promotes Experimental Allergic Encephalomyelitis in Mice. *PLoS ONE* **2013**, *8*, e80128. [CrossRef] [PubMed]

140. Nellaker, C.; Yao, Y.; Jones-Brando, L.; Mallet, F.; Yolken, R.H.; Karlsson, H. Transactivation of elements in the human endogenous retrovirus W family by viral infection. *Retrovirology* **2006**, *3*, 44. [CrossRef] [PubMed]

141. Laufer, G.; Mayer, J.; Mueller, B.F.; Mueller-Lantzsch, N.; Ruprecht, K. Analysis of transcribed human endogenous retrovirus W env loci clarifies the origin of multiple sclerosis-associated retrovirus env sequences. *Retrovirology* **2009**, *6*, 37. [CrossRef] [PubMed]

142. Roebke, C.; Wahl, S.; Laufer, G.; Stadelmann, C.; Sauter, M.; Mueller-Lantzsch, N.; Mayer, J.; Ruprecht, K. An N-terminally truncated envelope protein encoded by a human endogenous retrovirus W locus on chromosome Xq22.3. *Retrovirology* **2010**, *7*, 69. [CrossRef] [PubMed]

143. Do Olival, G.S.; Faria, T.S.; Nali, L.H.S.; de Oliveira, A.C.P.; Casseb, J.; Vidal, J.E.; Cavenaghi, V.B.; Tilbery, C.P.; Moraes, L.; Fink, M.C.S.; et al. Genomic analysis of ERVWE2 locus in patients with multiple sclerosis: Absence of genetic association but potential role of human endogenous retrovirus type W elements in molecular mimicry with myelin antigen. *Front. Microbiol.* **2013**, *4*, 1–7. [CrossRef] [PubMed]

144. García-Montojo, M.; de la Hera, B.; Varadé, J.; de la Encarnación, A.; Camacho, I.; Domínguez-Mozo, M.; Arias-Leal, A.; García-Martínez, A.; Casanova, I.; Izquierdo, G.; et al. HERV-W polymorphism in chromosome X is associated with multiple sclerosis risk and with differential expression of MSRV. *Retrovirology* **2014**, *11*, 2. [CrossRef] [PubMed]

145. Varadé, J.; García-Montojo, M.; de la Hera, B.; Camacho, I.; García-Martnez, M.Á.; Arroyo, R.; Álvarez-Lafuente, R.; Urcelay, E. Multiple sclerosis retrovirus-like envelope gene: Role of the chromosome 20 insertion. *BBA Clin.* **2015**, *3*, 162–167. [CrossRef] [PubMed]

146. Bhat, R.K.; Ellestad, K.K.; Wheatley, B.M.; Warren, R.; Holt, R.A.; Power, C. Age- and disease-dependent HERV-W envelope allelic variation in brain: Association with neuroimmune gene expression. *PLoS ONE* **2011**, *6*, e19176. [CrossRef] [PubMed]

147. Hon, G.M.; Erasmus, R.T.; Matsha, T. Multiple sclerosis-associated retrovirus and related human endogenous retrovirus-W in patients with multiple sclerosis: A literature review. *J. Neuroimmunol.* **2013**, *263*, 8–12. [CrossRef] [PubMed]

148. P. Ryan, F. Human Endogenous Retroviruses in Multiple Sclerosis: Potential for Novel Neuro-Pharmacological Research. *Curr. Neuropharmacol.* **2011**, *9*, 360–369.

149. Curtin, F.; Perron, H.; Kromminga, A.; Porchet, H.; Lang, A.B. Preclinical and early clinical development of GNbAC1, a humanized IgG4 monoclonal antibody targeting endogenous retroviral MSRV-Env protein. *MAbs* **2015**, *7*, 265–275. [CrossRef] [PubMed]

150. Curtin, F.; Perron, H.; Faucard, R.; Porchet, H.; Lang, A.B. Treatment Against Human Endogenous Retrovirus: A Possible Personalized Medicine Approach for Multiple Sclerosis. *Mol. Diagn. Ther.* **2015**, *19*, 255–265. [CrossRef] [PubMed]

151. Gaudin, P.; Ijaz, S.; Tuke, P.W.; Marcel, F.; Paraz, A.; Seigneurin, J.M.; Mandrand, B.; Perron, H.; Garson, J.A. Infrequency of detection of particle-associated MSRV/HERV-W RNA in the synovial fluid of patients with rheumatoid arthritis. *Rheumatology (Oxford)* **2000**, *39*, 950–954. [CrossRef] [PubMed]

152. Bendiksen, S.; Martinez-Zubiavrra, I.; Tümmler, C.; Knutsen, G.; Elvenes, J.; Olsen, E.; Olsen, R.; Moens, U. Human Endogenous Retrovirus W Activity in Cartilage of Osteoarthritis Patients. *BioMed Res. Int.* **2014**, *2014*, 1–14. [CrossRef] [PubMed]

153. Faucard, R.; Madeira, A.; Gehin, N.; Authier, F.-J.; Panaite, P.-A.; Lesage, C.; Burgelin, I.; Bertel, M.; Bernard, C.; Curtin, F.; et al. Human Endogenous Retrovirus and Neuroinflammation in Chronic Inflammatory Demyelinating Polyradiculoneuropathy. *EBioMedicine* **2016**, *6*, 190–198. [CrossRef] [PubMed]

154. Molès, J.P.; Tesniere, A.; Guilhou, J.J. A new endogenous retroviral sequence is expressed in skin of patients with psoriasis. *Br. J. Dermatol.* **2005**, *153*, 83–89. [CrossRef] [PubMed]

155. Hohenadl, C.; Germaier, H.; Walchner, M.; Hagenhofer, M.; Herrmann, M.; Stürzl, M.; Kind, P.; Hehlmann, R.; Erfle, V.; Leib-Mösch, C. Transcriptional activation of endogenous retroviral sequences in human epidermal keratinocytes by UVB irradiation. *J. Investig. Dermatol.* **1999**, *113*, 587–594. [CrossRef] [PubMed]

156. Schanab, O.; Humer, J.; Gleiss, A.; Mikula, M.; Sturlan, S.; Grunt, S.; Okamoto, I.; Muster, T.; Pehamberger, H.; Waltenberger, A. Expression of human endogenous retrovirus K is stimulated by ultraviolet radiation in melanoma. *Pigment Cell Melanoma Res.* **2011**, *24*, 656–665. [CrossRef] [PubMed]

157. De Sousa Nogueira, M.A.; Biancardi Gavioli, C.F.; Pereira, N.Z.; de Carvalho, G.C.; Domingues, R.; Aoki, V.; Sato, M.N. Human endogenous retrovirus expression is inversely related with the up-regulation of interferon-inducible genes in the skin of patients with lichen planus. *Arch. Dermatol. Res.* **2015**, *307*, 259–264. [CrossRef] [PubMed]

158. Weis, S.; Llenos, I.C.; Sabunciyan, S.; Dulay, J.R.; Isler, L.; Yolken, R.; Perron, H. Reduced expression of human endogenous retrovirus (HERV)-W GAG protein in the cingulate gyrus and hippocampus in schizophrenia, bipolar disorder, and depression. *J. Neural Transm.* **2007**, *114*, 645–655. [CrossRef] [PubMed]

159. Diem, O.; Schäffner, M.; Seifarth, W.; Leib-Mösch, C. Influence of antipsychotic drugs on human endogenous retrovirus (HERV) transcription in brain cells. *PLoS ONE* **2012**, *7*, e30054. [CrossRef] [PubMed]

160. Oluwole, S.O.A.; Yao, Y.; Conradi, S.; Kristensson, K.; Karlsson, H. Elevated levels of transcripts encoding a human retroviral envelope protein (syncytin) in muscles from patients with motor neuron disease. *Amyotroph. Lateral Scler.* **2007**, *8*, 67–72. [CrossRef] [PubMed]

161. Jeong, B.-H.; Lee, Y.-J.; Carp, R.I.; Kim, Y.-S. The prevalence of human endogenous retroviruses in cerebrospinal fluids from patients with sporadic Creutzfeldt-Jakob disease. *J. Clin. Virol.* **2010**, *47*, 136–142. [CrossRef] [PubMed]

162. Balestrieri, E.; Arpino, C.; Matteucci, C.; Sorrentino, R.; Pica, F.; Alessandrelli, R.; Coniglio, A.; Curatolo, P.; Rezza, G.; Macciardi, F.; et al. HERVs Expression in Autism Spectrum Disorders. *PLoS ONE* **2012**, *7*, e48831. [CrossRef] [PubMed]

163. Balestrieri, E.; Pitzianti, M.; Matteucci, C.; D'Agati, E.; Sorrentino, R.; Baratta, A.; Caterina, R.; Zenobi, R.; Curatolo, P.; Garaci, E.; et al. Human endogenous retroviruses and ADHD. *World J. Biol. Psychiatry* **2014**, *15*, 499–504. [CrossRef] [PubMed]

164. Deb-Rinker, P.; Klempan, T.A.; O'Reilly, R.L.; Torrey, E.F.; Singh, S.M. Molecular characterization of a MSRV-like sequence identified by RDA from monozygotic twin pairs discordant for schizophrenia. *Genomics* **1999**, *61*, 133–144. [CrossRef] [PubMed]

165. Karlsson, H.; Bachmann, S.; Schröder, J.; McArthur, J.; Torrey, E.F.; Yolken, R.H. Retroviral RNA identified in the cerebrospinal fluids and brains of individuals with schizophrenia. *Proc. Natl. Acad. Sci. USA* **2001**, *98*, 4634–4639. [CrossRef] [PubMed]

166. Karlsson, H.; Schröder, J.; Bachmann, S.; Bottmer, C.; Yolken, R.H. HERV-W-related RNA detected in plasma from individuals with recent-onset schizophrenia or schizoaffective disorder. *Mol. Psychiatry* **2003**, *9*, 12–13. [CrossRef] [PubMed]

167. Yao, Y.; Schröder, J.; Nellåker, C.; Bottmer, C.; Bachmann, S.; Yolken, R.H.; Karlsson, H. Elevated levels of human endogenous retrovirus-W transcripts in blood cells from patients with first episode schizophrenia. *Genes Brain Behav.* **2008**, *7*, 103–112. [CrossRef] [PubMed]

168. Huang, W.; Li, S.; Hu, Y.; Yu, H.; Luo, F.; Zhang, Q.; Zhu, F. Implication of the env gene of the human endogenous retrovirus W family in the expression of BDNF and DRD3 and development of recent-onset schizophrenia. *Schizophr. Bull.* **2011**, *37*, 988–1000. [CrossRef] [PubMed]

169. Qin, C.; Li, S.; Yan, Q.; Wang, X.; Chen, Y.; Zhou, P.; Lu, M.; Zhu, F. Elevation of Ser9 phosphorylation of GSK3β is required for HERV-W env-mediated BDNF signaling in human U251 cells. *Neurosci. Lett.* **2016**, *627*, 84–91. [CrossRef] [PubMed]

170. Perron, H.; Mekaoui, L.; Bernard, C.; Veas, F.; Stefas, I.; Leboyer, M. Endogenous Retrovirus Type W GAG and Envelope Protein Antigenemia in Serum of Schizophrenic Patients. *Biol. Psychiatry* **2008**, *64*, 1019–1023. [CrossRef] [PubMed]

171. Hegyi, H. GABBR1 has a HERV-W LTR in its regulatory region–a possible implication for schizophrenia. *Biol. Direct* **2013**, *8*, 5. [CrossRef] [PubMed]

172. Frank, O.; Giehl, M.; Zheng, C.; Hehlmann, R.; Leib-Mosch, C.; Seifarth, W. Human endogenous retrovirus expression profiles in samples from brains of patients with schizophrenia and bipolar disorders. *J. Virol.* **2005**, *79*, 10890–10901. [CrossRef] [PubMed]

173. Perron, H.; Hamdani, N.; Faucard, R.; Lajnef, M.; Jamain, S.; Daban-Huard, C.; Sarrazin, S.; LeGuen, E.; Houenou, J.; Delavest, M.; et al. Molecular characteristics of Human Endogenous Retrovirus type-W in schizophrenia and bipolar disorder. *Transl. Psychiatry* **2012**, *2*, e201. [CrossRef] [PubMed]

174. Gürtler, C.; Bowie, A.G. Innate immune detection of microbial nucleic acids. *Trends Microbiol.* **2013**, *21*, 413–420. [CrossRef] [PubMed]

175. Melder, D.C.; Pankratz, V.S.; Federspiel, M.J. Evolutionary Pressure of a Receptor Competitor Selects Different Subgroup A Avian Leukosis Virus Escape Variants with Altered Receptor Interactions. *J. Virol.* **2003**, *77*, 10504–10514. [CrossRef] [PubMed]

176. Spencer, T.E.; Mura, M.; Gray, C.A.; Griebel, P.J.; Palmarini, M. Receptor usage and fetal expression of ovine endogenous betaretroviruses: Implications for coevolution of endogenous and exogenous retroviruses. *J. Virol.* **2003**, *77*, 749–753. [CrossRef] [PubMed]

177. An, D.S.; Xie, Y.M.; Chen, I.S. Envelope gene of the human endogenous retrovirus HERV-W encodes a functional retrovirus envelope. *J. Virol.* **2001**, *75*, 3488–3489. [CrossRef] [PubMed]

178. Vincendeau, M.; Göttesdorfer, I.; Schreml, J.M.H.; Wetie, A.G.N.; Mayer, J.; Greenwood, A.D.; Helfer, M.; Kramer, S.; Seifarth, W.; Hadian, K.; et al. Modulation of human endogenous retrovirus (HERV) transcription during persistent and de novo HIV-1 infection. *Retrovirology* **2015**, *12*, 27. [CrossRef] [PubMed]

179. Hurst, T.; Pace, M.; Katzourakis, A.; Phillips, R.; Klenerman, P.; Frater, J.; Magiorkinis, G. Human endogenous retrovirus (HERV) expression is not induced by treatment with the histone deacetylase (HDAC) inhibitors in cellular models of HIV-1 latency. *Retrovirology* **2016**, *13*, 10. [CrossRef] [PubMed]

180. Uleri, E.; Mei, A.; Mameli, G.; Poddighe, L.; Serra, C.; Dolei, A. HIV Tat acts on endogenous retroviruses of the W family and this occurs via Toll-like receptor 4: Inference for neuroAIDS. *AIDS* **2014**, *28*, 2659–2670. [CrossRef] [PubMed]

181. Toufaily, C.; Landry, S.; Leib-Mosch, C.; Rassart, E.; Barbeau, B. Activation of LTRs from different human endogenous retrovirus (HERV) families by the HTLV-1 tax protein and T-cell activators. *Viruses* **2011**, *3*, 2146–2159. [CrossRef] [PubMed]

182. Garrison, K.E.; Jones, R.B.; Meiklejohn, D.A.; Anwar, N.; Ndhlovu, L.C.; Chapman, J.M.; Erickson, A.L.; Agrawal, A.; Spotts, G.; Hecht, F.M.; et al. T cell responses to human endogenous retroviruses in HIV-1 infection. *PLoS Pathog.* **2007**, *3*, 1617–1627. [CrossRef] [PubMed]

183. Perron, H.; Suh, M.; Lalande, B.; Gratacap, B.; Laurent, A.; Stoebner, P.; Seigneurin, J.R. Herpes simplex virus ICP0 and ICP4 immediate early proteins strongly enhance expression of a retrovirus harboured by a leptomeningeal cell line from a patient with multiple sclerosis. *J. Gen. Virol.* **1993**, *74*, 65–72. [CrossRef] [PubMed]

184. Ruprecht, K.; Obojes, K.; Wengel, V.; Gronen, F.; Kim, K.S.; Perron, H.; Schneider-Schaulies, J.; Rieckmann, P. Regulation of human endogenous retrovirus W protein expression by herpes simplex virus type 1: Implications for multiple sclerosis. *J. Neurovirol.* **2006**, *12*, 65–71. [CrossRef] [PubMed]

185. Lee, W.J.; Kwun, H.J.; Kim, H.S.; Jang, K.L. Activation of the human endogenous retrovirus W long terminal repeat by herpes simplex virus type 1 immediate early protein 1. *Mol. Cells* **2003**, *15*, 75–80. [PubMed]

186. Krone, B.; Oeffner, F.; Grange, J.M. Is the risk of multiple sclerosis related to the "biography" of the immune system? *J. Neurol.* **2009**, *256*, 1052–1060. [CrossRef] [PubMed]

187. Mameli, G.; Poddighe, L.; Mei, A.; Uleri, E.; Sotgiu, S.; Serra, C.; Manetti, R.; Dolei, A. Expression and activation by Epstein Barr virus of human endogenous retroviruses-W in blood cells and astrocytes: Inference for multiple sclerosis. *PLoS ONE* **2012**, *7*, e44991. [CrossRef] [PubMed]

188. Assinger, A.; Yaiw, K.-C.; Göttesdorfer, I.; Leib-Mösch, C.; Söderberg-Nauclér, C. Human cytomegalovirus (HCMV) induces human endogenous retrovirus (HERV) transcription. *Retrovirology* **2013**, *10*, 132. [CrossRef] [PubMed]

189. Bergallo, M.; Galliano, I.; Montanari, P.; Gambarino, S.; Mareschi, K.; Ferro, F.; Fagioli, F.; Tovo, P.A.; Ravanini, P. CMV induces HERV-K and HERV-W expression in kidney transplant recipients. *J. Clin. Virol.* **2015**, *68*, 28–31. [CrossRef] [PubMed]

190. Yu, C.; Shen, K.; Lin, M.; Chen, P.; Lin, C.; Chang, G.D.; Chen, H. GCMa regulates the syncytin-mediated trophoblastic fusion. *J. Biol. Chem.* **2002**, *277*, 50062–50068. [CrossRef] [PubMed]

191. Li, F.; Nellåker, C.; Sabunciyan, S.; Yolken, R.H.; Jones-Brando, L.; Johansson, A.-S.; Owe-Larsson, B.; Karlsson, H. Transcriptional derepression of the ERVWE1 locus following influenza A virus infection. *J. Virol.* **2014**, *88*, 4328–4337. [CrossRef] [PubMed]

192. Ponferrada, V.G.; Mauck, B.S.; Wooley, D.P. The envelope glycoprotein of human endogenous retrovirus HERV-W induces cellular resistance to spleen necrosis virus. *Arch. Virol.* **2003**, *148*, 659–675. [CrossRef] [PubMed]

193. Machnik, G.; Klimacka-Nawrot, E.; Sypniewski, D.; Matczyńska, D.; Gałka, S.; Bednarek, I.; Okopień, B. Porcine endogenous retrovirus (PERV) infection of HEK-293 cell line alters expression of human endogenous retrovirus (HERV-W) sequences. *Folia Biol. (Czech Republic)* **2014**, *60*, 35–46.

194. Sassaman, D.M.; Dombroski, B.A.; Moran, J.V.; Kimberland, M.L.; Naas, T.P.; DeBerardinis, R.J.; Gabriel, A.; Swergold, G.D.; Kazazian, H.H. Many human L1 elements are capable of retrotransposition. *Nat. Genet.* **1997**, *16*, 37–43. [CrossRef] [PubMed]

Review

The Role of Somatic L1 Retrotransposition in Human Cancers

Emma C. Scott [1,2] and Scott E. Devine [1,2,3,4,*]

1   Graduate Program in Molecular Medicine, University of Maryland, Baltimore, MD 21201, USA; escott@umaryland.edu

2   Institute for Genome Sciences, University of Maryland School of Medicine, Baltimore, MD 21201, USA

3   Department of Medicine, University of Maryland School of Medicine; Baltimore, MD 21201, USA

4   Greenebaum Comprehensive Cancer Center, University of Maryland School of Medicine, Baltimore, MD 21201, USA

*   Correspondence: sdevine@som.umaryland.edu; Tel.: +1-410-706-2343

Academic Editors: David J. Garfinkel and Katarzyna J. Purzycka
Received: 17 March 2017; Accepted: 22 May 2017; Published: 31 May 2017

Abstract: The human LINE-1 (or L1) element is a non-LTR retrotransposon that is mobilized through an RNA intermediate by an L1-encoded reverse transcriptase and other L1-encoded proteins. L1 elements remain actively mobile today and continue to mutagenize human genomes. Importantly, when new insertions disrupt gene function, they can cause diseases. Historically, L1s were thought to be active in the germline but silenced in adult somatic tissues. However, recent studies now show that L1 is active in at least some somatic tissues, including epithelial cancers. In this review, we provide an overview of these recent developments, and examine evidence that somatic L1 retrotransposition can initiate and drive tumorigenesis in humans. Recent studies have: (i) cataloged somatic L1 activity in many epithelial tumor types; (ii) identified specific full-length L1 source elements that give rise to somatic L1 insertions; and (iii) determined that L1 promoter hypomethylation likely plays an early role in the derepression of L1s in somatic tissues. A central challenge moving forward is to determine the extent to which L1 driver mutations can promote tumor initiation, evolution, and metastasis in humans.

Keywords: retrotransposon; somatic retrotransposition; cancer genomics; LINE-1, L1

1. Introduction

Transposable genetic elements constitute at least 45% of the human genome [1]. Some of these mobile elements or "jumping genes" have the ability to replicate themselves and insert new copies elsewhere in the genome [2]. The most prevalent mobile element in humans is the Long Interspersed Element 1, abbreviated LINE-1 or L1, which has expanded to over 500,000 copies in the human genome and makes up ~17% of the human genome sequence [1]. This massive copy number expansion is the result of over 150 million years of L1 propagation that began with the incorporation of these elements into ancestral genomes sometime before the mammalian radiation [1]. L1s belong to the non-long-terminal-repeat (non-LTR) class of retrotransposons, which have certain sequence attributes and move through an RNA intermediate. Non-LTR retrotransposons themselves originated over 600 million years ago (probably from eubacterial reverse transcriptases) during the Precambrian Era and likely predate multicellular eukaryotic life [3,4]. The end result of L1 expansion over these hundreds of millions of years is a human genome that is littered with L1 copies [1]. A small number of these endogenous L1 retrotransposons remain actively mobile today and continue to mutagenize human genomes. We present here a brief review of our current understanding of how these elements influence human genomes, and ultimately, human health. A major focus is to examine somatic L1 retrotransposition as a causative agent in human cancers.

2. Mobilization of L1 Retrotransposons

In order to examine how L1s influence human genomes and disease, we must first understand how these elements are mobilized (Figure 1).

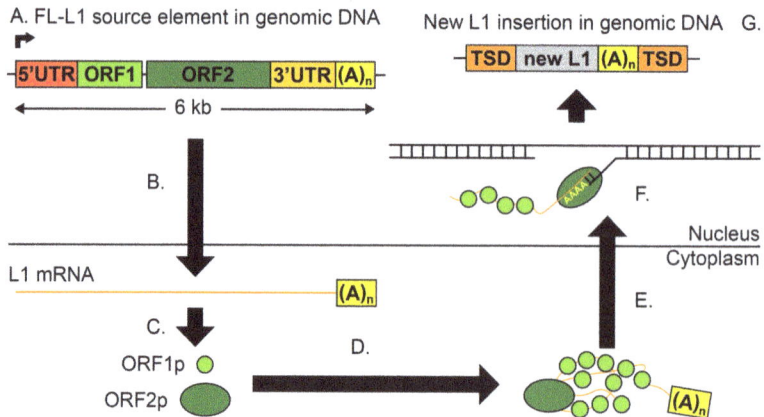

Figure 1. Mobilization of L1s. New L1 insertions are generated via the five step process depicted here. This process begins with a full-length (FL)-L1 source element in the genomic DNA (**A**; colored bar; L1 features are not to scale). This element is transcribed (**B**) and the resulting mRNA (orange) is exported into the cytoplasm. This mRNA is translated (**C**) into the open reading frame (ORF)1p (light green) and ORF2p (dark green) proteins, which bind the L1 mRNA to form a ribonucleoprotein complex (**D**). This complex is imported (**E**) into the nucleus. Finally, the new L1 insertion is generated by target-primed reverse transcription (**F**). The result of this mobilization process is another copy of L1 (grey) located somewhere in the genome, flanked by target site duplications (TSDs; orange) and with a poly(A) tail (yellow) (**G**). UTR; untranslated region.

This process begins with a canonical full-length L1 (FL-L1) source element that is ~6 kb long and consists of a promoter that is located within a 5′ untranslated region (UTR), two non-overlapping open reading frames (ORFs), a 3′ UTR, and a poly(A) tail (Figure 1A). Only a fraction of the L1 copies in the human genome have these features, as most copies are either 5′ truncated or have other deleterious mutations (see below). The first step in L1 mobilization is transcription of the FL-L1 source element from its internal promoter (Figure 1A), most likely by RNA Polymerase II [5]. This process arises only from FL-L1 source elements, since truncated L1s are missing important sequence features in the first 100 bp of the 5′ UTR that are critical for transcription initiation [6]. The resulting bicistronic mRNA is then exported from the nucleus into the cytoplasm, where it is translated to make two L1-encoded proteins, ORF1p and ORF2p (Figure 1B). ORF1p is a small RNA-binding protein [7]. ORF2p is a larger protein that encodes both endonuclease (EN) [8] and reverse transcriptase (RT) functions [9]. These proteins then bind the L1 mRNA that produced them to form a ribonucleoprotein complex [10,11] that enters the nucleus (Figure 1D,E). The ORF1p and ORF2p proteins have a strong *cis* preference for mobilizing the specific mRNA that gave rise to them, which allows for the preferential amplification of functional L1 copies [11–13]. Recently, a third ORF (named ORF0) was discovered in primate L1s (including humans); ORF0 is transcribed from an antisense promoter in the canonical L1 5′ UTR and can be translated into the ORF0 protein or alternative ORF0-fusion proteins that include the coding exons of neighboring genes [14]. Although ORF0 proteins can increase L1 activity (through an unknown mechanism), they are not necessary for retrotransposition [14].

The final step of L1 mobilization is known as target-primed reverse transcription (TPRT) (Figure 1F). This mechanism was first characterized in the non-LTR retrotransposon R2 from the silkworm *Bombyx mori* [15], and later was shown to accurately describe L1 integration as well [16]. Briefly, the process begins when the EN domain of ORF2p nicks the genomic DNA at its target site. The consensus recognition sequence for the EN domain is 5'-TTTT/A-3' [8,17,18], although there is considerable flexibility in the exact site that is bound and cut by EN [19]. This cleavage exposes a 3'OH group, which is then used by the ORF2p RT domain to prime the reverse transcription of the L1 mRNA, starting from the poly(A) tail [20,21] and extending towards the 5' end of the mRNA [15,16]. Finally, the complementary strand is synthesized and the junctions are repaired through mechanisms that are not well understood, likely involving host factors. The result of this process is a newly-inserted L1 copy, or "offspring" insertion, at a second genomic site (Figure 1G). It is important to note that the original FL-L1 source element remains intact in this process, and is capable of producing additional offspring insertions [22].

New L1 insertions have important hallmark features of retrotransposition: a poly(A) tail, flanking target site duplications (TSDs), frequent 5' truncation, and occasional 5' inversion [5,23–26]. Additionally, L1s can sometimes mobilize downstream sequences in a process that is known as 3' transduction; this is thought to occur when transcription continues through the L1 polyadenylation signal and terminates after another signal in the adjacent genomic DNA [27–29]. The L1 retrotransposition machinery also can be hijacked by the nonautonomous retrotransposons *Alu* [30] and SVA [31,32], as well as cellular mRNAs [12]. As a consequence, the L1-mediated TPRT mechanism ultimately is responsible for the mobilization of most, if not all, recently inserted mobile elements in human genomes. Likewise, all of these mobile elements have features in common with L1 (e.g., TSDs that are similar in length and sequence composition) because they are all mobilized by the same TPRT mechanism and L1 proteins. Polymorphic copies of HERV-K elements (endogenous retroviruses) also are found in human genomes, but evidence suggests that functional HERV-K copies are either very rare or no longer exist in human populations [33,34].

The large number of individual steps that are required for L1 mobilization present numerous opportunities for human cells to regulate this mutagenic activity, and many such L1 regulators have evolved. In fact, there are mechanisms of L1 repression that act at each step of the mobilization process. For example, one of the most well-studied mechanisms of L1 repression is the methylation of CpG dinucleotides in the L1 promoter [35,36]. Promoter methylation silences L1s by repressing transcription, which is the first step of the mobilization process [36]. Beyond methylation, there are numerous other activities that inhibit L1 mobilization. These include histone modifications and chromatin remodeling (which also regulate L1 transcription), small RNA-mediated mechanisms (including piwi-interacting RNAs and small-interfering RNAs), and numerous cellular proteins that inhibit L1 mobilization post-transcriptionally in both the cytoplasm and the nucleus (reviewed recently by JL Goodier [37]). These redundant mechanisms work together to repress L1 activity.

3. An Historical Perspective of L1 Activity

Mobile genetic elements were first discovered in maize genomes by Barbara McClintock in 1950 [2]. This work eventually won her the Nobel Prize in Physiology or Medicine in 1983 [38], once the ubiquitous nature of transposable elements was recognized and fully appreciated. Repetitive elements were first discovered in the human genome during the course of DNA renaturation experiments in the 1960s, but these elements remained a mystery for some time [39,40]. L1 elements specifically were found in the human genome in the early 1980s [41–43]. From their first description by Adams et al. in 1980 it was suggested that L1s could be "essentially parasitic" DNA without a function (page 6126); this speculation was based on the lack of measurable L1 transcription in bone marrow cells [41]. This concept of parasitic or selfish DNA was not new. The exact origin of the idea is difficult to determine, but it was popularized around this time by the publication of Richard Dawkins' book *The Selfish Gene* in 1976, and there were even ongoing reviews and debates about this notion in *Nature* in 1980 [44–46]. However, this classification of L1s as useless "junk DNA" was challenged as soon as it was introduced, both theoretically within the same article by Adams et al. [41] and by others as scientists began to recognize the mutagenic potential of these sequences.

During the 1980s there was a furious rush to describe the newly-discovered L1 repetitive elements. This included classifying L1s, determining their function, and beginning to understand their activity—all of which occurred in the short span of eight years. When L1s were first discovered in 1980, they were simply described as long repeated sequences of variable length that were dispersed across all human chromosomes [41]. Researchers in the field quickly determined that L1s had been discovered by multiple laboratories, and introduced standardized terminology by 1982 [42]. Long repetitive elements in general were referred to as LINEs, and the most common element was termed LINE-1 or L1. Next, researchers concluded (from many lines of evidence) that these repeats could potentially be mobile elements and speculated that they were moving through an RNA intermediate (summarized by Singer and Skowronski in 1985 [47]). In just the next year (1986), concrete evidence to confirm this hypothesis was found with the discovery that L1 ORF2 could potentially encode a protein with homology to reverse transcriptase proteins [43]. Around the same time, FL-L1 transcripts were shown to be expressed in a human teratocarcinoma cell line [48,49]. These studies provided additional support for the theory that L1s might represent mobile elements that could mutagenize the human genome, a hypothesis that would soon be confirmed.

The first evidence for ongoing mobility of L1s in human genomes came in 1988, when Kazazian et al. found two disease-causing de novo germline L1 insertions [25]. This L1 activity was discovered during a screen of hemophilia A patients for pathogenic mutations in the Factor VIII gene (F8) on Chromosome X. Two unrelated patients were found to have germline L1 insertions in the F8 gene that were absent from their parents' genomes, indicating that these L1 insertions probably occurred in gametes or during early embryogenesis [25]. The two most important findings of this study were that L1s are actively mobile in human genomes and that offspring insertions can cause disease by disrupting genes. Months after this initial discovery, a putative somatic L1 insertion was identified in a case of breast cancer [50]. This 7–8 kb insertion was located in an intron of the MYC proto-oncogene. However, only part of this L1 insertion was sequenced and the structure lacked the hallmarks of TPRT [51]. Thus, although this study offered the first suggestion of somatic L1 activity in humans, such activity was unlikely in this case.

The next bona fide instance of L1 retrotransposition was found in 1992, when a somatic L1 insertion was discovered by Miki et al. in a case of colorectal cancer (CRC) [52]. Similar to the L1 mutations in F8 [25], this insertion was discovered during a screen for mutations in the APC tumor suppressor gene (TSG) in tumors of CRC patients. This screen was performed because APC plays a pivotal role in this kind of cancer: the majority of CRC tumors are initiated by mutation or loss of function of both APC alleles in normal colon cells, which results in the formation of a precancerous lesion that can eventually progress to carcinoma [53–55]. This L1 insertion was absent from adjacent normal colon tissue, suggesting that it occurred in a cancerous or precancerous colon cell sometime during the tumorigenesis process. This was remarkable because it was the first somatic L1 retrotransposition event that was documented in humans. This 750 bp insertion was flanked by TSDs and had a poly(A) tail, indicating that it was a bona fide somatic insertion produced by the TPRT mechanism. Likewise, this insertion interrupted the last coding exon of APC, so it was expected to disrupt normal APC function. The authors emphasized that the insertion likely initiated the tumor, but in view of our current understanding of CRC, this interpretation is somewhat unclear (see Section 5 for discussion).

For many years this was the only available evidence that L1 elements were capable of retrotransposition in somatic tissues. There was another putative example of somatic L1 activity published in 1997 [56]. However, this insertion again lacked the expected hallmarks of TPRT, and therefore, most likely was not produced by somatic retrotransposition [51,56]. Thus, genuine somatic L1 insertions appeared to be rare in humans in these early studies. In contrast, germline L1 insertions have been linked to many additional human diseases throughout the 1990s, 2000s, and to the present. In fact, germline L1 insertions have caused at least 29 cases of human disease, and have contributed to another 94 cases indirectly by mobilizing *Alu*, SVA, or other mRNAs (reviewed in [57]). In most cases, these disease-causing insertions occurred within the coding exons of known genes and disrupted gene function (although insertions in the promoters, 5′ UTR, 3′ UTR, and introns of genes also have been observed).

At this point in time scientists were undertaking important mechanistic research on L1 retrotransposons, including studies aimed at understanding how L1s are mobilized (briefly reviewed in Section 2). Much of the work during this period was done with a clever cell culture-based retrotransposition assay [58]. Specifically, Moran et al. demonstrated in 1996 that a plasmid-borne FL-L1 source element was highly active in HeLa cells, and could generate numerous offspring insertions in its host's HeLa cell chromosomes [58]. Retrotransposition of the L1 element was dependent upon the two proteins that were encoded by the plasmid-borne L1 copy (ORF1p and ORF2p), as targeted mutations of these regions abolished L1 activity. Moreover, the new L1 insertions that were generated in HeLa chromosomes had the expected features of TPRT-mediated events. This study was remarkable because it confirmed that L1s are indeed active retrotransposons in humans [58]. However, the L1 retrotransposition assay itself is in some ways equally remarkable because it has fueled decades of productive research on L1 biology. For example, Brouha et al. [59] and Beck et al. [60] have used this assay to identify active FL-L1 source elements in human genomes. Others have used the assay to study the roles of the L1-encoded ORF0p, ORF1p, and ORF2p proteins in retrotransposition [8,14,58,61–63]. Conceptually similar *Alu* [30] and SVA [31,32] assays have confirmed that both of these nonautonomous elements hijack the L1 machinery for their own mobilization. The L1 assay also has been adapted for the creation of mouse models to study the timing and effects of L1 retrotransposition in the mouse [64–68]. Many other advances have leveraged these assays as well (reviewed in [69]).

4. Somatic L1 Activity in Human Genomes

As outlined above in Section 3, Miki et al. reported the earliest example of a somatic L1 insertion that might have helped to drive tumorigenesis in humans [52]. However, almost two decades passed before it became apparent that somatic L1 insertions occur frequently in human epithelial cancers. After a hiatus of 18 years, our laboratory demonstrated in 2010 that somatic L1 insertions occur frequently in human lung tumors [51]. A series of other studies subsequently revealed that somatic L1 retrotransposition is a hallmark feature of human epithelial cancers [70–81] (see Table 1). These observations have led to the suggestion that L1 might generate driver mutations in proto-oncogenes or TSGs that could fuel tumorigenesis in humans [51,52,70] (see Section 5). In the following sections, we review these studies and explore the role of somatic L1 retrotransposition in human cancers.

4.1. A Second Discovery of L1 Retrotransposition in Cancer

As mentioned above, we rediscovered somatic L1 retrotransposition in human tumors in a study that leveraged next-generation sequencing technologies to investigate L1 retrotransposition in human populations and cancers [51]. This study introduced the L1-Seq assay, which is a modified and updated version of L1 Display [82]. Similar to L1 Display, L1-Seq exploits sequence features of the youngest, most active L1 (L1-Ta) elements to selectively amplify the 3′ insertion junctions of these young L1 copies. In contrast to L1 Display assays, which use a gel electrophoresis step to visualize new L1 insertions, L1-Seq assays instead apply DNA sequencing technologies directly to the junction fragments to discover the chromosomal coordinates of new L1 insertions in a high-throughput manner.

After developing and optimizing the L1-Seq approach, we used it to discover new L1 insertions in 38 diverse humans, eight tumor-derived cell lines, 20 non-small cell lung tumors (with matched normal tissues), and 10 brain tumors (with matched blood leukocyte controls) [51]. This screen identified 802 novel L1 insertions, the majority of which were rare germline insertions. However, nine somatic L1 insertions also were identified in six of the 20 lung tumors, which were validated with PCR and Sanger sequencing. Somatic L1 insertions were not found in the brain tumors that were examined in this study, foreshadowing the fact that somatic L1 insertions do not occur in all tumor types (see Section 4.2). Taken together with the Miki et al. 1992 study, our study revealed that somatic L1 retrotransposition occurs in at least two types of human cancers (colon and lung), and suggested that somatic L1 retrotransposition might occur more broadly than previously appreciated [51,52].

4.2. Cataloguing Somatic Retrotransposition in Cancer

During the seven years that have elapsed since our 2010 study [51], many additional studies have documented somatic L1 activity in a range of human epithelial cancers. The main findings of these 14 research articles are summarized in Table 1. These papers have established many principles of somatic L1 activity, while at the same time posing important new questions.

Table 1. Studies of somatic L1 activity in cancer genomes. Column three gives the total number of somatic L1 insertions discovered in the total number of tumors assayed; these estimates take validation rates into account when applicable.

Reference	Tumor Type	Insertions (Tumors)	Important Findings
Miki et al. 1992 [52]	Colorectal	1 (1)	First genuine somatic L1 activity; L1 insertion in *APC* might have initiated colorectal cancer (CRC), but somewhat unclear
Iskow et al. 2010 [51]	Lung, brain	8, 0 (20, 10)	Introduced high-throughput L1-Seq assay; Established that somatic L1 activity occurs frequently in lung tumors, but not in brain tumors; Suggested that L1s might drive tumorigenesis; Found a hypomethylation signature that distinguishes L1-permissive lung tumors
Lee et al. 2012 [70]	Colorectal, prostate, ovarian, brain, blood	178 (43)	Somatic L1 activity only in epithelial tumors, absent from brain and blood; Genes with somatic L1 insertions typically had decreased expression; Compared features of somatic and germline L1s
Solyom et al. 2012 [71]	Colorectal	72 (16)	Positive correlation between patient age and number of somatic L1s; Most L1 insertions occurred after tumor initiation
Shukla et al. 2013 [72]	Liver	12 (19)	Intronic somatic L1 insertion into a regulatory element increased expression of candidate liver oncogene *ST18*; Suggested that L1s might be somatically active in normal liver cells
Pitkänen et al. 2014 [73]	Colorectal	83 (92)	All L1 insertions originated from one source element on Chromosome 22, in *TTC28*; These L1 insertions were previously mischaracterized as translocations
Helman et al. 2014 [74]	11 types	695 (976)	Somatic L1 insertion in an exon of the *PTEN* tumor suppressor gene (TSG); Lung, colorectal, head and neck, and uterine cancers had highest L1 mobilization levels
Tubio et al. 2014 [75]	12 types	2711 (290)	3′ transductions make up 24% of somatic L1 activity; A small number of source elements gave rise to most L1 insertions with transductions; Active sources had promoter hypomethylation; Activity of sources fluctuates over the course of tumor evolution
Paterson et al. 2015 [76]	Esophageal	5108 (43)	The majority of L1s were discovered by searching for somatic poly(A) insertions, so some probably represent L1-mediated transposition of non-L1 sequence; Identified active source elements using 3′ transductions

Table 1. *Cont.*

Reference	Tumor Type	Insertions (Tumors)	Important Findings
Rodić et al. 2015 [77]	Pancreatic	409 (20)	Inverse correlation between survival and both the number of somatic L1 insertions and ORF1p protein expression; Retrotransposition occurs throughout tumor development, but is discontinuous
Ewing et al. 2015 [78]	Colorectal, pancreatic, gastric, testicular	104 (18)	Frequent somatic L1 insertions in precancerous adenomas; Most somatic L1 insertions were clonal; Validated one somatic non-germline L1 insertion in normal colon; Suggested that L1 insertions are occurring in normal colon or very early in tumorigenesis
Doucet-O'Hare et al. 2015 [79]	Esophageal	118 (20)	Found somatic L1 insertions in patients with Barrett's Esophagus (a cancer-predisposing condition) and esophageal cancer; L1 activity seen in patients that did not develop cancer; Suggested that somatic L1 activity could occur in normal or metaplastic cells
Scott et al. 2016 [80]	Colorectal	27 (1)	An L1-initiated CRC caused by L1 mutagenesis of *APC* TSG; Tumor initiated by activity of a hot, population-specific FL-L1 source element, which was hypomethylated and expressed in normal colon tissue; Demonstrated that L1s can evade somatic repression and initiate tumorigenesis
Achanta et al. 2016 [83]	Brain	1 (10)	Found one somatic L1 insertion in a secondary glioblastoma; Cannot rule out that this occurred in normal brain because compared to DNA from blood
Carreira et al. 2016 [84]	Brain	0 (14)	Could only validate one TPRT-independent somatic L1 insertion and one likely *Alu-Alu* recombination event; Conclude that L1 retrotransposition does not occur in primary glioblastoma or glioma
Tang et al. 2017 [81]	Ovarian; pancreatic	35, 205 (8, 13)	Found one somatic L1 insertion in *BRCA1* TSG intron, in an ovarian cancer; Some pancreatic L1 insertions (76) were discovered in an earlier analysis of this same sequencing data [77] and used for methodological validation here

Although these studies employed several strategies to measure somatic L1 activity, most of the methods can be grouped into two basic categories: (1) targeted resequencing assays and (2) bioinformatics tools that use whole genome sequencing (WGS) or whole exome sequencing (WES) to discover somatic L1 insertions. Targeted resequencing tools exploit specific sequence features of young L1s to selectively amplify and sequence novel L1 insertion junctions. These assays are similar to the previously described L1 Display [82] and L1-Seq [51] assays, often with further improvements in multiplexing, genome coverage, and enrichment of L1 junction fragments prior to sequencing [71,72,77–79,81,83,84]. With the decreased price of WGS, and the availability of such data in public repositories, several groups have developed bioinformatics tools to discover somatic L1 insertions in silico using WGS or WES data [70,74,75,80]. One advantage of this approach is that existing WGS or WES data from large consortia (like The Cancer Genome Atlas project (TCGA) [70,74,75] and the International Cancer Genome Consortium (ICGC) [75]) can be directly screened for somatic L1 insertions. As the cost of genome sequencing continues to decrease, WGS is increasingly being used to discover somatic L1 insertions in smaller laboratories as well [80]. This approach also provides the opportunity to assess other somatic variants (without any additional sequencing) to understand how L1 insertions work together with other mutagenic processes to drive tumorigenesis [75,80]. It is important to note that all of these techniques require validation to confirm that putative somatic insertions have the expected features of TPRT-mediated events (e.g., poly(A) tails and flanking TSDs).

Somatic L1 activity in cancer genomes has been found to be almost entirely confined to tumors arising from epithelial tissues [70] (see Table 1 for details). The highest levels of L1 mobilization are found in lung [51,74,75] and colorectal [70,71,74,75,78,80] cancers. Moderate L1 activity is seen in esophageal [76,79], pancreatic [77,78,81], head and neck [74,75], uterine [74], ovarian [70,74,81], gastric [78], and prostate [70,75] cancers. Lower but detectable levels of L1 activity have been seen in breast [74,75], bone [75], liver [72], kidney [74], and testicular [78] cancers. Only one TPRT-mediated somatic L1 insertion has been found in a case of brain cancer, though it is unclear whether this event occurred in the normal brain or during tumorigenesis because the normal DNA that was used for comparison was isolated from blood instead of adjacent normal brain tissue [83]. One other somatic L1 insertion was identified in a glioma tumor-derived cell line [75], though the exact timing of this insertion is again not discernible. Numerous other studies have found no such events in 60 brain tumors that were examined, leading to the consensus that somatic L1 activity in brain cancers is either absent or extremely rare [51,70,74,84]. This is somewhat ironic, given that L1s are very active in normal brain tissues (see Section 4.5). A similar lack of L1 activity has been noted in 25 examined hematologic malignancies [70,74]. These results are perhaps not surprising in light of in vitro experiments that demonstrated very low levels of engineered L1 activity in precursor cells of these cancer types (astrocytes [85] and hematopoietic stem cells [86]); these data also further support the theory that somatic L1 activity is only found in tumors of epithelial origin [70]. Within each tumor type, the number of somatic L1 insertions per tumor also can vary substantially. This is illustrated most clearly in the lung tumors that were sequenced by the ICGC, where the number of somatic L1 insertions per tumor ranged from zero to over 800, with an average of ~63 insertions per tumor [75].

Additionally, L1s are responsible for the occasional somatic retrotransposition of other sequences in cancer genomes. The most commonly mobilized non-L1 sequences are 3' transductions, which are produced when the sequence downstream of a FL-L1 source element is mobilized (see Sections 2 and 4.3). In one large study, 3' transductions occurred in 24% (655/2756) of the somatic retrotransposition events that were discovered in tumors [75]; half of these transductions were so-called "orphans" consisting only of downstream DNA without any L1 sequence [75,87,88]. These 3' transductions can amplify exons, entire genes, and regulatory sequences, providing another mechanism by which L1s might alter the function of cancer cells [75].

Processed pseudogenes also are generated in somatic tissues via the retrotransposition of cellular mRNAs by the L1 machinery [89,90]. Such events occur at a low rate in human tumors (42 insertions in 629 tumors examined in an in-depth study of this phenomenon) [90]. Similar to L1 activity, somatic processed pseudogene formation is observed most frequently in lung and colorectal cancers, and less often in other epithelial cancers. Of note, one processed pseudogene insertion was recovered from a chondrosarcoma, which is a cartilage tumor of mesenchymal origin [90]; this finding contradicts the theory that L1 activity is restricted to epithelial tissues [70]. Infrequent somatic *Alu* insertions also have been observed in various epithelial cancers [70,71,74,75], and one somatic SVA insertion has been validated in a head and neck cancer [74]. Thus, in addition to simple L1 insertions, other non-L1 sequences are mobilized by the L1-mediated TPRT mechanism at lower frequencies in somatic tumor tissues.

4.3. Identification of Active Full-length (FL)-L1 Source Elements in Tumors

It is increasingly becoming possible to identify the specific FL-L1 source elements that produce somatic offspring insertions in tumors. A key development in this regard has been a much more extensive knowledge of the FL-L1 source elements that are harbored by human genomes. Based on the reference (REF) human genome sequence, Brouha et al. initially estimated that every individual has a collection of approximately 80 to 100 FL-L1 source elements that are retrotransposition-competent [59]. Later studies demonstrated that each individual also has non-reference (non-REF) FL-L1 source elements [60,80,91] (and our unpublished data). We refer to this collection of REF and non-REF FL-L1 source elements as an individual's FL-L1 source element profile [80]. Source element profiles appear to vary considerably from one person to the next, and such differences likely produce variation in the levels of germline and somatic L1 mutagenesis that are caused by L1s [59,60,80,91] (and our unpublished data). Thus, a key goal for the future will be to understand exactly how these FL-L1 profiles vary in human populations and how this variation affects L1-mediated diseases, including cancers.

Currently, there are two available methods to determine source/offspring relationships between L1 elements. The first uses 3′ transductions to identify offspring insertions that were generated from a specific source element; this technique is very effective and accurate, but is useful only for the subset of insertions that have 3′ transductions [22,73,75]. We recently developed a second method to track source/offspring relationships using interior mutations that are frequently found in FL-L1 source elements [80]. Each source element has its own unique set of interior mutations that is inherited by offspring insertions emanating from that source element, and this provides a means to track source/offspring relationships [80].

Both of these methods have been used to study the FL-L1 elements that are active in human tumors. One large study employed the 3′ transduction method and found that a relatively small number of source elements (72) were responsible for the bulk (95%) of measurable somatic L1 activity in 290 tumors of 12 types [75]. Even more surprising, two of these source elements were responsible for over a third of the somatic activity that could be tracked with 3′ transductions [75]. Activity from one of these elements (along with many others) has been mistaken for chromosomal translocations by multiple cancer genomics groups [73,75]. This is an extremely important distinction because translocations can create novel fusion genes (which are often oncogenic) whereas L1 insertions cannot [75]. It is important to note that only ~24% of somatic offspring can be attributed to specific source elements using 3′ transductions, and much of what we currently know about source elements in cancer is based upon such methods. Thus, additional studies will be necessary to identify all of the source elements that are somatically active in human cancers. As a step in this direction, we recently developed and used the interior mutation method to identify source elements that produced somatic L1 insertions in a case of CRC [80]. Three non-REF source elements (on Chromosomes 17, 14, and 12) produced the majority of somatic L1 insertions in this tumor, including an insertion that disrupted the *APC* TSG and initiated tumorigenesis [80].

These studies have uncovered some important features of active source elements. For instance, both REF and non-REF FL-L1 source elements can contribute to somatic L1 retrotransposition [75,80]. Source elements that are active in tumors have two intact ORFs, and most belong to the youngest L1 subfamily (Ta-1d) [80]. Some sources are "hot" L1s [75,80], which are particularly active in the cell culture-based retrotransposition assay [58] (see Section 3). The global distribution of source elements also can differ—some source elements are population-specific, while others are present in all 26 of the global populations that were examined by the 1000 Genomes Project [80]. The number of active source elements, and the amount of activity for each source element, varies considerably between tumors and tumor types. This activity can even change with the different stages of tumor evolution, with levels of L1 activity fluctuating during cancer development and progression [75,77]. In some instances, somatic FL-L1 insertions can even themselves give rise to further somatic L1 activity [75].

4.4. Mechanism of Reactivation of L1s in Cancer Genomes

The discovery of frequent somatic L1 activity in cancer has researchers asking why this phenomenon is occurring and why it is so variable between tumors and tissues. Promoter methylation is one of the earliest lines of defense against L1 activity, and therefore, hypomethylation is thought to be a necessary step for L1 reactivation in tumors. This hypothesis has been corroborated by many groups. These studies examined methylation at three different levels, using techniques that vary in scope. First, the Illumina Infinium platform assesses global genomic methylation within CpG islands [51]. Second, average L1 promoter methylation across all FL-L1 copies can be measured using general bisulfite sequencing PCR (with internal L1 primers that amplify most FL-L1 promoters [71,72]). Finally, methylation of specific FL-L1 source elements can be measured with targeted bisulfite sequencing PCR by amplifying the 5' junctions of these elements (using a primer in the upstream genomic DNA in combination with an internal L1 primer) [75,80].

Using the first method (the Illumina Infinium platform), we identified a hypomethylation signature at 59 genomic CpG sites that was associated with L1-permissive lung cancers [51]. Soon thereafter, Solyom et al. identified four CpG sites in L1 promoters that were hypomethylated in CRC tumors compared to normal colon tissue [71]. Hypomethylation of the entire CpG island was later observed in the L1 promoters of liver tumors compared to adjacent normal tissue [72]. Moreover, there was a strong positive correlation between the level of hypomethylation at these promoters and the number of somatic L1 insertions that were produced in all three tumors that were examined [72]. These last two studies used methods that measure the average methylation levels at many L1 promoters throughout the genome (i.e., the second, general method listed above). In contrast, two recent studies have inspected the promoters of specific FL-L1 source elements using the third (targeted) method outlined above. These studies found that the source elements (i) were hypomethylated in the tumors in which they caused somatic L1 insertions; (ii) were usually methylated in tumors that lacked activity from the source element; and (iii) were often methylated in normal tissue [75,80]. The functional consequences of L1 promoter hypomethylation also have been demonstrated—a source element that produced a somatic insertion early in tumorigenesis was hypomethylated and transcribed in both normal and tumor colon tissue [80]. Collectively, these results suggest that methylation represses L1 elements in normal somatic tissues, but is either absent or removed from elements that have become reanimated. Other host factors and mechanisms likely contribute to this process as well (see Section 2). Additional work is needed to better understand how these elements are silenced and derepressed in somatic human tissues, and how these processes impact tumorigenesis.

4.5. L1 Retrotransposition Contributes to Genomic Diversity in the Adult Brain

In addition to cancer genomes, somatic retrotransposition also occurs in neuronal cells [66,85] and normal brain tissues [66,85,92]. Using an array capture and sequencing-based technology to discover new L1 insertions, the Faulkner group documented high levels of L1 activity in the human brain (850 putative somatic L1 insertions in three individuals) [92]. Somatic retrotransposition was confirmed in the brain independently by the Walsh lab using single cell sequencing technology, although at much lower levels [93]. The Faulkner lab subsequently carried out their own single cell analyses in human neuronal brain tissues, and verified that somatic L1 retrotransposition indeed occurs at high frequencies in such tissues [94] (in agreement with their earlier 2011 study [92]). Although there is some disagreement about the absolute level of somatic retrotransposition that occurs in the human brain, these studies seem to agree that such mobilization occurs. While this is not directly related to cancer, it is important to consider that somatic mobilization is not limited to the germline and tumors, but also occurs in the brain and may occur in other normal somatic tissues as well (see Section 6.2). It is of interest to note that the first transposable element that was discovered by Barbara McClintock is mobilized exclusively in the somatic cells of maize [2].

5. L1 as a Driver of Tumorigenesis

Perhaps the most important discussion ongoing in the human retrotransposon field is the exact timing of L1 activity during tumor development and progression. This boils down to one central question: are somatic L1 insertions drivers of tumorigenesis or mere passengers along for the ride? Nearly every published paper in the field addresses this question, including the first documented case of somatic L1 activity from 1992 [52]. This somatic L1 insertion disrupted a coding exon of the *APC* TSG and likely was a driver of tumorigenesis [52] (Table 2). However, in view of our current understanding of CRC, the precise role of this insertion in tumorigenesis remains unclear. This is because the second *APC* allele was not examined in that study and the authors did not rule out a possible role for faulty DNA repair as an initiator of tumorigenesis [52] (see [80] for details). Thus, the precise stage of tumorigenesis that was affected by this L1 driver mutation (i.e., tumor initiation vs. a later stage) remains unclear. There have been additional findings supporting the theory that somatic L1 insertions participate in tumorigenesis. For example, a somatic L1 mutation has been identified in a uterine tumor in the known cancer gene *PTEN* (Table 2). However, a relatively small number of clear L1 driver mutations have been discovered to date (Table 2; see Section 6.1 for further discussion of this point).

Studies published in 2012 [70] and 2014 [74] confirmed that intronic L1 insertions in human tumors usually result in decreased expression of the mutated genes, and thus, could theoretically contribute to tumorigenesis by decreasing the expression of TSGs. However, the frequency of this occurrence is under debate: two additional studies had the contradictory finding that the vast majority of intronic somatic L1 insertions had no effect on gene expression in liver, lung, and colon tumors [72,75]. Thus, although L1 insertions can change gene expression in tumors at least occasionally, more work is needed to reach a consensus on how frequently this occurs. Interestingly, the unique type of mutation that is caused by L1 mobilization (i.e., structural variation) also can have unexpected effects on gene expression. For example, Shukla et al. [72] identified a somatic L1 insertion within an intron of the *ST18* gene that disrupted a *cis*-regulatory repressor element, which in turn led to increased expression of the *ST18* gene (Table 2). This group concluded that *ST18* is likely a proto-oncogene that has a role in driving liver cancer in this case [72]. It has since been demonstrated that *ST18* is important for the development and persistence of liver tumors by facilitating interactions with tumor associated macrophages [95]. Therefore, the somatic L1 insertion in *ST18* likely was integral for the formation and maintenance of the tumor in which it was discovered [72]. Both somatically-acquired and germline L1 elements in the genome also can have large effects on the transcriptome through a number of diverse mechanisms that are independent of L1 activity [96] (recently reviewed in [97]). It can be imagined that these mechanisms could impact the expression of proto-oncogenes and TSGs, adding another level of complexity of the potential roles of retrotransposons in cancer cells [72,75].

Table 2. Likely driver mutations caused by somatic L1 retrotransposon insertions in known proto-oncogenes and TSGs.

Gene	Location of Insertion	Tumor Type	Reference
APC	16th exon (coding)	Colorectal	Miki et al. 1992 [52]
APC	16th exon (coding)	Colorectal	Scott et al. 2016 [80]
PTEN	6th exon (coding)	Uterine	Helman et al. 2014 [74]
ST18	Intron (repressor)	Liver	Shukla et al. 2013 [72]

One major unresolved question is: how often do L1 driver mutations initiate tumorigenesis in normal cells? Insertions that occur after tumor growth is well underway would not appear to participate in tumorigenesis per se, but instead might play a role in tumor evolution or metastasis. Doucet-O'Hare et al. examined the evolution of esophageal cancer from a precancerous condition and demonstrated that L1s can be active very early during the tumorigenesis process, and even found somatic L1 insertions in precancerous lesions that never progressed to cancer (over a 15 year period) [79]. In a similar study, Ewing et al. also found somatic L1 insertions in precancerous lesions (adenomas) of the colon and in normal colon tissue [78]. These papers together strongly suggest that tumor-initiating L1 insertions could occur in a normal cell and then become amplified into a tumor through selection.

We recently demonstrated that L1 indeed can initiate tumorigenesis in normal colon cells [80]. *APC* is a gatekeeper TSG that is frequently mutated in patients with CRC—in fact, both copies of *APC* must be mutated to initiate most cases of CRC [53–55]. We discovered a somatic L1 insertion that disrupted the last coding exon of this gene [80], only 388 bp upstream of the somatic L1 insertion that was discovered by Miki et al. in 1992 [52] (Table 2). Importantly, we also established that the second *APC* allele was inactivated by a point mutation, and that this tumor had stable microsatellites, indicating that faulty DNA repair did not initiate tumorigenesis in this case [80]. Instead, we showed that the tumor developed in concordance with the well-established two-hit genetic model for CRC wherein both *APC* gatekeeper alleles are mutated in a normal cell [53–55]. L1 disrupted one of the two *APC* alleles and a stop codon disrupted the second allele. Thus, somatic L1 mutations can initiate tumorigenesis in normal colon cells.

6. Closing Remarks and Future Directions

This review has summarized the current literature documenting somatic L1 retrotransposition in human cancers. The wave of papers that has been published on this topic over the last seven years has provided a survey of which tumor types are permissive for somatic L1 activity; the field also has begun to address the important question of why this phenomenon is occurring and the effects it has on cancer development and progression. However, in the course of discovering important characteristics of somatic L1 activity, these papers also have raised several new questions that need to be addressed. To close this review, we summarize below some unsettled questions and future directions.

6.1. Why Don't We See L1-Initiated Cancers More Frequently?

As discussed in Section 5, there have only been a few reported instances of probable L1-initiated tumors (Table 2). Though somatically-mobilized L1s can indeed initiate and drive tumorigenesis [80], there have been a notably small number of such cases discovered, especially considering the total number of tumors that have been examined for somatic L1 activity across all the studies in Table 1. This raises the question: why don't we see this phenomenon more frequently? Although one possibility is that L1s initiate tumorigenesis in somatic cells only rarely, as suggested by multiple groups [71,75,77], there also are a few reasons that we may be underestimating the frequency L1-mediated cancers.

First and foremost, the genetic pathways for tumor development have not been thoroughly defined for most tumor types. Thus, we might be finding somatic L1 insertions in proto-oncogenes and TSGs that have not yet been discovered, and as a consequence, we cannot yet link these insertions to tumor development. In this regard, somatic L1 insertions may define a novel set of proto-oncogenes and TSGs that can only drive tumorigenesis when mutated by L1. In support of this idea, recurring L1 mutations have been identified in several novel genes that were not previously linked to tumorigenesis [70,71,74–76,78]. This phenomenon is reminiscent of studies in mice where tumors were induced by the Sleeping Beauty transposon [98]. Even when an L1 insertion occurs within a known proto-oncogene or TSG, it can be difficult to link the insertion unambiguously to the tumor in which it was discovered. For example, in some cases a known proto-oncogene or TSG might not have been linked to a specific tumor type (e.g., the role of *ST18* in liver cancer development was determined a few years after the L1 insertion in this gene was reported [72,95]). Even if the gene has been clearly implicated in the tumor type previously, it can be difficult to interpret the impact of some L1 insertions without extensive experimentation (e.g., it is difficult to predict the functional consequences of intronic insertions, such as the one that was discovered in *BRCA1* in a case of ovarian cancer [81]).

Another confounding factor is that the temporal order in which gene mutations occur during tumor initiation and evolution is unclear in most tumor types (recently reviewed in [99]). In many cases, the discovery of tumor progression pathways is hindered by the extraordinary mutational heterogeneity that is found within most cancer types [100]. As a result, there are only a few tumor types (e.g., CRC) in which we can currently determine whether tumors are actually initiated by L1 insertions in known driver genes. Thus, although projects such as TCGA and ICGC have begun to explore the mutational landscapes of many human cancers, much more work is needed to fully understand how L1 mutagenesis contributes to tumor formation.

6.2. To What Extent Are L1s Active in Normal Noncancerous Cells and Tissues?

Although L1 clearly is quite active in the normal somatic tissues of the brain and epithelial cancers, we are just beginning to explore the extent of somatic activity in other normal tissues. Several lines of evidence suggest that L1 can evade somatic repression in at least some other normal tissues. In one study, 21 putative somatic L1 insertions were identified in normal liver cells using a targeted sequencing assay [72]. In another study, two somatic L1 insertions were discovered in normal stomach cells [78]. Yet another study found nine putative somatic L1 insertions in the normal esophagus [79]. In all three of these studies, independent validation of somatic L1 activity in the normal tissue was either very limited or not possible. A confounding factor was that germline insertions could appear to be new somatic insertions in adjacent normal tissue if the tumor underwent chromosomal loss over the site. Regardless, these studies are likely to have reported true somatic insertions, since it can be reasonably expected that germline insertions would be detected by PCR in the tumor-infiltrating noncancerous cells that comprise the tumor stroma. Thorough validation might have been possible if a second normal tissue were available in these studies (however, this was not the case). In contrast, two additional somatic L1 insertions have been identified in normal tissues that were fully validated by PCR: the first in normal colon and absent from liver cells [78]; the second in normal esophagus (and a precancerous lesion) and absent from blood cells [79]. Finally, in our CRC study we determined that a tumor-forming L1 insertion in *APC* occurred at the earliest stages of tumorigenesis (most likely in a normal colon cell), providing further evidence that normal colon tissues can support somatic L1 activity [80]. Thus, there is a lot of evidence in the literature (albeit sometimes preliminary in nature) to suggest that normal adult tissues may broadly support somatic L1 retrotransposition.

On the basis of this limited evidence, is it possible that most (if not all) normal epithelial tissues support L1 activity? If so, this might have been largely missed for the same reason that it was initially overlooked in the brain: that each cell in a given tissue generates a unique collection of somatic L1 retrotransposition events that cannot be detected in bulk tissue. The solution to this problem is to adapt L1 discovery methods to single cell sequencing technologies. This approach has been pioneered in brain tissues by the Walsh [93] and Faulkner [94] labs, and should be adaptable to other normal tissues as well. Through whole genome amplification, this technique is able to both sequence the genome of a single cell and also provide material for validation [93,94]. Although this approach is still in its infancy, it likely will be useful for finding the somatic L1 insertions that the literature suggests are mutagenizing the genomes of normal cells throughout the human body. Clearly, more work will be necessary to determine whether this is the case.

6.3. How Does Inter-Individual Genomic Variation Affect Somatic L1 Activity?

Finally, we need to address how differences in FL-L1 source element profiles influence tumorigenesis. As discussed in Section 4.3, each individual inherits a different collection of FL-L1 source elements. The content of these profiles could have considerable effects on the risk of an individual developing L1-mediated diseases, including cancers. In this regard, many questions remain unanswered: How many hot FL-L1 source elements are present in each human's genome and how does this vary from one human to the next? How many elements can evade somatic repression and initiate human cancers in normal somatic tissues, and in which tissues does this occur? This is a particularly important question because it is quite possible that most of the somatic L1 insertions that have been discovered in tumors thus far were produced by source elements that only became derepressed after tumorigenesis was underway. If this were the case, it might help to explain why the community has identified only a handful of clearly recognizable L1 driver mutations in human cancers: most insertions were generated too late to initiate or drive tumorigenesis. However, several lines of evidence indicate that at least some FL-L1 source elements can evade somatic repression in normal tissues, and generate driver mutations sufficiently early to initiate tumorigenesis. We need to explore this class of events more carefully and determine how often source elements can generate tumor-initiating mutations in normal cells. Events that occur later in tumorigenesis also need to be explored further for roles in tumor evolution and metastasis.

We also need to gain a better understanding of how FL-L1 source element profiles vary in human populations. Although some FL-L1s are ubiquitously found in most or all human genomes, many others are found only in a subset of individuals and are inherited in a population-specific manner [80]. As a result, ancestry could play an important role in determining an individual's mutagenic burden from germline and somatic L1 activity. For example, in our CRC study, a population-specific FL-L1 source element on Chromosome 17 of the patient's genome initiated tumorigenesis [80]. Since this element is restricted to populations that are associated with the African diaspora, the cancer risk that is associated with this element also would be restricted to such populations. At the present time, very little is known about how source element profiles vary across diverse human demographies, and how these differences affect tumorigenesis.

6.4. Conclusions

We have presented a brief overview of research examining somatic L1 retrotransposition in human genomes, focusing on landmark studies outlining the activity of these mobile elements in human cancers. Researchers in this field have characterized many aspects of somatic L1 activity in the short span of only seven years. However, several major questions remain unresolved. One unsettled question is: How often does somatic L1 retrotransposition initiate and drive tumorigenesis in humans? Despite the fact that thousands of somatic L1 insertions have been recovered from many tumor types, only a handful of clearly-recognizable driver mutations have been discovered. We also have a very incomplete understanding of L1 activity in normal somatic tissues. If L1s are highly active in most adult somatic tissues, we clearly need to gain a better understanding of the consequences of this activity. Likewise, in our current era of human genomics and precision medicine, it is increasingly important to define inter-individual variation and determine how variation in FL-L1 source element profiles may impact human health and disease.

Acknowledgments: We thank Eugene Gardner and Nelson Chuang for thoughtful discussions on the topics of this review. This work was funded by the following National Institutes of Health grants: T32 CA154274 (E.C.S.), R01 CA166661 (S.E.D.), and R01 HG002898 (S.E.D.).

Conflicts of Interest: The authors declare no conflict of interest.

Abbreviations

The following abbreviations are used in this manuscript:

L1	Long Interspersed Element 1
LTR	long-terminal-repeat
FL-L1	full-length L1
UTR	untranslated region
ORF	open reading frame
EN	endonuclease
RT	reverse transcriptase
TSD	target site duplication
TPRT	target-primed reverse transcription
CRC	colorectal cancer
TSG	tumor suppressor gene
WGS	whole genome sequencing
WES	whole exome sequencing
TCGA	The Cancer Genome Atlas
ICGC	International Cancer Genome Consortium
REF	reference

References

1. Lander, E.S.; Linton, L.M.; Birren, B.; Nusbaum, C.; Zody, M.C.; Baldwin, J.; Devon, K.; Dewar, K.; Doyle, M.; FitzHugh, W.; et al. International Human Genome Sequencing Consortium. Initial sequencing and analysis of the human genome. *Nature* **2001**, *409*, 860–921.
2. McClintock, B. The origin and behavior of mutable loci in maize. *Proc. Natl. Acad. Sci. USA* **1950**, *36*, 344–355.

3.   Malik, H.S.; Burke, W.D.; Eickbush, T.H. The age and evolution of non-LTR retrotransposable elements. *Mol. Biol. Evol.* **1999**, *16*, 793–805.

4.   Eickbush, T.H.; Malik, H.S. Origins and Evolution of Retrotransposons. In *Mobile DNA II*; ASM Press: Washington, DC, USA, 2002.

5.   Ostertag, E.M.; Kazazian, H.H. Biology of mammalian L1 retrotransposons. *Ann. Rev. Genet.* **2001**, *35*, 501–538.

6.   Swergold, G.D. Identification, characterization, and cell specificity of a human LINE-1 promoter. *Mol. Cell. Biol.* **1990**, *10*, 6718–6729.

7.   Hohjoh, H.; Singer, M.F. Sequence-specific single-strand RNA binding protein encoded by the human LINE-1 retrotransposon. *EMBO J.* **1997**, *16*, 6034–6043.

8.   Feng, Q.; Moran, J.V.; Kazazian, H.H.; Boeke, J.D. Human L1 retrotransposon encodes a conserved endonuclease required for retrotransposition. *Cell* **1996**, *87*, 905–916.

9.   Mathias, S.L.; Scott, A.F.; Kazazian, H.H.; Boeke, J.D.; Gabriel, A. Reverse transcriptase encoded by a human transposable element. *Science* **1991**, *254*, 1808–1810.

10. Hohjoh, H.; Singer, M.F. Cytoplasmic ribonucleoprotein complexes containing human LINE-1 protein and RNA. *EMBO J.* **1996**, *15*, 630–639.

11. Kulpa, D.A.; Moran, J.V. Cis-preferential LINE-1 reverse transcriptase activity in ribonucleoprotein particles. *Nat. Struct. Mol. Biol.* **2006**, *13*, 655–660.

12. Esnault, C.; Maestre, J.; Heidmann, T. Human LINE retrotransposons generate processed pseudogenes. *Nat. Genet.* **2000**, *24*, 363–367.

13. Wei, W.; Gilbert, N.; Ooi, S.L.; Lawler, J.F.; Ostertag, E.M.; Kazazian, H.H.; Boeke, J.D.; Moran, J.V. Human L1 retrotransposition: Cis preference versus *trans* complementation. *Mol. Cell. Biol.* **2001**, *21*, 1429–1439.

14. Denli, A.M.; Narvaiza, I.; Kerman, B.E.; Pena, M.; Benner, C.; Marchetto, M.C.N.; Diedrich, J.K.; Aslanian, A.; Ma, J.; Moresco, J.J.; et al. Primate-Specific ORF0 Contributes to Retrotransposon-Mediated Diversity. *Cell* **2015**, *163*, 583–593.

15. Luan, D.D.; Korman, M.H.; Jakubczak, J.L.; Eickbush, T.H. Reverse transcription of R2Bm RNA is primed by a nick at the chromosomal target site: A mechanism for non-LTR retrotransposition. *Cell* **1993**, *72*, 595–605.

16. Cost, G.J.; Feng, Q.; Jacquier, A.; Boeke, J.D. Human L1 element target-primed reverse transcription in vitro. *EMBO J.* **2002**, *21*, 5899–5910.

17. Jurka, J. Sequence patterns indicate an enzymatic involvement in integration of mammalian retroposons. *Proc. Natl. Acad. Sci. USA* **1997**, *94*, 1872–1877.

18. Morrish, T.A.; Gilbert, N.; Myers, J.S.; Vincent, B.J.; Stamato, T.D.; Taccioli, G.E.; Batzer, M.A.; Moran, J.V. DNA repair mediated by endonuclease-independent LINE-1 retrotransposition. *Nat. Genet.* **2002**, *31*, 159–165.

19. Cost, G.J.; Boeke, J.D. Targeting of Human Retrotransposon Integration Is Directed by the Specificity of the L1 Endonuclease for Regions of Unusual DNA Structure. *Biochemistry* **1998**, *37*, 18081–18093.

20. Ahl, V.; Keller, H.; Schmidt, S.; Weichenrieder, O. Retrotransposition and Crystal Structure of an Alu RNP in the Ribosome-Stalling Conformation. *Mol. Cell* **2015**, *60*, 715–727.

21. Doucet, A.J.; Wilusz, J.E.; Miyoshi, T.; Liu, Y.; Moran, J.V. A 3′ Poly(A) Tract Is Required for LINE-1 Retrotransposition. *Mol. Cell* **2015**, *60*, 728–741.

22. Macfarlane, C.M.; Collier, P.; Rahbari, R.; Beck, C.R.; Wagstaff, J.F.; Igoe, S.; Moran, J.V.; Badge, R.M. Transduction-specific ATLAS reveals a cohort of highly active L1 retrotransposons in human populations. *Hum. Mutat.* **2013**, *34*, 974–985.

23. Thayer, R.E.; Singer, M.F. Interruption of an alpha-satellite array by a short member of the KpnI family of interspersed, highly repeated monkey DNA sequences. *Mol. Cell. Biol.* **1983**, *3*, 967–973.

24. Grimaldi, G.; Skowronski, J.; Singer, M.F. Defining the beginning and end of KpnI family segments. *EMBO J.* **1984**, *3*, 1753–1759.

25. Kazazian, H.H.; Wong, C.; Youssoufian, H.; Scott, A.F.; Phillips, D.G.; Antonarakis, S.E. Haemophilia A resulting from de novo insertion of L1 sequences represents a novel mechanism for mutation in man. *Nature* **1988**, *332*, 164–166.

26. Ostertag, E.M.; Kazazian, H.H. Twin priming: A proposed mechanism for the creation of inversions in L1 retrotransposition. *Genome Res.* **2001**, *11*, 2059–2065.

27. Moran, J.V.; DeBerardinis, R.J.; Kazazian, H.H. Exon shuffling by L1 retrotransposition. *Science* **1999**, *283*, 1530–1534.

28. Goodier, J.L.; Ostertag, E.M.; Kazazian, H.H. Transduction of 3′-flanking sequences is common in L1 retrotransposition. *Hum. Mol. Genet.* **2000**, *9*, 653–657.

29. Pickeral, O.K.; Makałowski, W.; Boguski, M.S.; Boeke, J.D. Frequent human genomic DNA transduction driven by LINE-1 retrotransposition. *Genome Res.* **2000**, *10*, 411–415.

30. Dewannieux, M.; Esnault, C.; Heidmann, T. LINE-mediated retrotransposition of marked Alu sequences. *Nat. Genet.* **2003**, *35*, 41–48.

31. Hancks, D.C.; Goodier, J.L.; Mandal, P.K.; Cheung, L.E.; Kazazian, H.H. Retrotransposition of marked SVA elements by human L1s in cultured cells. *Hum. Mol. Genet.* **2011**, *20*, 3386–3400.

32. Raiz, J.; Damert, A.; Chira, S.; Held, U.; Klawitter, S.; Hamdorf, M.; Löwer, J.; Strätling, W.H.; Löwer, R.; Schumann, G.G. The non-autonomous retrotransposon SVA is trans-mobilized by the human LINE-1 protein machinery. *Nucleic Acids Res.* **2012**, *40*, 1666–1683.

33. Dewannieux, M.; Harper, F.; Richaud, A.; Letzelter, C.; Ribet, D.; Pierron, G.; Heidmann, T. Identification of an infectious progenitor for the multiple-copy HERV-K human endogenous retroelements. *Genome Res.* **2006**, *16*, 1548–1556.

34. Wildschutte, J.H.; Williams, Z.H.; Montesion, M.; Subramanian, R.P.; Kidd, J.M.; Coffin, J.M. Discovery of unfixed endogenous retrovirus insertions in diverse human populations. *Proc. Natl. Acad. Sci. USA* **2016**, *113*, E2326–E2334.

35. Thayer, R.E.; Singer, M.F.; Fanning, T.G. Undermethylation of specific LINE-1 sequences in human cells producing a LINE-1-encoded protein. *Gene* **1993**, *133*, 273–277.

36. Hata, K.; Sakaki, Y. Identification of critical CpG sites for repression of L1 transcription by DNA methylation. *Gene* **1997**, *189*, 227–234.

37. Goodier, J.L. Restricting retrotransposons: A review. *Mob. DNA* **2016**, *7*, 1–30.

38. Comfort, N.C. From controlling elements to transposons: Barbara McClintock and the Nobel Prize. *Trends Biochem. Sci.* **2001**, *26*, 454–457.

39. Waring, M.; Britten, R.J. Nucleotide sequence repetition: A rapidly reassociating fraction of mouse DNA. *Science* **1966**, *154*, 791–794.

40. Britten, R.J.; Kohne, D.E. Repeated sequences in DNA. Hundreds of thousands of copies of DNA sequences have been incorporated into the genomes of higher organisms. *Science* **1968**, *161*, 529–540.

41. Adams, J.W.; Kaufman, R.E.; Kretschmer, P.J.; Harrison, M.; Nienhuis, A.W. A family of long reiterated DNA sequences, one copy of which is next to the human beta globin gene. *Nucleic Acids Res.* **1980**, *8*, 6113–6128.

42. Singer, M.F. SINEs and LINEs: Highly repeated short and long interspersed sequences in mammalian genomes. *Cell* **1982**, *28*, 433–434.

43. Hattori, M.; Kuhara, S.; Takenaka, O.; Sakaki, Y. L1 family of repetitive DNA sequences in primates may be derived from a sequence encoding a reverse transcriptase-related protein. *Nature* **1986**, *321*, 625–628.

44. Doolittle, W.F.; Sapienza, C. Selfish genes, the phenotype paradigm and genome evolution. *Nature* **1980**, *284*, 601–603.

45. Orgel, L.E.; Crick, F. Selfish DNA: The ultimate parasite. *Nature* **1980**, *284*, 604–607.

46. Orgel, L.E.; Crick, F.H.; Sapienza, C. Selfish DNA. *Nature* **1980**, *288*, 645–646.

47. Singer, M.F.; Skowronski, J. Making sense out of LINES: Long interspersed repeat sequences in mammalian genomes. *Trends Biochem. Sci.* **1985**, *10*, 119–122.

48. Skowronski, J.; Singer, M.F. Expression of a cytoplasmic LINE-1 transcript is regulated in a human teratocarcinoma cell line. *Proc. Natl. Acad. Sci. USA* **1985**, *82*, 6050–6054.

49. Skowronski, J.; Fanning, T.G.; Singer, M.F. Unit-length line-1 transcripts in human teratocarcinoma cells. *Mol. Cell. Biol.* **1988**, *8*, 1385–1397.

50. Morse, B.; Rotherg, P.G.; South, V.J.; Spandorfer, J.M.; Astrin, S.M. Insertional mutagenesis of the myc locus by a LINE-1 sequence in a human breast carcinoma. *Nature* **1988**, *333*, 87–90.

51. Iskow, R.C.; McCabe, M.T.; Mills, R.E.; Torene, S.; Pittard, W.S.; Neuwald, A.F.; Van Meir, E.G.; Vertino, P.M.; Devine, S.E. Natural mutagenesis of human genomes by endogenous retrotransposons. *Cell* **2010**, *141*, 1253–1261.

52. Miki, Y.; Nishisho, I.; Horii, A.; Miyoshi, Y.; Utsunomiya, J.; Kinzler, K.W.; Vogelstein, B.; Nakamura, Y. Disruption of the APC gene by a retrotransposal insertion of L1 sequence in a colon cancer. *Cancer Res.* **1992**, *52*, 643–645.

53. Fearon, E.R.; Vogelstein, B. A genetic model for colorectal tumorigenesis. *Cell* **1990**, *61*, 759–767.

54. Powell, S.M.; Zilz, N.; Beazer-Barclay, Y.; Bryan, T.M.; Hamilton, S.R.; Thibodeau, S.N.; Vogelstein, B.; Kinzler, K.W. APC mutations occur early during colorectal tumorigenesis. *Nature* **1992**, *359*, 235–237.

55. Fearon, E.R. Molecular genetics of colorectal cancer. *Ann. Rev. Pathol.* **2011**, *6*, 479–507.

56. Liu, J.; Nau, M.M.; Zucman-Rossi, J.; Powell, J.I.; Allegra, C.J.; Wright, J.J. LINE-I element insertion at the t(11;22) translocation breakpoint of a desmoplastic small round cell tumor. *Genes Chromosomes Cancer* **1997**, *18*, 232–239.

57. Hancks, D.C.; Kazazian, H.H. Roles for retrotransposon insertions in human disease. *Mob. DNA* **2016**, *7*, 9.

58. Moran, J.V.; Holmes, S.E.; Naas, T.P.; DeBerardinis, R.J.; Boeke, J.D.; Kazazian, H.H. High frequency retrotransposition in cultured mammalian cells. *Cell* **1996**, *87*, 917–927.

59. Brouha, B.; Schustak, J.; Badge, R.M.; Lutz-Prigge, S.; Farley, A.H.; Moran, J.V.; Kazazian, H.H. Hot L1s account for the bulk of retrotransposition in the human population. *Proc. Natl. Acad. Sci. USA* **2003**, *100*, 5280–5285.

60. Beck, C.R.; Collier, P.; Macfarlane, C.; Malig, M.; Kidd, J.M.; Eichler, E.E.; Badge, R.M.; Moran, J.V. LINE-1 retrotransposition activity in human genomes. *Cell* **2010**, *141*, 1159–1170.

61. Alisch, R.S.; Garcia-Perez, J.L.; Muotri, A.R.; Gage, F.H.; Moran, J.V. Unconventional translation of mammalian LINE-1 retrotransposons. *Genes Dev.* **2006**, *20*, 210–224.

62. Morrish, T.A.; Garcia-Perez, J.L.; Stamato, T.D.; Taccioli, G.E.; Sekiguchi, J.; Moran, J.V. Endonuclease-independent LINE-1 retrotransposition at mammalian telomeres. *Nature* **2007**, *446*, 208–212.

63. Cook, P.R.; Jones, C.E.; Furano, A.V. Phosphorylation of ORF1p is required for L1 retrotransposition. *Proc. Natl. Acad. Sci. USA* **2015**, *112*, 4298–4303.

64. Ostertag, E.M.; DeBerardinis, R.J.; Goodier, J.L.; Zhang, Y.; Yang, N.; Gerton, G.L.; Kazazian, H.H. A mouse model of human L1 retrotransposition. *Nat. Genet.* **2002**, *32*, 655–660.

65. Prak, E.T.L.; Dodson, A.W.; Farkash, E.A.; Kazazian, H.H. Tracking an embryonic L1 retrotransposition event. *Proc. Natl. Acad. Sci. USA* **2003**, *100*, 1832–1837.

66. Muotri, A.R.; Chu, V.T.; Marchetto, M.C.N.; Deng, W.; Moran, J.V.; Gage, F.H. Somatic mosaicism in neuronal precursor cells mediated by L1 retrotransposition. *Nature* **2005**, *435*, 903–910.

67. An, W.; Han, J.S.; Wheelan, S.J.; Davis, E.S.; Coombes, C.E.; Ye, P.; Triplett, C.; Boeke, J.D. Active retrotransposition by a synthetic L1 element in mice. *Proc. Natl. Acad. Sci. USA* **2006**, *103*, 18662–18667.

68. Kano, H.; Godoy, I.; Courtney, C.; Vetter, M.R.; Gerton, G.L.; Ostertag, E.M.; Kazazian, H.H. L1 retrotransposition occurs mainly in embryogenesis and creates somatic mosaicism. *Genes Dev.* **2009**, *23*, 1303–1312.

69. Richardson, S.R.; Doucet, A.J.; Kopera, H.C.; Moldovan, J.B.; Garcia-Perez, J.L.; Moran, J.V. The Influence of LINE-1 and SINE Retrotransposons on Mammalian Genomes. *Microbiol. Spectr.* **2015**, *3*, doi:10.1128/microbiolspec.MDNA3-0061-2014.

70. Lee, E.; Iskow, R.; Yang, L.; Gokcumen, O.; Haseley, P.; Luquette, L.J.; Lohr, J.G.; Harris, C.C.; Ding, L.; Wilson, R.K.; et al. Landscape of somatic retrotransposition in human cancers. *Science* **2012**, *337*, 967–971.

71. Solyom, S.; Ewing, A.D.; Rahrmann, E.P.; Doucet, T.; Nelson, H.H.; Burns, M.B.; Harris, R.S.; Sigmon, D.F.; Casella, A.; Erlanger, B.; et al. Extensive somatic L1 retrotransposition in colorectal tumors. *Genome Res.* **2012**, *22*, 2328–2338.

72. Shukla, R.; Upton, K.R.; Muñoz-Lopez, M.; Gerhardt, D.J.; Fisher, M.E.; Nguyen, T.; Brennan, P.M.; Baillie, J.K.; Collino, A.; Ghisletti, S.; et al. Endogenous retrotransposition activates oncogenic pathways in hepatocellular carcinoma. *Cell* **2013**, *153*, 101–111.

73. Pitkänen, E.; Cajuso, T.; Katainen, R.; Kaasinen, E.; Välimäki, N.; Palin, K.; Taipale, J.; Aaltonen, L.A.; Kilpivaara, O. Frequent L1 retrotranspositions originating from TTC28 in colorectal cancer. *Oncotarget* **2014**, *5*, 853–859.

74. Helman, E.; Lawrence, M.S.; Stewart, C.; Sougnez, C.; Getz, G.; Meyerson, M. Somatic retrotransposition in human cancer revealed by whole-genome and exome sequencing. *Genome Res.* **2014**, *24*, 1053–1063.

75. Tubio, J.M.C.; Li, Y.; Ju, Y.S.; Martincorena, I.; Cooke, S.L.; Tojo, M.; Gundem, G.; Pipinikas, C.P.; Zamora, J.; Raine, K.; et al. Mobile DNA in cancer. Extensive transduction of nonrepetitive DNA mediated by L1 retrotransposition in cancer genomes. *Science* **2014**, *345*, 1251343.

76. Paterson, A.L.; Weaver, J.M.J.; Eldridge, M.D.; Tavaré, S.; Fitzgerald, R.C.; Edwards, P.A.W. Mobile element insertions are frequent in oesophageal adenocarcinomas and can mislead paired-end sequencing analysis. *BMC Genom.* **2015**, *16*, 473.

77. Rodić, N.; Steranka, J.P.; Makohon-Moore, A.; Moyer, A.; Shen, P.; Sharma, R.; Kohutek, Z.A.; Huang, C.R.; Ahn, D.; Mita, P.; et al. Retrotransposon insertions in the clonal evolution of pancreatic ductal adenocarcinoma. *Nat. Med.* **2015**, *21*, 1060–1064.

78. Ewing, A.D.; Gacita, A.; Wood, L.D.; Ma, F.; Xing, D.; Kim, M.S.; Manda, S.S.; Abril, G.; Pereira, G.; Makohon-Moore, A.; et al. Widespread somatic L1 retrotransposition occurs early during gastrointestinal cancer evolution. *Genome Res.* **2015**, *25*, 1536–1545.

79. Doucet-O'Hare, T.T.; Rodić, N.; Sharma, R.; Darbari, I.; Abril, G.; Choi, J.A.; Young Ahn, J.; Cheng, Y.; Anders, R.A.; Burns, K.H.; et al. LINE-1 expression and retrotransposition in Barrett's esophagus and esophageal carcinoma. *Proc. Natl. Acad. Sci. USA* **2015**, *112*, E4894–E4900.

80. Scott, E.C.; Gardner, E.J.; Masood, A.; Chuang, N.T.; Vertino, P.M.; Devine, S.E. A hot L1 retrotransposon evades somatic repression and initiates human colorectal cancer. *Genome Res.* **2016**, *26*, 745–755.

81. Tang, Z.; Steranka, J.P.; Ma, S.; Grivainis, M.; Rodić, N.; Huang, C.R.L.; Shih, I.M.; Wang, T.L.; Boeke, J.D.; Fenyö, D.; et al. Human transposon insertion profiling: Analysis, visualization and identification of somatic LINE-1 insertions in ovarian cancer. *Proc. Natl. Acad. Sci. USA* **2017**, *114*, E733–E740.

82. Sheen, F.M.; Sherry, S.T.; Risch, G.M.; Robichaux, M.; Nasidze, I.; Stoneking, M.; Batzer, M.A.; Swergold, G.D. Reading between the LINEs: Human genomic variation induced by LINE-1 retrotransposition. *Genome Res.* **2000**, *10*, 1496–1508.

83. Achanta, P.; Steranka, J.P.; Tang, Z.; Rodić, N.; Sharma, R.; Yang, W.R.; Ma, S.; Grivainis, M.; Huang, C.R.L.; Schneider, A.M.; et al. Somatic retrotransposition is infrequent in glioblastomas. *Mob. DNA* **2016**, *7*, 22.

84. Carreira, P.E.; Ewing, A.D.; Li, G.; Schauer, S.N.; Upton, K.R.; Fagg, A.C.; Morell, S.; Kindlova, M.; Gerdes, P.; Richardson, S.R.; et al. Evidence for L1-associated DNA rearrangements and negligible L1 retrotransposition in glioblastoma multiforme. *Mob. DNA* **2016**, *7*, 21.

85. Coufal, N.G.; Garcia-Perez, J.L.; Peng, G.E.; Yeo, G.W.; Mu, Y.; Lovci, M.T.; Morell, M.; O'Shea, K.S.; Moran, J.V.; Gage, F.H. L1 retrotransposition in human neural progenitor cells. *Nature* **2009**, *460*, 1127–1131.

86. Macia, A.; Widmann, T.J.; Heras, S.R.; Ayllon, V.; Sanchez, L.; Benkaddour-Boumzaouad, M.; Muñoz-Lopez, M.; Rubio, A.; Amador-Cubero, S.; Blanco-Jimenez, E.; et al. Engineered LINE-1 retrotransposition in nondividing human neurons. *Genome Res.* **2017**, *27*, 335–348.

87. Ostertag, E.M.; Goodier, J.L.; Zhang, Y.; Kazazian, H.H. SVA elements are nonautonomous retrotransposons that cause disease in humans. *Am. J. Hum. Genet.* **2003**, *73*, 1444–1451.

88. Solyom, S.; Ewing, A.D.; Hancks, D.C.; Takeshima, Y.; Awano, H.; Matsuo, M.; Kazazian, H.H. Pathogenic orphan transduction created by a nonreference LINE-1 retrotransposon. *Hum. Mutat.* **2012**, *33*, 369–371.

89. Ewing, A.D.; Ballinger, T.J.; Earl, D.; Broad Institute Genome Sequencing and Analysis Program and Platform; Harris, C.C.; Ding, L.; Wilson, R.K.; Haussler, D. Retrotransposition of gene transcripts leads to structural variation in mammalian genomes. *Genome Biol.* **2013**, *14*, R22.

90. Cooke, S.L.; Shlien, A.; Marshall, J.; Pipinikas, C.P.; Martincorena, I.; Tubio, J.M.C.; Li, Y.; Menzies, A.; Mudie, L.; Ramakrishna, M.; et al. Processed pseudogenes acquired somatically during cancer development. *Nat. Commun.* **2014**, *5*, 3644.

91. Sudmant, P.H.; Rausch, T.; Gardner, E.J.; Handsaker, R.E.; Abyzov, A.; Huddleston, J.; Zhang, Y.; Ye, K.; Jun, G.; Hsi-Yang Fritz, M.; et al. An integrated map of structural variation in 2,504 human genomes. *Nature* **2015**, *526*, 75–81.

92. Baillie, J.K.; Barnett, M.W.; Upton, K.R.; Gerhardt, D.J.; Richmond, T.A.; De Sapio, F.; Brennan, P.M.; Rizzu, P.; Smith, S.; Fell, M.; et al. Somatic retrotransposition alters the genetic landscape of the human brain. *Nature* **2011**, *479*, 534–537.

93. Evrony, G.D.; Cai, X.; Lee, E.; Hills, L.B.; Elhosary, P.C.; Lehmann, H.S.; Parker, J.J.; Atabay, K.D.; Gilmore, E.C.; Poduri, A.; et al. Single-neuron sequencing analysis of L1 retrotransposition and somatic mutation in the human brain. *Cell* **2012**, *151*, 483–496.

94. Upton, K.R.; Gerhardt, D.J.; Jesuadian, J.S.; Richardson, S.R.; Sánchez-Luque, F.J.; Bodea, G.O.; Ewing, A.D.; Salvador-Palomeque, C.; van der Knaap, M.S.; Brennan, P.M.; et al. Ubiquitous L1 mosaicism in hippocampal neurons. *Cell* **2015**, *161*, 228–239.

95. Ravà, M.; D'Andrea, A.; Doni, M.; Kress, T.R.; Ostuni, R.; Bianchi, V.; Morelli, M.J.; Collino, A.; Ghisletti, S.; Nicoli, P.; et al. Mutual epithelium-macrophage dependency in liver carcinogenesis mediated by ST18. *Hepatology (Baltimore, Md.)* **2017**, *65*, 1708–1719.

96. Faulkner, G.J.; Kimura, Y.; Daub, C.O.; Wani, S.; Plessy, C.; Irvine, K.M.; Schroder, K.; Cloonan, N.; Steptoe, A.L.; Lassmann, T.; et al. The regulated retrotransposon transcriptome of mammalian cells. *Nat. Genet.* **2009**, *41*, 563–571.

97. Elbarbary, R.A.; Lucas, B.A.; Maquat, L.E. Retrotransposons as regulators of gene expression. *Science* **2016**, *351*, doi:10.1126/science.aac7247.

98. Collier, L.S.; Carlson, C.M.; Ravimohan, S.; Dupuy, A.J.; Largaespada, D.A. Cancer gene discovery in solid tumours using transposon-based somatic mutagenesis in the mouse. *Nature* **2005**, *436*, 272–276.

99. Turajlic, S.; McGranahan, N.; Swanton, C. Inferring mutational timing and reconstructing tumour evolutionary histories. *Biochim. Biophys. Acta* **2015**, *1855*, 264–275.

100. Lawrence, M.S.; Stojanov, P.; Mermel, C.H.; Robinson, J.T.; Garraway, L.A.; Golub, T.R.; Meyerson, M.; Gabriel, S.B.; Lander, E.S.; Getz, G. Discovery and saturation analysis of cancer genes across 21 tumour types. *Nature* **2014**, *505*, 495–501.

viruses

MDPI

Review

Protein-Coding Genes' Retrocopies and Their Functions

Magdalena Regina Kubiak and Izabela Makałowska *

Department of Integrative Genomics, Institute of Anthropology, Faculty of Biology, Adam Mickiewicz University in Poznan, 61-614 Poznan, Poland; magdalena.kubiak@amu.edu.pl
* Correspondence: izabel@amu.edu.pl; Tel.: +48-61-8295835

Academic Editors: David J. Garfinkel and Katarzyna J. Purzycka
Received: 25 February 2017; Accepted: 11 April 2017; Published: 13 April 2017

Abstract: Transposable elements, often considered to be not important for survival, significantly contribute to the evolution of transcriptomes, promoters, and proteomes. Reverse transcriptase, encoded by some transposable elements, can be used in *trans* to produce a DNA copy of any RNA molecule in the cell. The retrotransposition of protein-coding genes requires the presence of reverse transcriptase, which could be delivered by either non-long terminal repeat (non-LTR) or LTR transposons. The majority of these copies are in a state of "relaxed" selection and remain "dormant" because they are lacking regulatory regions; however, many become functional. In the course of evolution, they may undergo subfunctionalization, neofunctionalization, or replace their progenitors. Functional retrocopies (retrogenes) can encode proteins, novel or similar to those encoded by their progenitors, can be used as alternative exons or create chimeric transcripts, and can also be involved in transcriptional interference and participate in the epigenetic regulation of parental gene expression. They can also act in *trans* as natural antisense transcripts, microRNA (miRNA) sponges, or a source of various small RNAs. Moreover, many retrocopies of protein-coding genes are linked to human diseases, especially various types of cancer.

Keywords: retrotransposon; retrotransposition; retrocopy; retrogene; gene duplication; genome evolution

1. Introduction

A large fraction of human and other eukaryotic genomes consist of sequences that originated, directly or indirectly, as a result of transposable elements (TE) activities. Most of these genomic elements are considered to be nonessential to survival. However, TEs have a significant influence on genome evolution. TEs are probably most commonly known as recombination hotspots; however, they also contribute to the evolution of promoters and proteomes. Considering the direct contribution to proteomes, two scenarios exist: the coding potential of a TE is "domesticated" to perform host cellular function or TE-derived sequences are exapted into a coding portion of existing genes to generate novel protein variants [1]. One of the most impressive examples of a domesticated TE is the recombination-activating protein RAG1 [2]. This protein was derived from the transposase gene of a *Transib* DNA transposon 500 million years ago [3]. Another great example is the domestication of the *gag* gene. As many as 85 *gag*-like genes might exist in the human genome [4]. The second direct contribution to proteomes is exaptation, i.e., cooptation of different TE fragments to a new role. We call exapted those TE fragments that became a part of a coding sequence (CDS) but do not code for a protein domain attributed to their original function. In humans *Alu* sequences are major donors of exons, but exons acquired from other elements, such as LINEs, endogenous retroviruses, and DNA transposons, have also been reported [5]. Examples of the exaptation of an endogenous retrovirus envelope (*env*) gene are the primate genes *Syncytin-1* and *Syncytin-2*, which might be involved in the formation of the placenta [6].

166

Reverse transcriptase (RT), encoded by some TEs, can be used in *trans* to produce a DNA copy of any RNA molecule in the cell. This copy, reintegrated into the genome, will most likely be "dead on arrival" because none of the regulatory elements can be copied in RNA-mediated gene duplication. Therefore, these sequences are often called retropseudogenes or processed pseudogenes. Although the majority of these retrocopies are in a state of "relaxed" selection and remain "dormant" because they are lacking regulatory regions, many become functional. The evolutionary path of these functional retrocopies, called retrogenes, is not uniform. In the course of evolution, they may undergo subfunctionalization and share their function with their parent [7], develop a brand new function (neofunctionalization) [7], or replace their progenitors [8].

Retrogenes were long considered to be unimportant copies, but are currently called "seeds of evolution" since they have made a significant contribution to molecular evolution [9]. It has been shown that retrogenes play an important role in the diversification of transcriptomes and proteomes and may be responsible for a wealth of species-specific features. Some of these differences are highly important in medical research and may be the reason why results from animal studies cannot be transferred to humans. For example, the functional mouse retrogene *Rps23r1* reduces Alzheimer's beta-amyloid levels and tau phosphorylation [10]. This particular retrogene is rodent-specific and does not exist in the human genome. Another elegant example of the functional phenotypic effect of retroposition was demonstrated by *fgf4* retrogene studies. Insertion of this retrogene is responsible for chondrodysplasia in dogs. All breeds with short legs are carriers of the *fgf4* retrogene [11].

The discovery that retro sequences considered "junk DNA" may be functional and play a crucial role in shaping genome-specific features was one of the most surprising breakthroughs of human and other genome analyses. A large number of studies were recently performed to explore these unique sequences, yet our knowledge of retrogenes' evolution is exceptionally limited. In this review paper, we present recent studies aiming to decipher the functions of transcriptionally active retrocopies of protein-coding genes.

2. Retrotransposons as a Source of Cellular Reverse Transcriptase

The possibility of the reverse flow of genetic information from RNA to DNA was initially proposed in research conducted on the chicken Rous sarcoma virus [12]. The suggestion that the viral RNA genome can be transcribed into a DNA sequence and integrated into the host genome, together with the subsequent discovery of adequate enzymes [13,14], received the Nobel Prize in 1975. At that time, various mobile genomic elements, such as *Ty* in yeast [15] and LINE1 in human [16], were found to encode a reverse transcriptase, which was quickly associated with their mobilization abilities. This abundant group of "jumping genes", called retrotransposons, has been divided into two families characterized by the presence or absence of flanking long terminal repeats (LTRs) (Figure 1). The first group includes retroviral-like elements with LTRs, and the second consists mainly of long interspersed nuclear elements (LINEs) and short interspersed nuclear elements (SINEs) without LTRs. The LTR and non-LTR retrotransposons can be further subdivided into autonomous retrotransposons, which encode proteins required for mobilization, and nonautonomous retrotransposons, which utilize the retrotransposition machinery of the others.

The origin of retroelements and the evolutionary relation between them and retroviruses are not currently clear. The major hindrance in evolutionary analysis is the lack of unitary molecular characteristics across all retrosequences. Despite this limitation, several approaches have been used to portray the relations, for example, comparison of the reverse transcriptase [17] or ribonuclease H domains [18]. According to these analyses, non-LTR retrotransposons are the oldest group of retroelements and might be derived from various sequences, for example, having a common ancestor with RNA viruses [17] or originating from the prokaryotic group of II introns, also called retrointrons [18,19]. In the case of LTR retrotransposons, the evolutionary path is even more ambiguous. Doubtless, they are closely related with retroviruses, which is reflected in the common structural traits and notable similarities during the retrotransposition process. Autonomous LTR retrotransposons contain *gag* and *pol* genes encoding structural proteins required for the formation of the virus-like particles and enzymes involved in reverse transcription and incorporation of new copies into the genome. The lack of the envelope (*env*) gene, which enables recognition and infection of the host cell, is considered a main difference between retrotransposons and retroviruses. Although *env*-like genes have been found in some retrotransposons, their function is not fully understood [20,21]. Due to these facts, it was suggested that retrotransposons evolved from retroviruses that lost their infectious properties as a result of mutational inactivation of the genes responsible for intercellular movement. Alternatively, LTR retrotransposons could have arisen from ancestral non-LTR retrotransposable elements, with acquisition of the *env* gene providing an opportunity for retrovirus evolution [17–19].

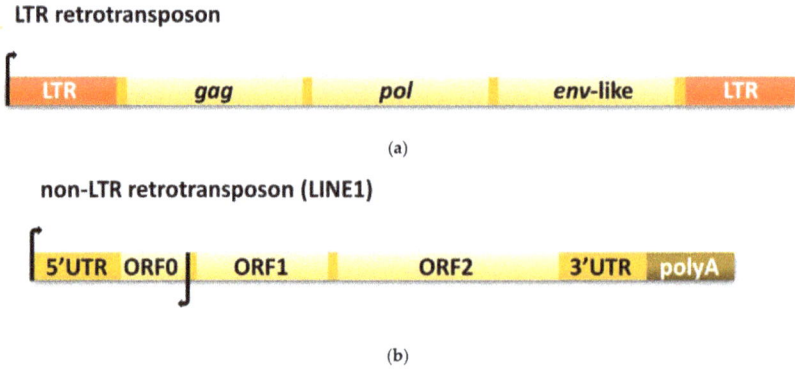

LTR retrotransposon

(a)

non-LTR retrotransposon (LINE1)

(b)

Figure 1. Schematic representation of the structural organization of (**a**) long terminal repeat (LTR) and (**b**) non-LTR retrotransposons. A detailed description can be found in Sections 2.2 and 2.3.

Both LTR and non-LTR retrotransposons are abundant in eukaryotic genomes (Figure 2). In plants, the most prominent fraction is composed of LTR retrotransposons and is frequently associated with an enlarged genome size. A good example is *Zea mays*, in which insertions of retrotransposons have doubled the genome size during the past three million years of evolution [22]. LTR retrotransposons occupy at least 55% of the *Z. mays* and 76% of the *Hordeum vulgare* genomes [23–25] and compose up to 58% and 91% of the *Allium cepa* and *Asparagus officinalis* DNA contigs, respectively [26]. Non-LTR retrotransposons are less abundant and usually cover only a few percent of the whole genome. Nevertheless, analysis of 23 plant genomes indicated that structurally intact (and therefore potentially active) LINEs are present from model species, such as *Arabidopsis thaliana*, to very complex genomes, such as *Picea abies* [27]. In metazoa, compared to in plants, retrotransposons usually constitute a smaller genome fraction (Figure 2). A remarkable example is one of the model organisms, *Drosophila melanogaster*, in which retrotransposons compose approximately 17% of the genome. In mammals, non-LTR retrotransposons are the most common; they represent approximately 28% and 35% of the mouse and human genomes, respectively. In contrast, LTR elements compose only 12% and 9% of these genomes [28]. However, not all retrotransposon families are still active. For example, only three non-LTR retrotransposon families—long interspersed elements 1 (L1s), *Alu* elements, and SVAs (named after their composite parts: SINE-R, VNTR (variable number of tandem repeats), and an *Alu*-like sequence)—are actively mobilized in the human genome [29].

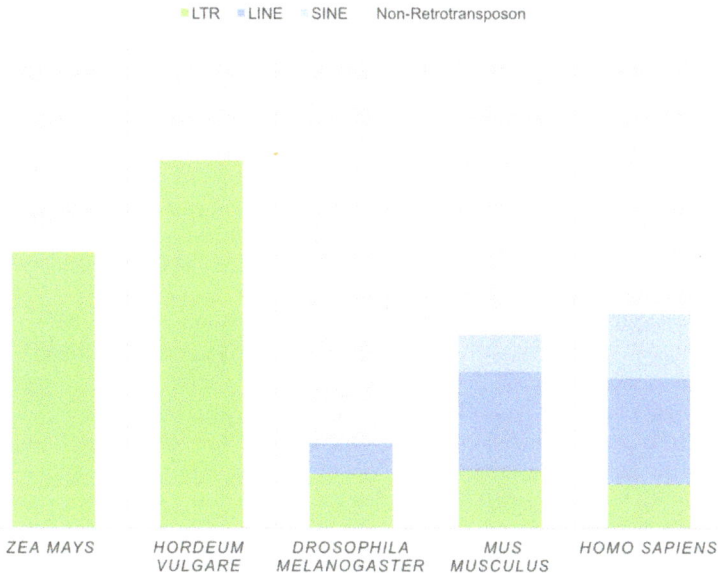

GENOMIC FRACTIONS OF RETROTRANSPOSONS

LTR   LINE   SINE   Non-Retrotransposon

Figure 2. Genomic fractions of retrotransposons in selected genomes. Based on: *Zea mays* [23], *Hordeum vulgare* [25], *Drosophila melanogaster*, *Mus musculus* (genome version mm10), *Homo sapiens* (genome version hg38)—Repeat Masker online dataset [28].

2.1. Retrotransposition of Nonautonomous Retrotransposons and Gene Copies

While autonomous retrotransposons encode their own mobilization machinery, the generation of nonautonomous retrotransposons and retrotransposed gene copies has long remained unclear. In the early 1980s, the analysis of repetitive sequences originating from small nuclear RNA and 7SL RNA showed that some are flanked by short direct repeats [30]. Because analogous repeats were found in combination with endogenous retroviruses and other mobile elements, a similar mechanism involving the insertion of reverse-transcribed RNA into the genome was suggested.

The endogenous reverse transcriptase activity and its ability to retrotranspose cellular non-retrotransposon mRNA were tested in a minigene system in HeLa cells [31]. Copies of a reporter gene with all characteristics indicating RNA-dependent duplication, including lack of introns, the presence of a polyA tract, and short repeats flanking the sequence, were found. Furthermore, the experiment showed that retrotransposition of mRNA in mammalian cells is still an active process. Further experimentation with mouse and human cells transfected by vectors containing reverse transcriptase revealed functional differences between the enzymes from retroviruses and those from LINE retrotransposon [32]. These results demonstrated that, in contrast to retroviral reverse transcriptase, LINE RT could generate reverse transcripts from RNAs, which do not show any sequence specificity or similarity to the LINE themselves. Therefore, autonomous non-LTR retrotransposons might be the source of endogenous RT involved in the formation of other retrotransposed genetic elements. Experiments performed by the same group [33] confirmed that LINE1 can act in *trans* and give rise to new retroposed gene copies. However, LINE1 elements are more effective in *cis*, which could be a consequence of the close proximity of LINE1-derived mRNA and proteins during translation or the limited life-time of protein in the lack of LINE1 mRNA [34]. This conclusion also explains the abundance of LINE1 elements in the genome compared to retrotransposed genes, as it was demonstrated that no more than 0.05% of the retrotransposition events conducted by LINE1 are related to retrotransposed gene formation [34].

Because the majority of experiments were conducted on animals, mostly mammalian models, the knowledge of retrotransposition in plants is restricted. Moreover, LINE elements constitute only a small fraction of the plant genome; their retrotransposition and creation of new gene copies is limited, but possible [35]. Since the first plant retrocopy was reported in *Solanum tuberosum* in 1987 [36], many retrotransposed genes have been found in various plant species [37–40]. For instance, a recent analysis of *A. thaliana* identified 251 retrocopies, of which 216 were described as novel [40]. In the RetrogeneDB2 repository, 1821 retrocopies in 37 plant species were identified [41]. Interestingly, analysis of plant retrocopies showed that LTR retrotransposons might also be involved in retrocopy formation. One of the best documented cases is the *Bs1* retrotransposon, which participated in the creation of retrocopies of three different cellular genes in the *Z. mays* genome [42,43]. Moreover, several additional retrocopies showing the signatures of LTR-mediated retroposition were recently found in both invertebrates and vertebrates [44]. Tan and coworkers [44] showed that LINE1-linked retrotransposition is dominant in mammals, whereas in mosquito, zebrafish, and chicken, retrocopies are created by both LTR- and non-LTR–mediated mechanisms. All polymorphic retrocopies found in *Drosophila* were formed by LTR retroposition. This type of retrotransposition is also more common in plants. Recent analysis of retrocopies derived from circular RNA (circRNA) in the mouse genome has shown LTR sequences localized in the flanking regions, which may suggest LTR-mediated retrotransposition [45].

Although the majority of retrotransposons have lost the capacity for mobilization—for instance, no more than 100 L1 in the human genome encode functional retrotranspositional machinery [46]—they are found to play various roles. They can shape genomes by acting as recombination hotspots or participating in exon-shuffling. They may also be used as new regulatory elements and alter the chromatin structure, thus influencing neighboring gene expression [47]. However, due to the integration of retrotransposons in random sites and the high likelihood of deleterious effects of insertion [48], different mechanisms of retrotransposition regulation evolved. Multilevel pathways are present in cells, starting from epigenetic silencing of retrotransposons by genomic DNA methylation and histone modification [49] to post-transcriptional positive and negative regulation [50].

2.2. LTR Retrotransposon-Based Transposition

The autonomous LTR retrotransposons are a diverse group; however, some common traits exist. They usually span several kilobases, are flanked by long terminal repeats, and contain promoters for RNA polymerase II localized in LTRs; however, only one RNA Pol II transcribes the retroelement. They encode at least two genes, which can overlap, be separated by terminal codons, or be fused into a single open reading frame [21,51]. One of the genes, *pol*, encodes various enzymatic domains, including reverse transcriptase and integrase, while the second, *gag*, produces structural proteins involved in the formation of virus-like particles (VLPs) (Figure 1a).

The mechanism of retrotransposition is similar to that observed in retroviruses. The process begins with transcription of the LTR retrotransposon by RNA polymerase II, after which the newly synthesized RNA is transported to the cytoplasm [20,51]. Next, translation and formation of VLPs occurs. Reverse transcriptase, integrase, and RNA molecules are typically packed in VLPs, and by chance cellular mRNA may also be encapsulated [20,44]. Reverse transcription usually starts by annealing tRNA to a primer binding site near the 5′ LTR. The microsimilarities between the LTR retrotransposon and cellular mRNA allow for template switching between these two molecules and therefore for incorporation of an mRNA sequence into the emerging cDNA [44]. Moreover, stretches of microsimilarity can appear in multiple places along the gene. Theoretically, template switching can occur several times, but frequently only a limited part of the parental mRNA is reverse transcribed. Furthermore, second strand synthesis occurs from a polypurine tract near the 3′ LTR. Next, the LTR ends of the new cDNA are bound by integrase (IN) and the VLP localizes a nearby nucleus. Finally, the cDNA–IN complex is transported to the nuclei, and integrase cuts the cellular DNA and joins the released ends with LTRs from the retrotransposed copy [51]. The complete cDNA is finally integrated into the genome, generally at a random site (Figure 3). In contrast to LINE1-mediated retroposition, new retrocopies acquire long tandem repeats and thus a promoter sequence, which enables further transcription and retrotransposition [44].

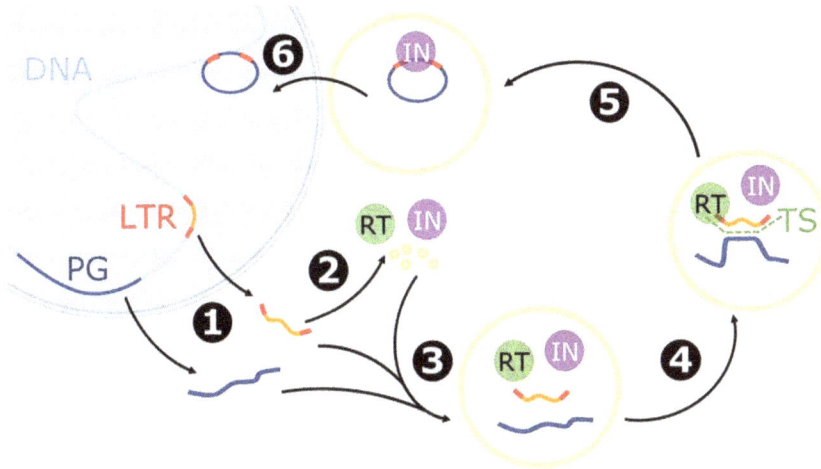

Figure 3. Model of mRNA retrotransposition mediated by the LTR retrotransposon (based on [20,44,51]). The blue sphere represents the nucleus. Stages: 1: Transcription of the LTR retrotransposon (LTR) and parental gene (PG). 2: Translation of LTR-encoded proteins, including reverse transcriptase (RT), integrase (IN), and proteins building virus-like particles (VLPs). 3: Formation of VLP with LTR-derived mRNA and parental gene mRNA. 4: Reverse transcription and template switch (TS). 5: Formation of the cDNA-IN complex. 6: Translocation and integration of chimeric cDNA into the genome.

2.3. Non-LTR Retrotransposition Mechanism

The most abundant non-LTR retrotransposon in mammals is LINE1 (L1); however, the set of the elements is quite variable between individuals [52]. For instance, a new insertion of L1 in humans occurs in between 1/95 and 1/270 of newborns [53]. Intact L1s are up to 6 kb in length and contain an internal promoter with sense and antisense activity localized in the 5' untranslated region [54]. We can distinguish two open reading frames: one encoding a smaller RNA-binding protein with chaperone activity [55] and the other encoding a larger protein with endonuclease and reverse transcriptase domains. Mutational analysis of these two open reading frames showed that both proteins are required for retrotransposition [56]. Additionally, a third, primate-specific open reading frame was recently found on the antisense strand [57], but the biological function of the transcript is not clear. The majority of L1 elements also contain a polyA tract in their 3' region (Figure 1b).

As experiments have shown, the LINE1 element can be transcribed by RNA polymerase II to mRNA, which is further used in translation and reverse transcription processes (Figure 4) [31,56]. The L1 mRNA is transported to the cytoplasm and translated, and the newly synthesized proteins interact with RNA in such a way that the one with enzymatic domains binds the 3' end, while the smaller ones with chaperone activity are attached along the entire RNA molecule [58]. The spatial distribution and proximity may influence the *cis* preference of L1 retrotransposition. However, small chaperone proteins encoded by L1 may also bind to other RNA molecules. For example, LINE1s are responsible for the *Alu* (from 7SL RNA) [59], snRNA [60] and mRNA retrocopies [33]. LINE1s can also mobilize primate-specific SVA retrotransposons [61,62], hY RNA [63] or even human endogenous retroviruses [64]. However, the observation that removal of the polyadenylation signal results in a loss of retrotransposition [65] indicates that the presence of a polyA tail may be one of the requirements for mobilizing RNA via a LINE1-derived mechanism.

The ribonucleoprotein particle formed by RNA and LINE1-encoded proteins needs to get close to the chromosome DNA, where the target-primed reverse transcription occurs. However, it is not clear how the import to the nucleus happens. It was proposed that retrotransposition occurs during cell division when the nuclear membrane is disrupted [66]. However, there is also evidence for retrotransposition in non-dividing cells [67,68] and, therefore, cell division may not be necessary for the ribonucleoprotein transport near genomic sequence. Nevertheless, the reverse transcriptase domain begins synthesis of a new DNA strand on an RNA template using the free 3′-OH resulting from endonuclease cleavage of the genomic sequence [69]. Alternatively, non-classic L1 insertion into pre-existing gaps in DNA can occur [70]. The subsequent steps have not been thoroughly studied, but it was proposed that a second nick is generated downstream, thereby enabling second DNA strand synthesis. The final step may also include creating target site duplication (TDS) of variable length. Additionally, in this process, the 5′ site of the newly arisen copy is often truncated.

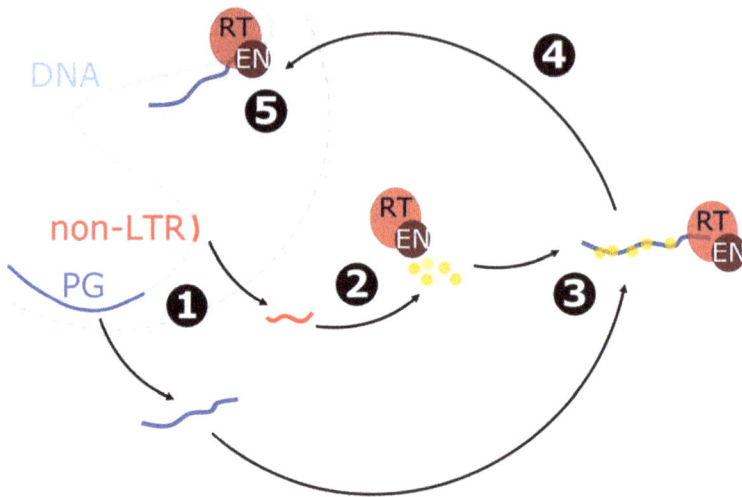

Figure 4. Model of mRNA retrotransposition mediated by non-LTR retrotransposon referring to LINE1. The blue sphere represents the nucleus; the dotted line shows the probability of nucleus membrane disruption. Description of the stages: 1: Transcription of the non-LTR retrotransposon (non-LTR) and parental gene (PG). 2: Translation of non-LTR–encoded proteins, including proteins with reverse transcriptase (RT) and endonuclease (EN) domains and RNA-binding proteins with chaperone activity. 3: Formation of ribonucleoprotein particle by binding of the proteins to the polyA tail of the parental gene mRNA. 4: Transport of ribonucleoprotein particle near the genome. 5: Reverse transcription and incorporation of the parental gene.

3. Number of Retrocopies across Genomes

The identification of retrocopies has been the subject of many studies and with the increasing number of sequenced genomes, as well as data from high-throughput experiments, new retrocopies are being discovered in various organisms. Examples of recent analyses include the comparative genomic study of the green algae retrogene repertoire [71] and inter-specific segregating retrocopies in cynomolgus and rhesus monkeys [72].

Basic annotations for retrotransposed gene copies, containing the localization and identified parental gene, are incorporated into several online databases. However, the most frequent description of these sequences is "processed pseudogene" or just "pseudogene", which could be misleading as it does not directly indicate origin by retrotransposition. The major difficulty in annotation is the lack of specific sequence motifs and the high rate of mutation accumulation in retrocopies. Even the most obvious feature, a lack of introns, does not have to be preserved, as retrogenes are known for intron incorporation [73]. Moreover, they may also acquire novel exons or be fused with another gene and used as an alternative exon [74]. On the other hand, intronless genes are not always an outcome of a retroposition event. For instance, single-exon histone-encoding genes are believed to have originated in prokaryota [75]. Other retrocopy traits, such as the polyA tails and insertion site repeats, are found only in evolutionarily young retrocopies. Regardless, the best approach to retrocopy identification so far is based on alignment of known protein-coding sequences to the genome.

The Ensembl genome browser is one of the major resources of publicly available genomic information. In addition to datasets analyzed in an automated way, manually curated data from the HAVANA project are included [76]. Annotation of retrocopies, called "processed pseudogenes", is based on an imperfect alignment between the genome and protein sequences, which enables multi-exon and single-exon gene models to be obtained. These single-exon annotations are interpreted as intronless, and therefore possibly retrotransposed, gene copies [77]. A similar method was applied in the PseudoPipe pipeline to predict all types of pseudogenes in the eukaryotic genomes stored on the Pseudogene.org server [78]. A slightly different approach was applied in UCSC Genome Browser, which besides Ensembl is the largest collection of genomic annotations. Here, the method for retrocopy identification was based on the RetroFinder pipeline, in which mRNA sequences were aligned to the genome [74,79]. Similar to the previous approaches, multi-exon and single-exon hits were obtained. The intronless candidates for retrocopies were then passed through multistage feature-based selection. For instance, a number of ancestral genes and putative retrocopy exons, as well as the presence and position of the polyA tail, were considered. PseudoPipe and RetroFinder, together with the HAVANA annotations, were used to produce a high-confidence dataset of pseudogenes, including retrocopies, in the human and mouse reference genomes in the GENCODE project [80].

More conservative approaches for retrotransposed gene copy annotation were applied in retroduplication-dedicated databases. The main purpose of utilizing more restrictive prediction criteria is the minimization of false positive results. On the other hand, they also provide expanded information enriched by potential function, interspecies conservation, and expression studies. There are three retrocopy-specific databases available. The first, HOPPSIGEN, is focused on human and mouse genomes [81]. A wider range of retrocopies, identified in six primate genomes, is available in RCPedia [82]. In addition, RetrogeneDB2 stores retrocopy information for 62 animal and 37 plant species [41]. Moreover, this database includes expression validation based not only on RNA-seq experiments but also on expressed sequence tag (EST), transcription start sites (TSS), and chromatin immunoprecipitation sequencing (ChIP-Seq) data. Another distinguishing feature of RetrogeneDB2 is the inclusion of data from retrocopy number variation studies that show retrocopies' indel frequencies across human populations [83].

The numbers of predicted retroposed pseudogenes across databases are summarized in Table 1. Although all the described methods rely on the alignment of known multi-exon coding genes (in the form of nucleotide or amino acid sequences) to the genome, the number of retrocopies differs because of the distinct filtering strategies, as well as the applied tools and parameters. For instance, the HOPPSIGEN dataset was obtained using the BLAST alignment tool [84], while in RetrogeneDB2, database retrocopies were identified on the basis of sequence alignments generated by LAST [85] (Table 2). Additional information about putative retrogenes can also be acquired from intronless gene databases. For instance, the IGD database [86] and SinEx DB [87] contain sets of single-exon coding genes for human and 10 mammalian genomes, respectively.

Table 1. Comparison of retrocopy sets available in the described databases. In addition, the number of retrogenes annotated in the *Homo sapiens*, *Pan troglodytes*, *Macaca mulatta*, and *Mus musculus* genomes is shown.

Database	Plants	Animals	Number of Retrocopies			
			Homo sapiens	*Pan troglodytes*	*Macaca mulatta*	*Mus musculus*
Non-specific databases						
Ensembl [1]	+	+	10,815	69	182	6999
UCSC [2]	+	+	13,742	–	–	18,456
GENCODE [3]	–	+	9074	–	–	6151
Pseudogene.org [4]	+	+	8739	7505	–	9809
Retrogene-dedicated databases						
HOPPSIGEN	–	+	5206	–	–	3428
RCPedia	–	+	7831	7733	7544	–
RetrogeneDB2	+	+	4611	3285	2377	4148

[1] Ensembl genome browser release 86 dataset filtered by "processed pseudogene" in BioMart; [2] UCSC RetroGenes v9 (*H. sapiens*) and v6 (*M. musculus*) dataset statistics; [3] GENECODE human (Release 25, GRCh38.p7), Mouse (Release M11, GRCm38.p4), "processed pseudogenes" with "gene" and "level 1" statuses were chosen from the comprehensive gene annotation GTF file; [4] class "processed" was selected for all organisms: chimp build 50 (CHIMP2), mouse build 84 (GRCm38), human build 83 (NCBI38).

Table 2. Approaches used in retrocopy-specific databases.

Database	Sequence Aligned to the Genome	Tool
HOPPSIGEN	gene coding sequence	TBLASTX
RCPedia	entire transcript	BLAT
RetrogeneDB2	protein	LAST

4. Molecular Functions of Genes Retrocopies

In the retroposition process, the parental regulatory elements are usually not inherited, and the new copy often slowly decays and is silenced by the accumulation of degenerative mutations in a process called pseudogenization or nonfunctionalization. However, a large number of transcripts originating from retrocopies were found in cells, which suggests the acquisition of active promoters. Retrogenes may use regulatory machinery of nearby genes or utilize distant CpG-rich sequences [88] and occasionally parts of their own sequence [89] to promote transcription. For instance, when the parental gene has multiple transcription start sites and the one located upstream of the promoter region is used, the retrotransposed transcript may contain a prominent part of the core promoter. A good example of such a case is the *PABP3* retrogene. Analysis of the sequence similarity between the abovementioned retrocopy and the parental gene showed high conservation of the 5′ upstream region, suggesting that the retrogene arose from a gene transcript containing a fragment of the promoter [89].

After gene duplication, the genome contains two similar genes, which may potentially have the same function. This initial functional redundancy, depending on the deleterious or beneficial impact on the organism, may be eliminated or preserved during evolution. Signatures of purifying selection, like intact open reading frames or lower rates of non-synonymous to synonymous mutation, are frequently used as evidence to support the putative functionality of the new copy [7]. However, sequence conservation is not direct evidence of functionality. Experiments focused on the molecular characteristics of retrogene products are usually necessary to confirm expression and to assess the function of the analyzed molecule.

The development of high-throughput sequencing methods has enabled wide characterization of genomes, transcriptomes, and even epigenomes of various organisms, as well as particular organs and tissues. Sequencing experiments provide information about gene expression, methylation patterns, and DNA–protein interactions, which may be used to create a complex description of retrogene functionality and regulation. However, the analysis of short reads produced by next-generation sequencers is challenging because of the difficulty in assigning them to one of two or more highly similar sequences, such as the parental gene and its retrocopies. Nevertheless, new technologies that produce long reads from a single molecule have improved, and retrocopy expression analyses should eventually become less problematic. While analytical problems may always exist, our current knowledge about retrocopies and their function is quite extensive, and many different examples are well documented (Figure 5).

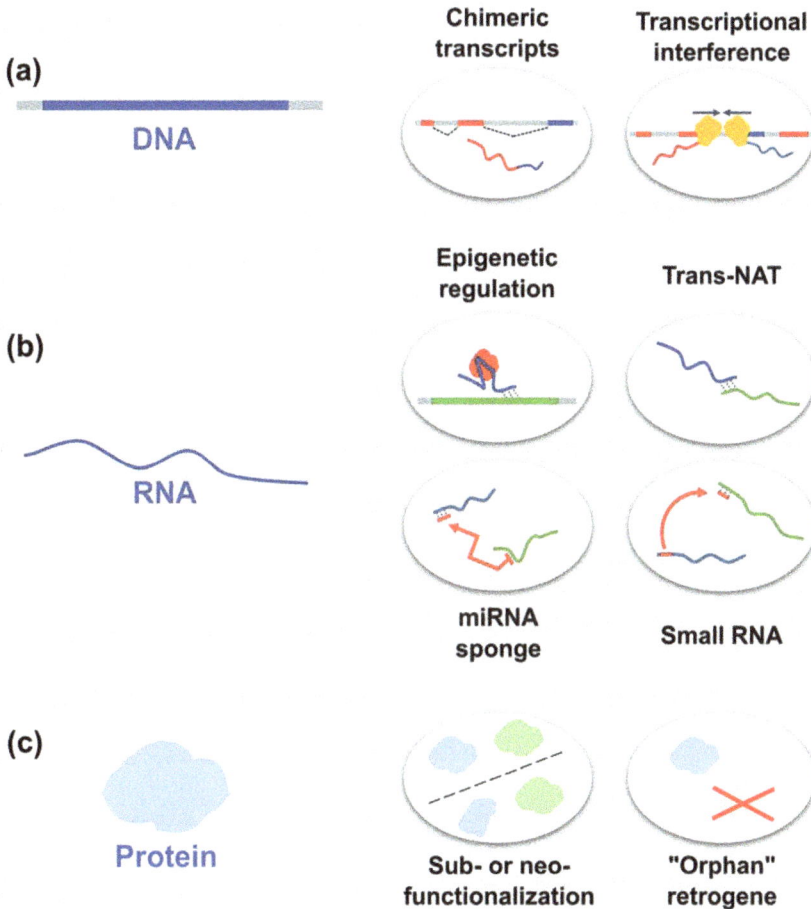

Figure 5. Selected functions of retrogenes. (**a**) On the DNA level, for example, retrogenes may be used as alternative exons and create chimeric transcripts or be involved in transcriptional interference; (**b**) as RNAs, they can participate in epigenetic regulation of parental gene expression and act as trans-natural antisense transcripts, microRNA (miRNA) sponges or a source of various small RNAs; (**c**) retrogenes can also encode proteins, which might retain the parental gene function (subfunctionalization), evolve a new function (neofunctionalization), or even functionally replace the parental gene ("orphan" retrogenes).

4.1. Protein-Coding Retrogenes

Retention of highly similar expressed sequences is often disadvantageous; therefore, conservation of the same gene function in its retrocopy is rare. Zhang [90] suggested that duplicates could be possible only in cases of highly demanded genes, such as rRNAs and histones. Although sharing the same function appears to be a natural consequence of gene duplication, retrogenes are often regulated in a different way than their ancestor genes because of distinct regulatory mechanisms. The main consequence of differences in regulatory machinery is spatio-temporal division of expression. In the most popular duplication-degeneration-complementation (DDC) model, the parental gene and retrocopy subdivide the ancestral function [91]. This mode of retrocopy evolution is called subfunctionalization.

One interesting but complex example of retrogene evolution is illustrated by a pair of retrocopies in *A. thaliana* [92]. The parental gene *CYP98A3* encodes the meta-hydroxylase engaged in the plant phenolic pathway and lignin biosynthesis. Two retrocopies of this gene, *CYP98A8* and *CYP98A9*, encode similar enzymes, which specialize in $3'$- and $5'$-hydroxylation of derivatives of spermidine localized in the pollen coat and wall. Analysis of the evolutionary history of the *CYP98* family shows that in *Brassicaceae*, the parental gene *CYP98A3* was retrotransposed and the *CYP98A8/9* ancestor retrogene went through tandem duplication [92]. Through further evolution, the loss of one copy was observed in some lineages; therefore, the remaining copy, *CYP98A8*, preserved the $3'$- and $5'$-hydroxylase activity. However, in the *A. thaliana* lineage, two copies were conserved, and subdivision of function occurred. *CYP98A9* acts as the $3'$-hydroxylase, while *CYP98A8* acts as the $5'$-hydroxylase. Moreover, the authors suggest that CYP89A9 retroprotein may have developed an additional function and play a role in flavonoid metabolism [92].

An example of subfunctionalization in humans is the cell cycle gene *CDC14B* and its retrocopy, *CDC14Bretro* [93]. Retrotransposition of the *CDC14Bpar* transcript occurred approximately 18–25 million years ago, and the ancestral function was probably conserved until the separation of African and Asian apes. However, in the African apes' ancestor genome, several mutations in the $5'$ end of retrogene sequence were fixed, and subcellular localization of the encoded protein was shifted from the microtubule to endoplasmic reticulum [93]. The authors suggest that the relocalization of the retrogene protein was due to a change in substrate and/or interaction partners and was related to the novel function development and specific testis/brain expression. Thus, this example may not be subfunctionalization but rather acquisition of a novel function that was previously not reported for the parental gene, i.e., neofunctionalization [94]. Another example where it is difficult to differentiate subfunctionalization from neofunctionalization is the *RAB6C* retrogene, which was retrotransposed approximately 21–25 million years ago in primates from the *RAB6A* gene [95]. In contrast to the parental gene, it is expressed in the centrosome, whereas the ancestor protein is found in the Golgi apparatus. This subcellular shift is probably a result of C-terminal extension impeding interactions with the Golgi. A *RAB6C* depletion experiment resulted in tetraploidization and duplication of the centrosome. Therefore, the retrogene-encoded protein developed a new function and is perhaps responsible for controlling cell cycle progression [95].

A retrogene-encoded protein may participate in an ancestral or new metabolic pathway, as shown above, but the process of retrocopy translation itself can also have an impact on the parental gene. An interesting example of this phenomenon is the connexin 43 (*Cx43*) gene and its retrocopy. Both encode proteins; however, retrogene expression is limited to breast cancer [96,97]. The *Cx43* retrogene has an intact open reading frame and encodes a protein of the same size as the parental gene, yet the ancestral function is not fully retained. While the parental *Cx43* gene is involved in cell growth control and intercellular communication, the retrogene seems to be not engaged in the second one [96]. Interestingly, further research indicated that the translational machinery preferentially binds to the retrogene, causing a shift in the parental gene mRNA from a polyribosome to monoribosome fraction, resulting in decreased expression of *Cx43* [97]. Silencing of the retrogene resulted in increased *Cx43* RNA and protein levels, supporting the regulatory role of the retrogene [97].

Another study of tumor-suppressor gene *TP53* suggested that gene duplicates, which arose via retrotransposition, play a role in the reduction of cancer risk in elephants [98]. Two of 19 retrogenes of the *TP53* gene, *TP53RTG12* and *TP53RTG19*, can be translated and enhance the DNA-damage response. In comparison to the parental TP53 protein, there are three significant differences: truncation in the DNA binding domain, lack of a nuclear localization signal, and lack of an oligomerization domain.

However, the interaction motif, which enables binding to a negative regulator, is conserved. Therefore, the researchers suggested two putative models of function for this protein. The retrogene-encoded proteins may bind to and block a TP53 negative regulator or may directly bind to the parental gene protein and prevent its ubiquitination [98].

A retrogene may not only share a function with the parental gene but may become a functional replacement after pseudogenization or deletion of its progenitor. The first so-called "orphan" retrogenes were discovered as a result of a comparative analysis of worm, chicken, and human genes. All 25 such cases identified in the human genome represent known and well-studied genes that were not previously recognized as retrocopies. Moreover, seven of them are associated with various human diseases, including diabetes, attention-deficit/hyperactivity disorder, congestive heart failure, and Huntington's disease. One of them, linked to hereditary spastic paraplegia, is the CHMP1B retrogene encoding chromatin-modifying protein 1B. The parental gene was pseudogenized in the ancestor of Old World and New World monkeys, but it is still active in rodents [8]. Another study identified a partially overlapping set of 10 "orphan" retrogenes [99]. Retroduplication and loss of parental genes was also found as a major mechanism involved in genome evolution and the generation of intronless genes in tunicates [100].

Another scenario of retrocopy function is dosage compensation when the level of the parental gene product is, for whatever reason, insufficient. For example, two testis-specific retrogenes, RPL10L and RPL39L, may compensate for their parental genes, which are inactivated during spermatogenesis [101]. A similar mechanism was proposed for the HNRNP G-T retrogene, which may functionally replace the parental protein in the course of meiosis [102].

Evidence of retrogene translation was also found during high-throughput data analysis. Because protein levels correlate with the levels of mRNA associated with polyribosomes, Mascarenhas and colleagues [103] analyzed the polyribosome loading of all RNA classes. An RNA sequencing experiment performed on cytosolic extracts and polyribosomal fractions showed that 18 pseudogenes exhibit significant polyribosome enrichment, which may suggest protein-coding potential [103]. Sixteen of these pseudogenes were found to be retrocopies.

Direct evidence of retrogene translation may be obtained via mass spectrometry. One interesting example of proteomic data utilization was an attempt to improve and further refine mouse genome annotations [104]. Analysis of 10.5 million tandem mass spectra enabled confirmation of the translation of known genes as well as identification of new protein-coding genes. Unique peptide hits were reported for nine retrocopies. One of them, retrotransposed from peptidylprolyl isomerase A (PPIA) gene, has two protein-coding variants that differ in the 5' region. Intriguingly, none of these translated retrogenes have orthologs in the human genome [104]. Tandem mass spectra were also used in the proteomic profiling of 30 human adult and fetal tissues and primary hematopoietic cells [105]. In these studies, translation of more than 17,000 known protein-coding genes and 808 novel coding regions, including 140 pseudogenes, was confirmed. Although the authors did not discriminate between pseudogene subtypes, retrogenes can be identified in this dataset. For instance, nine peptide sequences were matched uniquely with the fibrillarin-like 1 (FBLL1) retrogene. Some of the identified pseudogenes, like MAGE family member B6 pseudogene 1 (MAGEB6P1), had common for retrocopies testes-specific expression, while others, like voltage-dependent anion channel 1 pseudogene 7 (VDAC1P7), had broad expression patterns [105].

4.2. Consequences of Retrogene Insertion for the Host and nearby Genes

Across the 84,483 retrogenes annotated in 62 animal species deposited in RetrogeneDB, approximately 20% (18,468) are inserted into the intron of another gene. In the case of the human genome, this proportion is even larger, and so-called nested retrocopies constitute 44% of the total. As mentioned previously, a retrocopy may use a promoter of the host gene to become transcriptionally active. Retrocopies are also frequently found in gene-rich and actively transcribed chromatin regions [88]. Depending on the position of the retrocopy, different destinies for the emerging transcripts are proposed.

In the "hitchhike" scenario, a retrocopy inserted close to the 5' end of the host gene uses a 5' untranslated region (UTR) for its own transcription without a disruption of the host gene functions [7,88]. However, the production of resulting chimeric transcripts may theoretically reduce the normal host gene transcription level. Alternatively, a retrocopy insertion near the 3' end of the host gene may produce a new exon for a new splice isoform [88].

The first reported chimeric transcript generated by retrocopy insertion into a previously existing gene was *jingwei* observed in *Drosophila* [106]. A decade later, a fusion gene was found in a vertebrate during a study focused on resistance to human immunodeficiency virus type 1 (HIV-1) [107]. Interestingly, retroposition of cyclophilin A (*PPIA*) into the *TRIM5* gene, which occurred after the divergence of New World and Old World monkeys, resulted in the origin of a novel protein that was probably able to attach ubiquitin to HIV-1 virion proteins [107]. However, chimeric transcript creation is not a widespread phenomenon across the human and mouse genomes. Baertsch and colleagues [74] analyzed 726 highly expressed retrocopies and identified only 34 cases as potential gene fusions. In another study, new chimeric transcripts of 13 human and 14 mouse retrocopies, together with the upstream exons of the host gene, were identified [99]. As an example, the authors present the mouse *Taf9* retrogene, which uses the first two exons of the *Ak6* gene to become active.

The retrogene or part of it may also be used as an alternatively spliced exon of the host gene or a new 3' exon of a nearby gene. The *BRCA1* gene, for example, has an internal retrogene-derived exon, which generates a 22 amino acid cassette. In *SCP2*, *HLA-F*, and *KIAA0415*, alternatively spliced 3' end exons arising from antisense retrocopy insertions were found [74]. Chimeric transcripts were also found in an analysis focused on human-specific retrocopies harboring 5' CpG islands [108]. One is composed of exon 8 of the *RNF13* gene and *TMEM183A-r* retrogene and another is composed of the *HSF2BP* gene and *H2BFS* retrogene. In both cases the retrogene is located antisense to the parental gene orientation; therefore, these chimeric transcripts may potentially form RNA–RNA duplexes with their progenitors and participate in their regulation. The third chimeric transcript identified in these studies, created between the *VRK2* gene and *EIF3P3* retrogene, was expressed only in malignant prostate cancer cell lines [108]. Retrocopies expressed as a part of chimeric transcripts were also observed in plants. For instance, analysis of the retrogene repertoire in the genome of rice showed that more than one-third of the identified retrocopies recruit additional coding exons from nearby genes [38]. Retrogene-derived exons are found in many other plants' proteins; for example, a polygalacturonase-inhibiting protein encoded by a chimeric gene acting against *Aspergillus niger* polygalacutonase [38,109].

As already mentioned, the origination of chimeric transcripts may influence the expression level of a host gene. However, this is not the only way in which retrocopies regulate the expression of other genes. A group of *cis*-regulatory transcription-related functions was proposed for neighboring genes [110]. First, two genes localized in the same genomic locus and transcribed from independent promoters may directly impede each other's transcriptional processes [110]. This suppressive influence, called transcriptional interference, is a result of interactions between the dominant and sensitive promoters of overlapping genes. According to the promoter orientation and arrangement, several mechanisms of this process were proposed, including promoter competition, sitting duck interference, occlusion by another transient promoter occupation, collision of elongation complexes, and roadblocks precluding transcription [110]. Although transcriptional interference appears to be less frequent in higher eukaryotes, examples of this process have been reported; for instance, the results of an analysis focused on mouse and human genes that overlap in antisense orientation were consistent with the transcriptional collision model [111]. Transcriptional interference was also analyzed in the context of intronic retroelements and single-exon nested genes [112]. Using a minigene system and different deletion constructs, the *KTI12* retrogene located in the intron of the *TXNDC12* gene was analyzed. This particular case was selected based on the identification of three different ESTs, suggesting forced exonization of a portion of the *TXNDC12* intron upstream of the retrogene. The experimental results confirmed that expression of the *KTI12* retrocopy imposes utilization of cryptic acceptor splice sites and premature termination of *TXNDC12* gene transcription.

Host genes may also be affected by intronic retrogene methylation. Epigenetic regulation of expression is one of the mechanisms proposed for retrotransposon activity suppression; thus, a link between retrotransposition and methylation is strongly suggested. Recent analysis of retrocopy-associated CpG islands showed that 68% of them are methylated, which in comparison to the whole human genome is a significant proportion [113]. DNA methylation of cytosine bases at the CpG dinucleotides is a basic modification that occurs during genomic imprinting. Few examples of imprinted retrogenes have been reported. For instance, a systematic screen of known genes in mouse led to the identification of 11 imprinted retrogenes, of which three (*Mcts2*, *Dnajb3*, and *Oxct2a*) were nested in an intron of another gene. Interestingly, imprinting of these retrogenes is conserved in humans [114]. Detailed studies of the transcriptionally active *Mcts2* retrogene, inserted into the fourth intron of the *H13* gene, revealed that the retrogene's promoter is silenced by methylation in the female germline [115]. Surprisingly, the choice of polyA signal by the host gene depends on this epigenetic promoter modification. It was shown that expression of the retrogene from the paternal allele forces utilization of the upstream polyA site, resulting in a truncated transcript of the host gene [115]. Similar observations were reported for the *Nap1 l5* retrogene and *Herc3* host gene [116] and other retrogenes (for review, see: [117]).

4.3. Retrocopy Impact on Parental DNA

The main implication for the coexistence of highly similar sequences in a genome is the possibility of direct exchange of DNA fragments in a homologous recombination event. Retrogenes, like other genomic duplicates, can participate in gene conversion. However, this process was shown to be less frequent for genes localized in distant genomic regions [118]; therefore, it is less likely to occur between retrocopy–parental gene pairs, as they are in most cases localized on different chromosomes [119]. The retrotransposition process is also proposed as a possible mechanism in the RNA-mediated intron loss observed in Eukaryota [120–122]. In this model, precise deletion of the parental gene intron occurs as a result of recombination between genomic DNA and spliced, reverse-transcribed mRNA. Intron loss analysis conducted on 684 groups of orthologous genes from seven eukaryotic species supported the proposed mechanism [123]. Another study suggested that higher reverse transcriptase activity is connected to higher frequencies of intron loss and larger numbers of retrocopies. A correlation between these events was observed in mammals [124].

Retrogenes may have an impact on parental genes by contributing to epigenetic regulation of their expression. Nuclear antisense transcripts can act as a scaffold for chromatin remodeling complexes and therefore guide them to genomic loci on the basis of sequence complementarity [125]. It was also demonstrated that retrogenes may participate in this mechanism. A well-known example is the tumor-suppressor gene *PTEN*, whose expression is regulated at multiple levels by the *PTENpg1* retrogene [126,127]. *PTENpg1* has three different transcripts, of which two are antisense to the parental gene. One isoform, called *PTENpg1 asRNA alpha*, recruits the DNA methyl transferase 3A to the parental promoter [127]. As a result, repression of parental gene expression occurs by the addition of three methyl groups to histone H3 Lys27.

4.4. Retrogene Regulatory Functions on RNA Level

4.4.1. *Trans*-Natural Antisense Transcripts

The vast majority of transcribed retrogenes do not have conserved open reading frames and therefore have roles other than protein-coding functionality. Retrogenes may be expressed independently from both DNA strands, which increases the range of possible interactions on the RNA level. For instance, bioinformatic analysis of ESTs showed that retrogenes constitute 15% of the 87 pseudogenes that were found to be expressed from an antisense strand [128]. Moreover, recent studies focused on long non-coding RNAs that overlap retrocopies across the human genome identified three retrocopy-derived antisense RNAs (asRNAs). These retrocopies are potentially capable of forming RNA:RNA duplexes with their parental genes [129]. One of these long non-coding RNAs (lncRNAs) derived from the *HNRNPA1P7* retrogene may contribute to heterogeneous nuclear ribonucleoprotein A1 (*hnRNPA1*) pre-mRNA processing. The bioinformatic analysis strongly suggested that asRNA might mask the $5'$ splice site in the sixth intron of parental gene transcript and therefore enable the expression of isoform with a longer sixth exon. Another identified lncRNA, antisense to a ribosomal protein L23a (*RPL23A*) retrocopy, is potentially able to mask microRNA target sites in seven splice forms of the parental gene, thus controlling their stability [129]. An interesting example was found in the snail *Lymnaea stagnalis*. A retrocopy of a gene encoding nitric oxide synthase (NOS) includes a region of antisense homology to its progenitor's transcript. The antisense region of the pseudogene transcript forms an RNA:RNA duplex with the NOS-encoding mRNA and prevents its translation [130]. Antisense and sense transcripts can also interact. *PTENpg1*, described earlier in the context of epigenetic modification, has a second antisense transcript, *PTENpg1 asRNA beta*, which stabilizes the expression of the sense transcript [127].

4.4.2. MicroRNA Sponges

Continuing with the example of *PTENpg1*, another gene regulation mechanism should be mentioned. The sense transcript of *PTENpg1* shares many microRNA binding sites with the parental gene and consequently exhibits "sponging" activity, manifested by releasing miRNA-mediated repression of *PTEN* [126]. *PTEN* was identified as a tumor-suppressor gene that negatively regulates the phosphatidylinositol 3-kinase signaling pathway involved in cell proliferation control [131]. Downregulation of *PTEN*, which occurs when the retrogene transcript is absent, enhances proliferation and cell growth and decreases sensitivity to cell death, which promotes tumorigenesis [126,131]. The *PTEN-PTENpg1* endogenous competition for shared microRNA molecules has an oncosuppressive effect on different human cancers, including prostate and colon cancers [126], melanoma [132], renal cell carcinoma [133], and hepatocellular carcinoma [134]. Since the identification of *PTENpg1* as a microRNA sponge and functional antisense RNA and its role in tumorigenesis, many retrogene-derived non-coding RNAs have been analyzed in this context [135–138].

One interesting example is the high mobility group A1 (*HMGA1*) gene and its two retrogenes, *HMGA1P6* and *HMGA1P7* [139]. The parental gene encodes proteins involved in chromatin architecture organization and therefore gene expression regulation. Whereas in adult normal cells *HMGA1* proteins are expressed at a very low level, they are overexpressed in cancers. Analysis of the retrogene–parental gene expression pattern in thyroid and ovarian carcinomas showed a positive correlation and suggested gene co-regulation [139]. Both investigated retrogenes conserved the miRNA target site of the parental gene. The abilities of miRNA–retrogene interactions were experimentally evaluated by transfection of miRNAs into human breast adenocarcinoma cells. A significant reduction in the *HMGA1*, *HMGA1P6*, and *HMGA1P7* mRNA levels was observed, which strongly supports the retrogene miRNA "sponging" activity. These retrogenes may also regulate the expression of other genes, including cancer-related ones, such as *HMGA2*, *VEGF*, and *EZH2* [139]. Another research group found that the *HMGA1P7* retrogene is also involved in the regulation of *H19* non-coding gene and *IGF2* gene expression in human breast cancer [140]. In contrast to the oncosuppressive function of *PTENpg1*, *HMGA1P6* and *HMGA1P7* show oncogenic activity and contribute to cancer progression.

Retrogenes may also compete with parental genes for other molecules. For instance, the *Cx43* retrogene transcript shows higher affinity to translational machinery than the parental gene mRNA and therefore decreases the parental protein level [97].

4.4.3. Small RNA

Retrogenes have also been proposed as a source of several classes of small RNAs. For example, they may provide the sequence for novel miRNA genes. Primate-specific *miR-492* may be expressed from both *KRT19* gene and retrogene loci [141,142]. Interestingly, *hs-miR-492* was proposed to play a role in the progression of hepatoblastoma [142] and was found to be a proto-oncogenic miRNA, acting as a cell proliferation promoter in breast cancer [143]. Several other miRNAs stored in miRBase [144], the main source of miRNA annotations, lie within retrogenes. For instance, hsa-*miR-622*, which acts as a suppressor of tumorigenesis in cancers, including hepatocellular carcinoma [145], overlaps with the genomic locus of the *KRT18P27* retrocopy. Other examples include *hsa-miR-7161*, located in the *TATDN2P2* retrocopy, and *hsa-miR-4788*, overlapping *HMGB3P13*. However, experimental analysis must be conducted to verify the actual miRNA coding potential.

Piwi-interacting RNA (piRNA) involved in the formation of silencing complexes in animal germ lines can also be derived from retrocopies. Total small RNA profiling of the marmoset testis showed abundant expression of piRNAs [146]. Across various clusters, four antisense-oriented piRNAs localized within retrocopies were found, and the authors suggested that the piRNAs originating from these clusters might regulate the parental gene expression by cleaving mRNAs. Moreover, retrogene-derived piRNAs appear to be species-specific because clusters found in the marmoset were absent in the mouse [146]. Recent bioinformatic analysis focused on the small RNA of human sperm showed that piRNA clusters also contain non-coding genes, one of which overlaps with the *NPAP1P6* retrocopy [147]. Interestingly, recent evolutionary analysis showed that the parental gene of the *NPAP1P6* retrocopy, *NPAP1*, was created via duplication of a retrotransposed ancestral paralog derived from the vertebrate nucleoporin gene *POM121* [148]. *NPAP1* is a primate-specific imprinted gene that encodes a nuclear pore-associated protein associated with Prader–Willi syndrome. As the authors emphasized, this syndrome is linked with testis dysfunction, which supports the possible relation between many sperm piRNAs and the analyzed retrocopy [147]. Several additional piRNAs from retrocopies, including *IMPDH1P5*, *TMX2P1*, and *RP11-545A16.3*, which may potentially interact with the protein-coding genes due to high sequence complementarity, were identified [147]. Six retrogene-derived piRNAs that potentially regulate parental genes were found in late mouse spermatocytes [149]. For one of them, experimental evidence of post-transcriptional regulation of *Stambp* gene was provided. By generating two mouse strains with gene-trap insertions upstream of the retrocopy, the authors demonstrated relationships between the piRNA precursor, piRNA, and the parental gene levels, which clearly suggests a role of the retrocopy in *Stambp* regulation [149]. PiRNA clusters overlapping retrocopies were also found in other recent studies (e.g., [150,151]).

Antisense transcripts from retrocopies can pair with other transcripts, including those of the parental gene. This double-stranded RNA may be processed into endogenous small interfering RNA (endo-siRNA). Analysis of the small RNA profiles from wild-type and seven RNA-silencing mutants of *A. thaliana* showed overrepresentation of siRNAs in the transposons, retroelements, and pseudogenes, which may suggest that these sequences are regulated by siRNA-generating systems [152]. A year later, two scientific reports published in parallel showed that endo-siRNAs annotated in the retrocopy locus may regulate their parental genes' expression in mouse oocytes [153,154]. For instance, 77 small RNAs were mapped to an expressed retrocopy (*Gm15681*) of the protein phosphatase 4 regulatory subunit 1 gene (*Ppp4r1*). Moreover, almost all were oriented antisense to the parental gene, which strongly suggested that the siRNAs originated from the retrocopy-parental gene double stranded RNA (dsRNA) region [154]. In another analysis focused on developing rice grains [155], among 145 pseudogenes identified as good candidates for generating antisense small RNAs, 16.6% were retrocopies. However, their *cis*-activity was questioned due to a low rate of identified gene–retrogene complementary regions from which siRNA can be produced [155]. A cluster of siRNAs derived from pseudogenes was also identified in African *Typanosoma brucei* [156]. More detailed studies were conducted for endo-siRNAs derived from human pseudogenes in hepatocellular carcinoma. A well-documented example is the human endo-siRNA from the retrogene of the protein phosphatase 1K (*PPM1K*) gene. The retrocopy can fold into a hairpin structure due to inverted repeats and can be processed in at least two endo-siRNAs, one of which downregulates the parental gene and NIMA-related Kinase 8 (*NEK8*) gene, inhibiting cell proliferation in hepatocellular carcinoma. Thus, the retrogene could be considered a tumor-suppressor gene [157].

Retrocopy-derived small RNAs were also found in the human transcriptome during characterization of the non-coding RNA repertoire. A higher density of small RNA was observed in retrocopies than in duplicated pseudogenes or coding genes. Interestingly, transcription-dependent H3K9me3 enrichment was observed in some cases, suggesting that pseudogene-derived small RNAs, including retrocopy-derived RNAs, may play a role in modulating the epigenetic suppression of those pseudogenes, as well as neighboring gene expression [158].

5. Retrogenes in Diseases

Many retrocopies of protein-coding genes are linked to human diseases, especially various types of cancer (for review, see: [159–162]). It was recently suggested that expression of evolutionarily young non-coding genes in tumors might be considered a new biological phenomenon [163]. A good example of a cancer-related retrogene is the *RHOB* gene, a tumor suppressor of the Rho GTPases family, which arose via retroposition in the early stage of vertebrate evolution [164]. Another retrogene, *UTP14c*, was linked to ovarian cancer predisposition [165]. An analysis of 293 samples representing 13 cancer and normal tissue types revealed 218 pseudogenes expressed only in cancer samples. Out of them, 178 were observed in multiple cancers and 40 were identified in a single cancer type only [166].

As many reports have shown, retrocopies can be used as diagnostic biomarkers, such as the *INTS6P1* retrogene, for which a low expression level in plasma is linked with hepatocellular carcinoma [167], or as prognostic markers, such as tumor-suppressive *PTENpg1* [126,132–134] and the oncogenic *HMGA1P6* and *HMGA1P7* [139,140] retrogenes mentioned in previous sections. Another retrocopy of the *HMGA1* gene is also linked with disease; its overexpression has been found in human type 2 diabetes. Further analysis indicated that this retrocopy post-transcriptionally regulates parental gene expression by competing for critical RNA stability factor and, as a result, suppresses the expression of the insulin receptor gene. Therefore, this retrocopy contributes to insulin resistance [168].

Due to the reactivation of retrotransposons in somatic cells during cancer development, the formation of new retrocopies was observed [169,170]. Somatically acquired retrocopies are present in lung and colorectal cancers. Moreover, insertions occur not only in intragenic regions but also in other genes, which may have implications for their expression. For instance, a *KRT6A* retrocopy replaced the 3′UTR of the *MLL* gene transcript and a *PTPN12* retrocopy caused deletion of the promoter and first exon of the *MGA* gene [170].

Retrogenes may also play active roles in the regulation of signaling pathways involved in inflammation [171]. A mouse retrocopy of ribosomal protein S15A gene, called *Lethe* (*Rps15a-ps4*), is induced by inflammatory cytokines TNFα and IL-1β. Moreover, it can bind and block RelA homodimers, which are required for NF-κB activation; therefore, *Lethe* plays the role of a negative inflammatory response regulator. Interestingly, the retrogene is expressed in an age-dependent manner [171]. Mutation in another retrogene, *TACSTD2* (tumor-associated calcium signal transducer 2) causes gelatinous drop-like corneal dystrophy, leading to blindness [172]. An insertion of a retrocopy, similar to insertion of *L1* or *Alu* elements, may disrupt a gene structure. An example of such event is insertion of a retrocopy of *TMF1* gene into the *CYBB* gene on the X chromosome. This insertion induced aberrant *CYBB* mRNA splicing and introduced a premature stop codon that resulted in chronic granulomatous disease [173].

Retrocopies were also incorporated into analyses conducted in the context of neurodegenerative disorders, including Alzheimer's, Huntington's, and Parkinson's diseases [174], as well as muscular dystrophy [175].

As described above, the chimeric transcript resulting from cyclophilin A (*PPIA*) retrogene insertion into the *TRIM5* host gene in the owl monkey is involved in HIV-1 resistance [107]. Retrocopies may also be associated with the host response during pathogen infection in humans, as significant expression changes were observed after HIV-1 and human type 2 adenovirus infection [176,177]. For instance, expression pattern analysis after HIV-1 infection of human T-cells showed that the most upregulated pseudogene group was a group of retrocopies. Additionally, retrocopies accounted for eight out of 13 cases of underexpressed pseudogenes [176]. These results suggest that both tandemly duplicated and retroposed pseudogenes may be involved in host–pathogen interaction pathways.

6. Retroposition and Genetic Variation

Retroposition gives rise to considerable genetic variation between individuals. Recent developments in sequencing technology allow researchers to move beyond the analysis of individual genomes from model organisms to the study of retrocopies within a population. The 1000 Genomes Project [178] could be mentioned as an example of a large-scale sequencing project that enables the exploration of differences in copy-number variation within human populations. Recent studies [66,169,179], focused on the retrocopy repertoire in human populations, uncovered a total of 208 polymorphic retrocopies [180] called retroduplication variations (RDVs). Moreover, in two of them [66,169], RDV polymorphisms were used as genomic markers for the reconstruction of human population history. In another study, concentrated on retrocopies deletions, 214 indels that affected 190 retrocopies were identified. Out of them, 68 were found to be ancestral (i.e., their orthologs were found in at least one another Hominidae species) and the polymorphism of these retrocopies clearly resulted from a deletion. This study also showed a variation in the retrocopies' expression level [83].

7. Conclusions

Retrocopies were long considered non-functional pseudogenes or even "junk DNA"; however, current studies show that they contribute significantly to molecular evolution. Retrocopies have been found as factors shaping differences between species, individuals, or even tissues and cell types; therefore, they are considered a source of genetic polymorphism. They are also important players in complex cellular pathways, including immune response and tumorigenesis. Moreover, a large and increasing range of putative retrocopy functions make them an interesting subject of molecular and medical studies. The numerous studies performed to date have enriched our comprehension of the course and dynamics of retrocopy–gene interactions at the DNA, RNA, and protein levels. Nevertheless, many questions remain unsolved, and further analyses are necessary to accurately describe the plant and animal retrocopy repertoire, evolution, and functions.

Acknowledgments: This work was supported by the National Science Centre [grant No. 2013/11/B/NZ2/02598] and the KNOW Poznan RNA Centre [grant No. 01/KNOW2/2014].

Author Contributions: M.R.K. drafted the manuscript. Both authors edited, corrected, and approved the manuscript.

Conflicts of Interest: The authors declare no conflict of interest.

References

1. Makalowski, W.; Kischka, T.; Makałowska, I. Contribution of transposable elements to human proteins. In *Encyclopedia of Life Sciences*; John Wiley & Sons, Ltd.: Chichester, UK, 2017.
2. Agrawal, A.; Eastman, Q.M.; Schatz, D.G. Transposition mediated by RAG1 and RAG2 and its implications for the evolution of the immune system. *Nature* **1998**, *394*, 744–751. [PubMed]
3. Kapitonov, V.V.; Jurka, J. Harbinger transposons and an ancient HARBI1 gene derived from a transposase. *DNA Cell Biol.* **2004**, *23*, 311–324. [CrossRef] [PubMed]
4. Campillos, M.; Doerks, T.; Shah, P.K.; Bork, P. Computational characterization of multiple Gag-like human proteins. *Trends Genet.* **2006**, *22*, 585–589. [CrossRef] [PubMed]
5. Gotea, V.; Makałowski, W. Do transposable elements really contribute to proteomes? *Trends Genet.* **2006**, *22*, 260–267. [CrossRef] [PubMed]
6. Mi, S.; Lee, X.; Li, X.; Veldman, G.M.; Finnerty, H.; Racie, L.; LaVallie, E.; Tang, X.Y.; Edouard, P.; Howes, S.; et al. Syncytin is a captive retroviral envelope protein involved in human placental morphogenesis. *Nature* **2000**, *403*, 785–789. [PubMed]
7. Kaessmann, H.; Vinckenbosch, N.; Long, M. RNA-based gene duplication: Mechanistic and evolutionary insights. *Nat. Rev. Genet.* **2009**, *10*, 19–31. [CrossRef] [PubMed]
8. Ciomborowska, J.; Rosikiewicz, W.; Szklarczyk, D.; Makałowski, W.; Makałowska, I. "Orphan" retrogenes in the human genome. *Mol. Biol. Evol.* **2013**, *30*, 384–396. [CrossRef] [PubMed]
9. Brosius, J. Retroposons—Seeds of evolution. *Science* **1991**, *251*, 753. [CrossRef] [PubMed]

10. Zhang, Y.; Liu, S.; Zhang, X.; Li, W.-B.; Chen, Y.; Huang, X.; Sun, L.; Luo, W.; Netzer, W.J.; Threadgill, R.; et al. A functional mouse retroposed gene *Rps23r1* reduces Alzheimer's beta-amyloid levels and Tau phosphorylation. *Neuron* **2009**, *64*, 328–340. [CrossRef] [PubMed]

11. Parker, H.G.; VonHoldt, B.M.; Quignon, P.; Margulies, E.H.; Shao, S.; Mosher, D.S.; Spady, T.C.; Elkahloun, A.; Cargill, M.; Jones, P.G.; et al. An expressed *Fgf4* retrogene is associated with breed-defining chondrodysplasia in domestic dogs. *Science* **2009**, *325*, 995–998. [CrossRef] [PubMed]

12. Temin, H.M. The nature of the provirus of Rous sarcoma. *Natl. Cancer Inst. Monogr.* **1964**, *17*, 557–570.

13. Baltimore, D. RNA-dependent DNA polymerase in virions of RNA tumour viruses. *Nature* **1970**, *226*, 1209–1211. [CrossRef] [PubMed]

14. Temin, H.M.; Mizutani, S. RNA-dependent DNA polymerase in virions of Rous sarcoma virus. *Nature* **1970**, *226*, 1211–1213. [CrossRef] [PubMed]

15. Garfinkel, D.J.; Boeke, J.D.; Fink, G.R. Ty element transposition: Reverse transcriptase and virus-like particles. *Cell* **1985**, *42*, 507–517. [CrossRef]

16. Mathias, S.L.; Scott, A.F.; Kazazian, H.H.; Boeke, J.D.; Gabriel, A. Reverse transcriptase encoded by a human transposable element. *Science* **1991**, *254*, 1808–1810. [CrossRef] [PubMed]

17. Xiong, Y.; Eickbush, T.H. Origin and evolution of retroelements based upon their reverse transcriptase sequences. *EMBO J.* **1990**, *9*, 3353–3362. [PubMed]

18. Malik, H.S.; Eickbush, T.H. Phylogenetic analysis of ribonuclease H domains suggests a late, chimeric origin of LTR retrotransposable elements and retroviruses. *Genome Res.* **2001**, *11*, 1187–1197. [CrossRef] [PubMed]

19. Boeke, J.D. The unusual phylogenetic distribution of retrotransposons: A hypothesis. *Genome Res.* **2003**, *13*, 1975–1983. [CrossRef] [PubMed]

20. Havecker, E.R.; Gao, X.; Voytas, D.F. The diversity of LTR retrotransposons. *Genome Biol.* **2004**, *5*, 225. [CrossRef] [PubMed]

21. Eickbush, T.H.; Jamburuthugoda, V.K. The diversity of retrotransposons and the properties of their reverse transcriptases. *Virus Res.* **2008**, *134*, 221–234. [CrossRef] [PubMed]

22. SanMiguel, P.; Gaut, B.S.; Tikhonov, A.; Nakajima, Y.; Bennetzen, J.L. The paleontology of intergene retrotransposons of maize. *Nat. Genet.* **1998**, *20*, 43–45. [PubMed]

23. Meyers, B.C.; Tingey, S.V.; Morgante, M. Abundance, distribution, and transcriptional activity of repetitive elements in the maize genome. *Genome Res.* **2001**, *11*, 1660–1676. [CrossRef] [PubMed]

24. Messing, J.; Bharti, A.K.; Karlowski, W.M.; Gundlach, H.; Kim, H.R.; Yu, Y.; Wei, F.; Fuks, G.; Soderlund, C.A.; Mayer, K.F.X.; et al. Sequence composition and genome organization of maize. *Proc. Natl. Acad. Sci. USA* **2004**, *101*, 14349–14354. [CrossRef] [PubMed]

25. International Barley Genome Sequencing Consortium; Mayer, K.F.X.; Waugh, R.; Brown, J.W.S.; Schulman, A.; Langridge, P.; Platzer, M.; Fincher, G.B.; Muehlbauer, G.J.; Sato, K.; et al. A physical, genetic and functional sequence assembly of the barley genome. *Nature* **2012**, *491*, 711–716. [CrossRef] [PubMed]

26. Vitte, C.; Estep, M.C.; Leebens-Mack, J.; Bennetzen, J.L. Young, intact and nested retrotransposons are abundant in the onion and asparagus genomes. *Ann. Bot.* **2013**, *112*, 881–889. [CrossRef] [PubMed]

27. Heitkam, T.; Holtgräwe, D.; Dohm, J.C.; Minoche, A.E.; Himmelbauer, H.; Weisshaar, B.; Schmidt, T. Profiling of extensively diversified plant LINEs reveals distinct plant-specific subclades. *Plant J.* **2014**, *79*, 385–397. [CrossRef] [PubMed]

28. RepeatMasker Home Page. Available online: http://www.repeatmasker.org/ (accessed on 16 January 2017).

29. Konkel, M.K.; Batzer, M.A. A mobile threat to genome stability: The impact of non-LTR retrotransposons upon the human genome. *Semin. Cancer Biol.* **2010**, *20*, 211–221. [CrossRef] [PubMed]

30. Van Arsdell, S.W.; Denison, R.A.; Bernstein, L.B.; Weiner, A.M.; Manser, T.; Gesteland, R.F. Direct repeats flank three small nuclear RNA pseudogenes in the human genome. *Cell* **1981**, *26*, 11–17. [CrossRef]

31. Maestre, J.; Tchénio, T.; Dhellin, O.; Heidmann, T. mRNA retroposition in human cells: Processed pseudogene formation. *EMBO J.* **1995**, *14*, 6333–6338. [PubMed]

32. Dhellin, O.; Maestre, J.; Heidmann, T. Functional differences between the human LINE retrotransposon and retroviral reverse transcriptases for in vivo mRNA reverse transcription. *EMBO J.* **1997**, *16*, 6590–6602. [CrossRef] [PubMed]

33. Esnault, C.; Maestre, J.; Heidmann, T. Human LINE retrotransposons generate processed pseudogenes. *Nat. Genet.* **2000**, *24*, 363–367. [PubMed]

34. Wei, W.; Gilbert, N.; Ooi, S.L.; Lawler, J.F.; Ostertag, E.M.; Kazazian, H.H.; Boeke, J.D.; Moran, J.V. Human L1 retrotransposition: *Cis* preference versus *trans* complementation. *Mol. Cell. Biol.* **2001**, *21*, 1429–1439. [CrossRef] [PubMed]

35. Zhu, Z.; Tan, S.; Zhang, Y.; Zhang, Y.E. LINE-1-like retrotransposons contribute to RNA-based gene duplication in dicots. *Sci. Rep.* **2016**, *6*, 24755. [CrossRef] [PubMed]

36. Drouin, G.; Dover, G.A. A plant processed pseudogene. *Nature* **1987**, *328*, 557–558. [CrossRef]

37. Benovoy, D.; Drouin, G. Processed pseudogenes, processed genes, and spontaneous mutations in the *Arabidopsis* genome. *J. Mol. Evol.* **2006**, *62*, 511–522. [CrossRef] [PubMed]

38. Wang, W.; Zheng, H.; Fan, C.; Li, J.; Shi, J.; Cai, Z.; Zhang, G.; Liu, D.; Zhang, J.; Vang, S.; et al. High rate of chimeric gene origination by retroposition in plant genomes. *Plant Cell* **2006**, *18*, 1791–1802. [CrossRef] [PubMed]

39. Sakai, H.; Mizuno, H.; Kawahara, Y.; Wakimoto, H.; Ikawa, H.; Kawahigashi, H.; Kanamori, H.; Matsumoto, T.; Itoh, T.; Gaut, B.S. Retrogenes in rice (*Oryza sativa* L. ssp. japonica) exhibit correlated expression with their source genes. *Genome Biol. Evol.* **2011**, *3*, 1357–1368. [CrossRef] [PubMed]

40. Abdelsamad, A.; Pecinka, A. Pollen-specific activation of *Arabidopsis retrogenes* is associated with global transcriptional reprogramming. *Plant Cell* **2014**, *26*, 3299–3313. [CrossRef] [PubMed]

41. Rosikiewicz, W.; Kabza, M.; Kosiński, J.; Ciomborowska, J.; Kubiak, M.R.; Makałowska, I. RetrogeneDB—A database of plant and animal retrocopies. *Database (Oxford)*, under review.

42. Jin, Y.K.; Bennetzen, J.L. Integration and nonrandom mutation of a plasma membrane proton ATPase gene fragment within the Bs1 retroelement of maize. *Plant Cell* **1994**, *6*, 1177–1186. [CrossRef] [PubMed]

43. Elrouby, N.; Bureau, T.E. A novel hybrid open reading frame formed by multiple cellular gene transductions by a plant long terminal repeat retroelement. *J. Biol. Chem.* **2001**, *276*, 41963–41968. [CrossRef] [PubMed]

44. Tan, S.; Cardoso-Moreira, M.; Shi, W.; Zhang, D.; Huang, J.; Mao, Y.; Jia, H.; Zhang, Y.; Chen, C.; Shao, Y.; et al. LTR-mediated retroposition as a mechanism of RNA-based duplication in metazoans. *Genome Res.* **2016**, *26*, 1663–1675. [CrossRef] [PubMed]

45. Dong, R.; Zhang, X.-O.; Zhang, Y.; Ma, X.-K.; Chen, L.-L.; Yang, L. CircRNA-derived pseudogenes. *Cell Res.* **2016**, *26*, 747–750. [CrossRef] [PubMed]

46. Brouha, B.; Schustak, J.; Badge, R.M.; Lutz-Prigge, S.; Farley, A.H.; Moran, J.V.; Kazazian, H.H. Hot L1s account for the bulk of retrotransposition in the human population. *Proc. Natl. Acad. Sci. USA* **2003**, *100*, 5280–5285. [CrossRef] [PubMed]

47. Elbarbary, R.A.; Lucas, B.A.; Maquat, L.E. Retrotransposons as regulators of gene expression. *Science* **2016**, *351*, aac7247. [CrossRef] [PubMed]

48. Goodier, J.L.; Kazazian, H.H. Retrotransposons revisited: The restraint and rehabilitation of parasites. *Cell* **2008**, *135*, 23–35. [CrossRef] [PubMed]

49. Crichton, J.H.; Dunican, D.S.; Maclennan, M.; Meehan, R.R.; Adams, I.R. Defending the genome from the enemy within: Mechanisms of retrotransposon suppression in the mouse germline. *Cell. Mol. Life Sci.* **2014**, *71*, 1581–1605. [CrossRef] [PubMed]

50. Pizarro, J.G.; Cristofari, G. Post-Transcriptional Control of LINE-1 Retrotransposition by Cellular Host Factors in Somatic Cells. *Front. Cell Dev. Biol.* **2016**, *4*, 14. [CrossRef] [PubMed]

51. Schulman, A.H. Retrotransposon replication in plants. *Curr. Opin. Virol.* **2013**, *3*, 604–614. [CrossRef] [PubMed]

52. Beck, C.R.; Collier, P.; Macfarlane, C.; Malig, M.; Kidd, J.M.; Eichler, E.E.; Badge, R.M.; Moran, J.V. LINE-1 retrotransposition activity in human genomes. *Cell* **2010**, *141*, 1159–1170. [CrossRef] [PubMed]

53. Ewing, A.D.; Kazazian, H.H. High-throughput sequencing reveals extensive variation in human-specific L1 content in individual human genomes. *Genome Res.* **2010**, *20*, 1262–1270. [CrossRef] [PubMed]

54. Speek, M. Antisense promoter of human L1 retrotransposon drives transcription of adjacent cellular genes. *Mol. Cell. Biol.* **2001**, *21*, 1973–1985. [CrossRef] [PubMed]

55. Martin, S.L.; Bushman, F.D. Nucleic acid chaperone activity of the ORF1 protein from the mouse LINE-1 retrotransposon. *Mol. Cell. Biol.* **2001**, *21*, 467–475. [CrossRef] [PubMed]

56. Moran, J.V.; Holmes, S.E.; Naas, T.P.; DeBerardinis, R.J.; Boeke, J.D.; Kazazian, H.H. High frequency retrotransposition in cultured mammalian cells. *Cell* **1996**, *87*, 917–927. [CrossRef]

57. Denli, A.M.; Narvaiza, I.; Kerman, B.E.; Pena, M.; Benner, C.; Marchetto, M.C.N.; Diedrich, J.K.; Aslanian, A.; Ma, J.; Moresco, J.J.; et al. Primate-specific ORF0 contributes to retrotransposon-mediated diversity. *Cell* **2015**, *163*, 583–593. [CrossRef] [PubMed]

58. Khazina, E.; Truffault, V.; Büttner, R.; Schmidt, S.; Coles, M.; Weichenrieder, O. Trimeric structure and flexibility of the L1ORF1 protein in human L1 retrotransposition. *Nat. Struct. Mol. Biol.* **2011**, *18*, 1006–1014. [CrossRef] [PubMed]

59. Dewannieux, M.; Esnault, C.; Heidmann, T. LINE-mediated retrotransposition of marked Alu sequences. *Nat. Genet.* **2003**, *35*, 41–48. [CrossRef] [PubMed]

60. Doucet, A.J.; Droc, G.; Siol, O.; Audoux, J.; Gilbert, N. U6 snRNA Pseudogenes: Markers of Retrotransposition Dynamics in Mammals. *Mol. Biol. Evol.* **2015**, *32*, 1815–1832. [CrossRef] [PubMed]

61. Hancks, D.C.; Goodier, J.L.; Mandal, P.K.; Cheung, L.E.; Kazazian, H.H. Retrotransposition of marked SVA elements by human L1s in cultured cells. *Hum. Mol. Genet.* **2011**, *20*, 3386–3400. [CrossRef] [PubMed]

62. Raiz, J.; Damert, A.; Chira, S.; Held, U.; Klawitter, S.; Hamdorf, M.; Löwer, J.; Strätling, W.H.; Löwer, R.; Schumann, G.G. The non-autonomous retrotransposon SVA is *trans*-mobilized by the human LINE-1 protein machinery. *Nucleic Acids Res.* **2012**, *40*, 1666–1683. [CrossRef] [PubMed]

63. Perreault, J.; Noël, J.-F.; Brière, F.; Cousineau, B.; Lucier, J.-F.; Perreault, J.-P.; Boire, G. Retropseudogenes derived from the human Ro/SS-A autoantigen-associated hY RNAs. *Nucleic Acids Res.* **2005**, *33*, 2032–2041. [CrossRef] [PubMed]

64. Pavlícek, A.; Paces, J.; Elleder, D.; Hejnar, J. Processed pseudogenes of human endogenous retroviruses generated by LINEs: Their integration, stability, and distribution. *Genome Res.* **2002**, *12*, 391–399. [CrossRef] [PubMed]

65. Doucet, A.J.; Wilusz, J.E.; Miyoshi, T.; Liu, Y.; Moran, J.V. A 3′ Poly(A) Tract Is Required for LINE-1 Retrotransposition. *Mol. Cell* **2015**, *60*, 728–741. [CrossRef] [PubMed]

66. Abyzov, A.; Iskow, R.; Gokcumen, O.; Radke, D.W.; Balasubramanian, S.; Pei, B.; Habegger, L.; 1000 Genomes Project Consortium; Lee, C.; Gerstein, M. Analysis of variable retroduplications in human populations suggests coupling of retrotransposition to cell division. *Genome Res.* **2013**, *23*, 2042–2052. [CrossRef] [PubMed]

67. Kubo, S.; del Seleme, M.C.; Soifer, H.S.; Perez, J.L.G.; Moran, J.V.; Kazazian, H.H.; Kasahara, N. L1 retrotransposition in nondividing and primary human somatic cells. *Proc. Natl. Acad. Sci. USA* **2006**, *103*, 8036–8041. [CrossRef] [PubMed]

68. Macia, A.; Widmann, T.J.; Heras, S.R.; Ayllon, V.; Sanchez, L.; Benkaddour-Boumzaouad, M.; Muñoz-Lopez, M.; Rubio, A.; Amador-Cubero, S.; Blanco-Jimenez, E.; et al. Engineered LINE-1 retrotransposition in nondividing human neurons. *Genome Res.* **2017**, *27*, 335–348. [CrossRef] [PubMed]

69. Cost, G.J.; Feng, Q.; Jacquier, A.; Boeke, J.D. Human L1 element target-primed reverse transcription in vitro. *EMBO J.* **2002**, *21*, 5899–5910. [CrossRef] [PubMed]

70. Sen, S.K.; Huang, C.T.; Han, K.; Batzer, M.A. Endonuclease-independent insertion provides an alternative pathway for L1 retrotransposition in the human genome. *Nucleic Acids Res.* **2007**, *35*, 3741–3751. [CrossRef] [PubMed]

71. Jąkalski, M.; Takeshita, K.; Deblieck, M.; Koyanagi, K.O.; Makałowska, I.; Watanabe, H.; Makałowski, W. Comparative genomic analysis of retrogene repertoire in two green algae *Volvox carteri* and *Chlamydomonas reinhardtii*. *Biol. Direct* **2016**, *11*, 35. [CrossRef] [PubMed]

72. Zhang, X.; Zhang, Q.; Su, B. Emergence and evolution of inter-specific segregating retrocopies in cynomolgus monkey (*Macaca fascicularis*) and rhesus macaque (*Macaca mulatta*). *Sci. Rep.* **2016**, *6*, 32598. [CrossRef] [PubMed]

73. Szcześniak, M.W.; Ciomborowska, J.; Nowak, W.; Rogozin, I.B.; Makałowska, I. Primate and rodent specific intron gains and the origin of retrogenes with splice variants. *Mol. Biol. Evol.* **2011**, *28*, 33–37. [CrossRef] [PubMed]

74. Baertsch, R.; Diekhans, M.; Kent, W.J.; Haussler, D.; Brosius, J. Retrocopy contributions to the evolution of the human genome. *BMC Genom.* **2008**, *9*, 466. [CrossRef] [PubMed]

75. Slesarev, A.I.; Belova, G.I.; Kozyavkin, S.A.; Lake, J.A. Evidence for an early prokaryotic origin of histones H2A and H4 prior to the emergence of eukaryotes. *Nucleic Acids Res.* **1998**, *26*, 427–430. [CrossRef] [PubMed]

76. Ashurst, J.L.; Chen, C.-K.; Gilbert, J.G.R.; Jekosch, K.; Keenan, S.; Meidl, P.; Searle, S.M.; Stalker, J.; Storey, R.; Trevanion, S.; et al. The Vertebrate Genome Annotation (Vega) database. *Nucleic Acids Res.* **2005**, *33*, D459–D465. [CrossRef] [PubMed]

77. Aken, B.L.; Ayling, S.; Barrell, D.; Clarke, L.; Curwen, V.; Fairley, S.; Fernandez Banet, J.; Billis, K.; García Girón, C.; Hourlier, T.; et al. The Ensembl gene annotation system. *Database* **2016**, *2016*, baw093. [CrossRef] [PubMed]

78. Zhang, Z.; Carriero, N.; Zheng, D.; Karro, J.; Harrison, P.M.; Gerstein, M. PseudoPipe: An automated pseudogene identification pipeline. *Bioinformatics* **2006**, *22*, 1437–1439. [CrossRef] [PubMed]

79. Speir, M.L.; Zweig, A.S.; Rosenbloom, K.R.; Raney, B.J.; Paten, B.; Nejad, P.; Lee, B.T.; Learned, K.; Karolchik, D.; Hinrichs, A.S.; et al. The UCSC Genome Browser database: 2016 update. *Nucleic Acids Res.* **2016**, *44*, D717–D725. [CrossRef] [PubMed]

80. Pei, B.; Sisu, C.; Frankish, A.; Howald, C.; Habegger, L.; Mu, X.J.; Harte, R.; Balasubramanian, S.; Tanzer, A.; Diekhans, M.; et al. The GENCODE pseudogene resource. *Genome Biol.* **2012**, *13*, R51. [CrossRef] [PubMed]

81. Khelifi, A.; Adel, K.; Duret, L.; Laurent, D.; Mouchiroud, D.; Dominique, M. HOPPSIGEN: A database of human and mouse processed pseudogenes. *Nucleic Acids Res.* **2005**, *33*, D59–D66. [PubMed]

82. Navarro, F.C.P.; Galante, P.A.F. RCPedia: A database of retrocopied genes. *Bioinformatics* **2013**, *29*, 1235–1237. [CrossRef] [PubMed]

83. Kabza, M.; Kubiak, M.R.; Danek, A.; Rosikiewicz, W.; Deorowicz, S.; Polański, A.; Makałowska, I. Inter-population Differences in Retrogene Loss and Expression in Humans. *PLoS Genet.* **2015**, *11*, e1005579. [CrossRef] [PubMed]

84. Altschul, S.F.; Gish, W.; Miller, W.; Myers, E.W.; Lipman, D.J. Basic local alignment search tool. *J. Mol. Biol.* **1990**, *215*, 403–410. [CrossRef]

85. Kiełbasa, S.M.; Wan, R.; Sato, K.; Horton, P.; Frith, M.C. Adaptive seeds tame genomic sequence comparison. *Genome Res.* **2011**, *21*, 487–493. [CrossRef] [PubMed]

86. Louhichi, A.; Fourati, A.; Rebaï, A. IGD: A resource for intronless genes in the human genome. *Gene* **2011**, *488*, 35–40. [CrossRef] [PubMed]

87. Jorquera, R.; Ortiz, R.; Ossandon, F.; Cárdenas, J.P.; Sepúlveda, R.; González, C.; Holmes, D.S. SinEx DB: A database for single exon coding sequences in mammalian genomes. *Database* **2016**, *2016*, baw095. [CrossRef] [PubMed]

88. Vinckenbosch, N.; Dupanloup, I.; Kaessmann, H. Evolutionary fate of retroposed gene copies in the human genome. *Proc. Natl. Acad. Sci. USA* **2006**, *103*, 3220–3225. [CrossRef] [PubMed]

89. Okamura, K.; Nakai, K. Retrotransposition as a source of new promoters. *Mol. Biol. Evol.* **2008**, *25*, 1231–1238. [CrossRef] [PubMed]

90. Zhang, J. Evolution by gene duplication: An update. *Trends Ecol. Evol.* **2003**, *18*, 292–298. [CrossRef]

91. Force, A.; Lynch, M.; Pickett, F.B.; Amores, A.; Yan, Y.L.; Postlethwait, J. Preservation of duplicate genes by complementary, degenerate mutations. *Genetics* **1999**, *151*, 1531–1545. [PubMed]

92. Liu, Z.; Tavares, R.; Forsythe, E.S.; André, F.; Lugan, R.; Jonasson, G.; Boutet-Mercey, S.; Tohge, T.; Beilstein, M.A.; Werck-Reichhart, D.; et al. Evolutionary interplay between sister cytochrome P450 genes shapes plasticity in plant metabolism. *Nat. Commun.* **2016**, *7*, 13026. [CrossRef] [PubMed]

93. Rosso, L.; Marques, A.C.; Weier, M.; Lambert, N.; Lambot, M.-A.; Vanderhaeghen, P.; Kaessmann, H. Birth and Rapid Subcellular Adaptation of a Hominoid-Specific CDC14 Protein. *PLoS Biol.* **2008**, *6*, e140. [CrossRef] [PubMed]

94. Rastogi, S.; Liberles, D.A. Subfunctionalization of duplicated genes as a transition state to neofunctionalization. *BMC Evol. Biol.* **2005**, *5*, 28. [CrossRef] [PubMed]

95. Young, J.; Ménétrey, J.; Goud, B. RAB6C is a retrogene that encodes a centrosomal protein involved in cell cycle progression. *J. Mol. Biol.* **2010**, *397*, 69–88. [CrossRef] [PubMed]

96. Kandouz, M.; Bier, A.; Carystinos, G.D.; Alaoui-Jamali, M.A.; Batist, G. Connexin43 pseudogene is expressed in tumor cells and inhibits growth. *Oncogene* **2004**, *23*, 4763–4770. [CrossRef] [PubMed]

97. Bier, A.; Oviedo-Landaverde, I.; Zhao, J.; Mamane, Y.; Kandouz, M.; Batist, G. Connexin43 pseudogene in breast cancer cells offers a novel therapeutic target. *Mol. Cancer Ther.* **2009**, *8*, 786–793. [CrossRef] [PubMed]

98. Sulak, M.; Fong, L.; Mika, K.; Chigurupati, S.; Yon, L.; Mongan, N.P.; Emes, R.D.; Lynch, V.J. TP53 copy number expansion is associated with the evolution of increased body size and an enhanced DNA damage response in elephants. *Elife* **2016**, *5*, e11994. [PubMed]

99. Carelli, F.N.; Hayakawa, T.; Go, Y.; Imai, H.; Warnefors, M.; Kaessmann, H. The life history of retrocopies illuminates the evolution of new mammalian genes. *Genome Res.* **2016**, *26*, 301–314. [CrossRef] [PubMed]

100. Kim, D.S.; Wang, Y.; Oh, H.J.; Choi, D.; Lee, K.; Hahn, Y. Retroduplication and loss of parental genes is a mechanism for the generation of intronless genes in *Ciona intestinalis* and *Ciona savignyi*. *Dev. Genes Evol.* **2014**, *224*, 255–260. [CrossRef] [PubMed]

101. Uechi, T.; Maeda, N.; Tanaka, T.; Kenmochi, N. Functional second genes generated by retrotransposition of the X-linked ribosomal protein genes. *Nucleic Acids Res.* **2002**, *30*, 5369–5375. [CrossRef] [PubMed]

102. Elliott, D.J.; Venables, J.P.; Newton, C.S.; Lawson, D.; Boyle, S.; Eperon, I.C.; Cooke, H.J. An evolutionarily conserved germ cell-specific hnRNP is encoded by a retrotransposed gene. *Hum. Mol. Genet.* **2000**, *9*, 2117–2124. [CrossRef] [PubMed]

103. Mascarenhas, R.; Pietrzak, M.; Smith, R.M.; Webb, A.; Wang, D.; Papp, A.C.; Pinsonneault, J.K.; Seweryn, M.; Rempala, G.; Sadee, W. Allele-Selective Transcriptome Recruitment to Polysomes Primed for Translation: Protein-Coding and Noncoding RNAs, and RNA Isoforms. *PLoS ONE* **2015**, *10*, e0136798. [CrossRef] [PubMed]

104. Brosch, M.; Saunders, G.I.; Frankish, A.; Collins, M.O.; Yu, L.; Wright, J.; Verstraten, R.; Adams, D.J.; Harrow, J.; Choudhary, J.S.; et al. Shotgun proteomics aids discovery of novel protein-coding genes, alternative splicing, and "resurrected" pseudogenes in the mouse genome. *Genome Res.* **2011**, *21*, 756–767. [CrossRef] [PubMed]

105. Kim, M.-S.; Pinto, S.M.; Getnet, D.; Nirujogi, R.S.; Manda, S.S.; Chaerkady, R.; Madugundu, A.K.; Kelkar, D.S.; Isserlin, R.; Jain, S.; et al. A draft map of the human proteome. *Nature* **2014**, *509*, 575–581. [CrossRef] [PubMed]

106. Long, M.; Langley, C.H. Natural selection and the origin of jingwei, a chimeric processed functional gene in *Drosophila*. *Science* **1993**, *260*, 91–95. [CrossRef] [PubMed]

107. Sayah, D.M.; Sokolskaja, E.; Berthoux, L.; Luban, J. Cyclophilin A retrotransposition into TRIM5 explains owl monkey resistance to HIV-1. *Nature* **2004**, *430*, 569–573. [CrossRef] [PubMed]

108. Mori, S.; Hayashi, M.; Inagaki, S.; Oshima, T.; Tateishi, K.; Fujii, H.; Suzuki, S. Identification of Multiple Forms of RNA Transcripts Associated with Human-Specific Retrotransposed Gene Copies. *Genome Biol. Evol.* **2016**, *8*, 2288–2296. [CrossRef] [PubMed]

109. Jang, S.; Lee, B.; Kim, C.; Kim, S.-J.; Yim, J.; Han, J.-J.; Lee, S.; Kim, S.-R.; An, G. The OsFOR1 gene encodes a polygalacturonase-inhibiting protein (PGIP) that regulates floral organ number in rice. *Plant Mol. Biol.* **2003**, *53*, 357–369. [CrossRef] [PubMed]

110. Shearwin, K.E.; Callen, B.P.; Egan, J.B. Transcriptional interference—A crash course. *Trends Genet.* **2005**, *21*, 339–345. [CrossRef] [PubMed]

111. Osato, N.; Suzuki, Y.; Ikeo, K.; Gojobori, T. Transcriptional interferences in *cis* natural antisense transcripts of humans and mice. *Genetics* **2007**, *176*, 1299–1306. [CrossRef] [PubMed]

112. Kaer, K.; Branovets, J.; Hallikma, A.; Nigumann, P.; Speek, M. Intronic L1 retrotransposons and nested genes cause transcriptional interference by inducing intron retention, exonization and cryptic polyadenylation. *PLoS ONE* **2011**, *6*, e26099. [CrossRef] [PubMed]

113. Grothaus, K.; Kanber, D.; Gellhaus, A.; Mikat, B.; Kolarova, J.; Siebert, R.; Wieczorek, D.; Horsthemke, B. Genome-wide methylation analysis of retrocopy-associated CpG islands and their genomic environment. *Epigenetics* **2016**, *11*, 216–226. [CrossRef] [PubMed]

114. Wood, A.J.; Roberts, R.G.; Monk, D.; Moore, G.E.; Schulz, R.; Oakey, R.J. A screen for retrotransposed imprinted genes reveals an association between X chromosome homology and maternal germ-line methylation. *PLoS Genet.* **2007**, *3*, e20. [CrossRef] [PubMed]

115. Wood, A.J.; Schulz, R.; Woodfine, K.; Koltowska, K.; Beechey, C.V.; Peters, J.; Bourc'his, D.; Oakey, R.J. Regulation of alternative polyadenylation by genomic imprinting. *Genes Dev.* **2008**, *22*, 1141–1146. [CrossRef] [PubMed]

116. Cowley, M.; Wood, A.J.; Böhm, S.; Schulz, R.; Oakey, R.J. Epigenetic control of alternative mRNA processing at the imprinted Herc3/Nap1l5 locus. *Nucleic Acids Res.* **2012**, *40*, 8917–8926. [CrossRef] [PubMed]

117. McCole, R.B.; Oakey, R.J. Unwitting hosts fall victim to imprinting. *Epigenetics* **2008**, *3*, 258–260. [CrossRef] [PubMed]

118. Ezawa, K.; OOta, S.; Saitou, N. Genome-wide search of gene conversions in duplicated genes of mouse and rat. *Mol. Biol. Evol.* **2006**, *23*, 927–940. [CrossRef] [PubMed]

119. Pan, D.; Zhang, L. Quantifying the major mechanisms of recent gene duplications in the human and mouse genomes: A novel strategy to estimate gene duplication rates. *Genome Biol.* **2007**, *8*, R158. [CrossRef] [PubMed]

120. Nishioka, Y.; Leder, A.; Leder, P. Unusual alpha-globin-like gene that has cleanly lost both globin intervening sequences. *Proc. Natl. Acad. Sci. USA* **1980**, *77*, 2806–2809. [CrossRef] [PubMed]

121. Derr, L.K. The involvement of cellular recombination and repair genes in RNA-mediated recombination in Saccharomyces cerevisiae. *Genetics* **1998**, *148*, 937–945. [PubMed]

122. Mourier, T.; Jeffares, D.C. Eukaryotic intron loss. *Science* **2003**, *300*, 1393. [CrossRef] [PubMed]

123. Roy, S.W.; Gilbert, W. The pattern of intron loss. *Proc. Natl. Acad. Sci. USA* **2005**, *102*, 713–718. [CrossRef] [PubMed]

124. Zhu, T.; Niu, D.-K. Frequency of intron loss correlates with processed pseudogene abundance: A novel strategy to test the reverse transcriptase model of intron loss. *BMC Biol.* **2013**, *11*, 23. [CrossRef] [PubMed]

125. Rinn, J.L.; Kertesz, M.; Wang, J.K.; Squazzo, S.L.; Xu, X.; Brugmann, S.A.; Goodnough, L.H.; Helms, J.A.; Farnham, P.J.; Segal, E.; et al. Functional demarcation of active and silent chromatin domains in human HOX loci by noncoding RNAs. *Cell* **2007**, *129*, 1311–1323. [CrossRef] [PubMed]

126. Poliseno, L.; Salmena, L.; Zhang, J.; Carver, B.; Haveman, W.J.; Pandolfi, P.P. A coding-independent function of gene and pseudogene mRNAs regulates tumour biology. *Nature* **2010**, *465*, 1033–1038. [CrossRef] [PubMed]

127. Johnsson, P.; Ackley, A.; Vidarsdottir, L.; Lui, W.-O.; Corcoran, M.; Grandér, D.; Morris, K.V. A pseudogene long noncoding RNA network regulates PTEN transcription and translation in human cells. *Nat. Struct. Mol. Biol.* **2013**, *20*, 440–446. [CrossRef] [PubMed]

128. Muro, E.M.; Andrade-Navarro, M.A. Pseudogenes as an alternative source of natural antisense transcripts. *BMC Evol. Biol.* **2010**, *10*, 338. [CrossRef] [PubMed]

129. Bryzghalov, O.; Szcześniak, M.W.; Makałowska, I. Retroposition as a source of antisense long non-coding RNAs with possible regulatory functions. *Acta Biochim. Pol.* **2016**, *63*, 825–833. [CrossRef] [PubMed]

130. Korneev, S.A.; Park, J.H.; O'Shea, M. Neuronal expression of neural nitric oxide synthase (nNOS) protein is suppressed by an antisense RNA transcribed from an NOS pseudogene. *J. Neurosci.* **1999**, *19*, 7711–7720. [PubMed]

131. Stambolic, V.; Suzuki, A.; de la Pompa, J.L.; Brothers, G.M.; Mirtsos, C.; Sasaki, T.; Ruland, J.; Penninger, J.M.; Siderovski, D.P.; Mak, T.W. Negative regulation of PKB/Akt-dependent cell survival by the tumor suppressor PTEN. *Cell* **1998**, *95*, 29–39. [CrossRef]

132. Poliseno, L.; Haimovic, A.; Christos, P.J.; Vega, Y.; Saenz de Miera, E.C.; Shapiro, R.; Pavlick, A.; Berman, R.S.; Darvishian, F.; Osman, I. Deletion of PTENP1 pseudogene in human melanoma. *J. Investig. Dermatol.* **2011**, *131*, 2497–2500. [CrossRef] [PubMed]

133. Yu, G.; Yao, W.; Gumireddy, K.; Li, A.; Wang, J.; Xiao, W.; Chen, K.; Xiao, H.; Li, H.; Tang, K.; et al. Pseudogene PTENP1 functions as a competing endogenous RNA to suppress clear-cell renal cell carcinoma progression. *Mol. Cancer Ther.* **2014**, *13*, 3086–3097. [CrossRef] [PubMed]

134. Chen, C.-L.; Tseng, Y.-W.; Wu, J.-C.; Chen, G.-Y.; Lin, K.-C.; Hwang, S.-M.; Hu, Y.-C. Suppression of hepatocellular carcinoma by baculovirus-mediated expression of long non-coding RNA PTENP1 and microRNA regulation. *Biomaterials* **2015**, *44*, 71–81. [CrossRef] [PubMed]

135. Wang, L.; Guo, Z.-Y.; Zhang, R.; Xin, B.; Chen, R.; Zhao, J.; Wang, T.; Wen, W.-H.; Jia, L.-T.; Yao, L.-B.; et al. Pseudogene OCT4-pg4 functions as a natural micro RNA sponge to regulate OCT4 expression by competing for miR-145 in hepatocellular carcinoma. *Carcinogenesis* **2013**, *34*, 1773–1781. [CrossRef] [PubMed]

136. Peng, H.; Ishida, M.; Li, L.; Saito, A.; Kamiya, A.; Hamilton, J.P.; Fu, R.; Olaru, A.V.; An, F.; Popescu, I.; et al. Pseudogene INTS6P1 regulates its cognate gene INTS6 through competitive binding of miR-17-5p in hepatocellular carcinoma. *Oncotarget* **2015**, *6*, 5666–5677. [CrossRef] [PubMed]

137. Ye, X.; Fan, F.; Bhattacharya, R.; Bellister, S.; Boulbes, D.R.; Wang, R.; Xia, L.; Ivan, C.; Zheng, X.; Calin, G.A.; et al. VEGFR-1 Pseudogene Expression and Regulatory Function in Human Colorectal Cancer Cells. *Mol. Cancer Res.* **2015**, *13*, 1274–1282. [CrossRef] [PubMed]

138. Di Sanzo, M.; Aversa, I.; Santamaria, G.; Gagliardi, M.; Panebianco, M.; Biamonte, F.; Zolea, F.; Faniello, M.C.; Cuda, G.; Costanzo, F. FTH1P3, a Novel H-Ferritin Pseudogene Transcriptionally Active, Is Ubiquitously Expressed and Regulated during Cell Differentiation. *PLoS ONE* **2016**, *11*, e0151359. [CrossRef] [PubMed]

139. Esposito, F.; De Martino, M.; Petti, M.G.; Forzati, F.; Tornincasa, M.; Federico, A.; Arra, C.; Pierantoni, G.M.; Fusco, A. HMGA1 pseudogenes as candidate proto-oncogenic competitive endogenous RNAs. *Oncotarget* **2014**, *5*, 8341–8354. [CrossRef] [PubMed]

140. De Martino, M.; Forzati, F.; Marfella, M.; Pellecchia, S.; Arra, C.; Terracciano, L.; Fusco, A.; Esposito, F. HMGA1P7-pseudogene regulates H19 and Igf2 expression by a competitive endogenous RNA mechanism. *Sci. Rep.* **2016**, *6*, 37622. [CrossRef] [PubMed]

141. Devor, E.J. Primate microRNAs miR-220 and miR-492 lie within processed pseudogenes. *J. Hered.* **2006**, *97*, 186–190. [CrossRef] [PubMed]

142. Von Frowein, J.; Pagel, P.; Kappler, R.; von Schweinitz, D.; Roscher, A.; Schmid, I. MicroRNA-492 is processed from the keratin 19 gene and up-regulated in metastatic hepatoblastoma. *Hepatology* **2011**, *53*, 833–842. [CrossRef] [PubMed]

143. Shen, F.; Cai, W.-S.; Feng, Z.; Li, J.-L.; Chen, J.-W.; Cao, J.; Xu, B. MiR-492 contributes to cell proliferation and cell cycle of human breast cancer cells by suppressing SOX7 expression. *Tumour Biol.* **2015**, *36*, 1913–1921. [CrossRef] [PubMed]

144. Griffiths-Jones, S.; Grocock, R.J.; van Dongen, S.; Bateman, A.; Enright, A.J. miRBase: MicroRNA sequences, targets and gene nomenclature. *Nucleic Acids Res.* **2006**, *34*, D140–D144. [CrossRef] [PubMed]

145. Song, W.-H.; Feng, X.-J.; Gong, S.-J.; Chen, J.-M.; Wang, S.-M.; Xing, D.-J.; Zhu, M.-H.; Zhang, S.-H.; Xu, A.-M. MicroRNA-622 acts as a tumor suppressor in hepatocellular carcinoma. *Cancer Biol. Ther.* **2015**, *16*, 1754–1763. [CrossRef] [PubMed]

146. Hirano, T.; Iwasaki, Y.W.; Lin, Z.Y.-C.; Imamura, M.; Seki, N.M.; Sasaki, E.; Saito, K.; Okano, H.; Siomi, M.C.; Siomi, H. Small RNA profiling and characterization of piRNA clusters in the adult testes of the common marmoset, a model primate. *RNA* **2014**, *20*, 1223–1237. [CrossRef] [PubMed]

147. Pantano, L.; Jodar, M.; Bak, M.; Ballescà, J.L.; Tommerup, N.; Oliva, R.; Vavouri, T. The small RNA content of human sperm reveals pseudogene-derived piRNAs complementary to protein-coding genes. *RNA* **2015**, *21*, 1085–1095. [CrossRef] [PubMed]

148. Neumann, L.C.; Feiner, N.; Meyer, A.; Buiting, K.; Horsthemke, B. The imprinted *NPAP1* gene in the Prader-Willi syndrome region belongs to a *POM121*-related family of retrogenes. *Genome Biol. Evol.* **2014**, *6*, 344–351. [CrossRef] [PubMed]

149. Watanabe, T.; Cheng, E.; Zhong, M.; Lin, H. Retrotransposons and pseudogenes regulate mRNAs and lncRNAs via the piRNA pathway in the germline. *Genome Res.* **2015**, *25*, 368–380. [CrossRef] [PubMed]

150. Gebert, D.; Ketting, R.F.; Zischler, H.; Rosenkranz, D. piRNAs from Pig Testis Provide Evidence for a Conserved Role of the Piwi Pathway in Post-Transcriptional Gene Regulation in Mammals. *PLoS ONE* **2015**, *10*, e0124860. [CrossRef] [PubMed]

151. Milligan, M.J.; Harvey, E.; Yu, A.; Morgan, A.L.; Smith, D.L.; Zhang, E.; Berengut, J.; Sivananthan, J.; Subramaniam, R.; Skoric, A.; et al. Global Intersection of Long Non-Coding RNAs with Processed and Unprocessed Pseudogenes in the Human Genome. *Front. Genet.* **2016**, *7*, 26. [CrossRef] [PubMed]

152. Kasschau, K.D.; Fahlgren, N.; Chapman, E.J.; Sullivan, C.M.; Cumbie, J.S.; Givan, S.A.; Carrington, J.C. Genome-wide profiling and analysis of *Arabidopsis* siRNAs. *PLoS Biol.* **2007**, *5*, e57. [CrossRef] [PubMed]

153. Tam, O.H.; Aravin, A.A.; Stein, P.; Girard, A.; Murchison, E.P.; Cheloufi, S.; Hodges, E.; Anger, M.; Sachidanandam, R.; Schultz, R.M.; et al. Pseudogene-derived small interfering RNAs regulate gene expression in mouse oocytes. *Nature* **2008**, *453*, 534–538. [CrossRef] [PubMed]

154. Watanabe, T.; Totoki, Y.; Toyoda, A.; Kaneda, M.; Kuramochi-Miyagawa, S.; Obata, Y.; Chiba, H.; Kohara, Y.; Kono, T.; Nakano, T.; et al. Endogenous siRNAs from naturally formed dsRNAs regulate transcripts in mouse oocytes. *Nature* **2008**, *453*, 539–543. [CrossRef] [PubMed]

155. Guo, X.; Zhang, Z.; Gerstein, M.B.; Zheng, D. Small RNAs originated from pseudogenes: *cis*- or *trans*-acting? *PLoS Comput. Biol.* **2009**, *5*, e1000449. [CrossRef] [PubMed]

156. Wen, Y.-Z.; Zheng, L.-L.; Liao, J.-Y.; Wang, M.-H.; Wei, Y.; Guo, X.-M.; Qu, L.-H.; Ayala, F.J.; Lun, Z.-R. Pseudogene-derived small interference RNAs regulate gene expression in African *Trypanosoma brucei*. *Proc. Natl. Acad. Sci. USA* **2011**, *108*, 8345–8350. [CrossRef] [PubMed]

157. Chan, W.-L.; Yuo, C.-Y.; Yang, W.-K.; Hung, S.-Y.; Chang, Y.-S.; Chiu, C.-C.; Yeh, K.-T.; Huang, H.-D.; Chang, J.-G. Transcribed pseudogene ψPPM1K generates endogenous siRNA to suppress oncogenic cell growth in hepatocellular carcinoma. *Nucleic Acids Res.* **2013**, *41*, 3734–3747. [CrossRef] [PubMed]

158. Guo, X.; Lin, M.; Rockowitz, S.; Lachman, H.M.; Zheng, D. Characterization of Human Pseudogene-Derived Non-Coding RNAs for Functional Potential. *PLoS ONE* **2014**, *9*, e93972. [CrossRef] [PubMed]

159. Fatima, R.; Akhade, V.S.; Pal, D.; Rao, S.M. Long noncoding RNAs in development and cancer: Potential biomarkers and therapeutic targets. *Mol. Cell. Ther.* **2015**, *3*. [CrossRef] [PubMed]

160. Poliseno, L.; Marranci, A.; Pandolfi, P.P. Pseudogenes in Human Cancer. *Front. Med.* **2015**, *2*, 68. [CrossRef] [PubMed]

161. Zhang, X.; Zhang, J.; Ping, X.; Wang, Q.-L.; Lu, X. Pseudogene transcripts: Participants in tumorigenicity and promising therapeutic targets. *Leuk. Res.* **2016**, *42*, 105–106. [CrossRef] [PubMed]

162. Shi, X.; Nie, F.; Wang, Z.; Sun, M. Pseudogene-expressed RNAs: A new frontier in cancers. *Tumour Biol.* **2016**, *37*, 1471–1478. [CrossRef] [PubMed]

163. Kozlov, A.P. Expression of evolutionarily novel genes in tumors. *Infect. Agents Cancer* **2016**, *11*. [CrossRef] [PubMed]

164. Prendergast, G.C. Actin' up: RhoB in cancer and apoptosis. *Nat. Rev. Cancer* **2001**, *1*, 162–168. [CrossRef] [PubMed]

165. Rohozinski, J.; Edwards, C.L.; Anderson, M.L. Does expression of the retrogene UTP14c in the ovary pre-dispose women to ovarian cancer? *Med. Hypotheses* **2012**, *78*, 446–449. [CrossRef] [PubMed]

166. Kalyana-Sundaram, S.; Kumar-Sinha, C.; Shankar, S.; Robinson, D.R.; Wu, Y.-M.; Cao, X.; Asangani, I.A.; Kothari, V.; Prensner, J.R.; Lonigro, R.J.; et al. Expressed pseudogenes in the transcriptional landscape of human cancers. *Cell* **2012**, *149*, 1622–1634. [CrossRef] [PubMed]

167. Lui, K.Y.; Peng, H.-R.; Lin, J.-R.; Qiu, C.-H.; Chen, H.-A.; Fu, R.-D.; Cai, C.-J.; Lu, M.-Q. Pseudogene integrator complex subunit 6 pseudogene 1 (INTS6P1) as a novel plasma-based biomarker for hepatocellular carcinoma screening. *Tumour Biol.* **2016**, *37*, 1253–1260. [CrossRef] [PubMed]

168. Chiefari, E.; Iiritano, S.; Paonessa, F.; Le Pera, I.; Arcidiacono, B.; Filocamo, M.; Foti, D.; Liebhaber, S.A.; Brunetti, A. Pseudogene-mediated posttranscriptional silencing of HMGA1 can result in insulin resistance and type 2 diabetes. *Nat. Commun.* **2010**, *1*, 40. [CrossRef] [PubMed]

169. Ewing, A.D.; Ballinger, T.J.; Earl, D.; Broad Institute Genome Sequencing and Analysis Program and Platform; Harris, C.C.; Ding, L.; Wilson, R.K.; Haussler, D. Retrotransposition of gene transcripts leads to structural variation in mammalian genomes. *Genome Biol.* **2013**, *14*, R22. [CrossRef] [PubMed]

170. Cooke, S.L.; Shlien, A.; Marshall, J.; Pipinikas, C.P.; Martincorena, I.; Tubio, J.M.C.; Li, Y.; Menzies, A.; Mudie, L.; Ramakrishna, M.; et al. Processed pseudogenes acquired somatically during cancer development. *Nat. Commun.* **2014**, *5*, 3644. [CrossRef] [PubMed]

171. Rapicavoli, N.A.; Qu, K.; Zhang, J.; Mikhail, M.; Laberge, R.-M.; Chang, H.Y. A mammalian pseudogene lncRNA at the interface of inflammation and anti-inflammatory therapeutics. *eLife* **2013**, *2*, e00762. [CrossRef] [PubMed]

172. Tsujikawa, M.; Kurahashi, H.; Tanaka, T.; Nishida, K.; Shimomura, Y.; Tano, Y.; Nakamura, Y. Identification of the gene responsible for gelatinous drop-like corneal dystrophy. *Nat. Genet.* **1999**, *21*, 420–423. [PubMed]

173. De Boer, M.; van Leeuwen, K.; Geissler, J.; Weemaes, C.M.; van den Berg, T.K.; Kuijpers, T.W.; Warris, A.; Roos, D. Primary immunodeficiency caused by an exonized retroposed gene copy inserted in the CYBB gene. *Hum. Mutat.* **2014**, *35*, 486–496. [CrossRef] [PubMed]

174. Costa, V.; Esposito, R.; Aprile, M.; Ciccodicola, A. Non-coding RNA and pseudogenes in neurodegenerative diseases: "The (un)Usual Suspects". *Front. Genet.* **2012**, *3*, 231. [CrossRef] [PubMed]

175. Feng, Q.; Snider, L.; Jagannathan, S.; Tawil, R.; van der Maarel, S.M.; Tapscott, S.J.; Bradley, R.K. A feedback loop between nonsense-mediated decay and the retrogene DUX4 in facioscapulohumeral muscular dystrophy. *eLife* **2015**, *4*. [CrossRef] [PubMed]

176. Gupta, A.; Brown, C.T.; Zheng, Y.-H.; Adami, C. Differentially-Expressed Pseudogenes in HIV-1 Infection. *Viruses.* **2015**, *7*, 5191–5205. [CrossRef] [PubMed]

177. Zhao, H.; Chen, M.; Lind, S.B.; Pettersson, U. Distinct temporal changes in host cell lncRNA expression during the course of an adenovirus infection. *Virology* **2016**, *492*, 242–250. [CrossRef] [PubMed]

178. 1000 Genomes Project Consortium; Abecasis, G.R.; Auton, A.; Brooks, L.D.; DePristo, M.A.; Durbin, R.M.; Handsaker, R.E.; Kang, H.M.; Marth, G.T.; McVean, G.A. An integrated map of genetic variation from 1092 human genomes. *Nature* **2012**, *491*, 56–65. [PubMed]

179. Schrider, D.R.; Navarro, F.C.P.; Galante, P.A.F.; Parmigiani, R.B.; Camargo, A.A.; Hahn, M.W.; de Souza, S.J. Gene copy-number polymorphism caused by retrotransposition in humans. *PLoS Genet.* **2013**, *9*, e1003242. [CrossRef] [PubMed]
180. Richardson, S.R.; Salvador-Palomeque, C.; Faulkner, G.J. Diversity through duplication: Whole-genome sequencing reveals novel gene retrocopies in the human population. *Bioessays* **2014**, *36*, 475–481. [CrossRef] [PubMed]

MDPI AG

St. Alban-Anlage 66

4052 Basel, Switzerland

Tel. +41 61 683 77 34

Fax +41 61 302 89 18

http://www.mdpi.com

Viruses Editorial Office

E-mail: viruses@mdpi.com

http://www.mdpi.com/journal/viruses

www.ingramcontent.com/pod-product-compliance
Lightning Source LLC
Chambersburg PA
CBHW051850210326

41597CB00033B/5850